FINITE ELEMENTS:
A Second Course

VOLUME II

The Texas Finite Element Series

FINITE ELEMENTS:
A Second Course

VOLUME II

GRAHAM F. CAREY and **J. TINSLEY ODEN**

Texas Institute for Computational Mechanics
The University of Texas at Austin

PRENTICE-HALL, INC., *Englewood Cliffs, New Jersey 07632*

Library of Congress Cataloging in Publication Data

—————.

Finite elements.

([Texas finite element series])
Vol. 2 by Graham F. Carey and J. Tinsley Oden.
Vols. 2-   have bibliography.
Contents: v. 1.   An introduction—v. 2.   A second
course.
   Includes indexes.
   1.   Finite element method   I.   Carey, Graham F.
II.   Oden, J. Tinsley.   III.   Title.
IV.   Series.
TA347.F5B4   515.3'53   80-25153
ISBN 0-13-317057-8 (v. 1)   AACR2

Editorial/production supervision by Lori Opre
Cover design by: Edsal Enterprises
Manufacturing buyer: Gordon Osbourne

©1983 by Graham F. Carey and J. Tinsley Oden

Printed in the United States of America

10 9 8 7 6 5 4 3 2 1

ISBN 0-13-317065-9

PRENTICE-HALL INTERNATIONAL, INC., *London*
PRENTICE-HALL OF AUSTRALIA PTY. LIMITED, *Sydney*
EDITORA PRENTICE-HALL DO BRASIL, LTDA., *Rio de Janeiro*
PRENTICE-HALL CANADA INC., *Toronto*
PRENTICE-HALL OF INDIA PRIVATE LIMITED, *New Delhi*
PRENTICE-HALL OF JAPAN, INC., *Tokyo*
PRENTICE-HALL OF SOUTHEAST ASIA PTE. LTD., *Singapore*
WHITEHALL BOOKS LIMITED, *Wellington, New Zealand*

CONTENTS

PREFACE

This volume, the second in our series on finite elements, should not be viewed as a continuation of Volume I. We assume that the reader has not only had a first course in finite elements of comparable level to that given in Volume I but also has acquired a certain mathematical maturity that will allow him or her to appreciate some of the finer points of the method. Still, this book is directed to the student of engineering or applied science. Emphasis is placed on concepts and methods, and although we do consider several theoretical issues, our mission is one of revealing intrinsic properties of the method and not of burdening the development with too many technical details. We aim at a second course in the subject, as the title indicates, in which we give a greater insight of how the method works and wherein we discuss a variety of special topics that can be treated only lightly, if at all, in a first course.

As the text has been designed to be essentially self-contained, some of the ideas introduced in Volume I are again encountered here as part of a more comprehensive treatment. The major mathematical results are stated, but detailed proofs are not included here. Some of these theoretical aspects of the subject are covered in Volume IV. Similarly, the technical details concerning implementation and the numerical performance of the method have been left to Volume III. The numerous exercises are designed to corroborate and extend the formulations in the text.

We owe special thanks to several colleagues who have read earlier versions of parts of this manuscript and made many helpful suggestions; in particular, we are grateful to Professors T.J.R. Hughes, D.S. Malkus, J.N. Reddy, and M. Stern. We also register a special note of thanks to Barbara Brandt, Amy

Dominguez, Linda McClelland, and Nancy Webster, who typed the manuscript.

G. F. CAREY

J. T. ODEN

Austin, Texas

FINITE ELEMENTS:
A Second Course

VOLUME II

1

SOME GENERAL PROPERTIES

OF FINITE ELEMENT METHODS

1.1 THE FINITE ELEMENT METHOD: A BRIEF REVIEW

The reader who has some familiarity with finite element methods will recall the following concepts underlying these methods:

Weak Formulation: A variational or weak statement of the boundary-value problem is assumed to be given. For example, a variational statement of a typical two-point boundary-value problem* consists of seeking a function u in a suitable class H of admissible functions such that

$$\int_0^1 (u'v' + uv)\,dx = \int_0^1 fv\,dx + 3v(1) \qquad \forall v \in H \qquad (1.1.1)$$

Here $u' = du/dx$, the domain of the solution is the open inverval $(0, 1) = \{x \mid 0 < x < 1\}$, f is a function given as part of the data, the symbol $\forall$ is read "for every," and $\in$ is read "belonging to."

If the solution u is sufficiently *regular* [e.g., if u has continuous derivatives of order 2 on $(0, 1)$], and if each admissible function $v \in H$ has the property that $v(0) = 0$, any solution u of (1.1.1) is also a solution of the "classical"

---

* This problem is similar to the model problem discussed in detail in Chapter 1 of Volume I. We shall discuss some broad generalizations later in this chapter, but here we confine our attention to linear elliptic two-point boundary-value problems.

two-point boundary-value problem

$$\left.\begin{array}{r} -u''(x) + u(x) = f(x), \qquad x \in (0, 1) \\ u(0) = 0 \\ u'(1) = 3 \end{array}\right\} \qquad (1.1.2)$$

Conversely, any solution of (1.1.2) is also a solution of (1.1.1).

Essential and Natural Boundary Conditions: Boundary conditions fall naturally into two categories: essential boundary conditions, which enter into the definition of the space H of admissible functions, and natural boundary conditions, which affect the actual form of the variational statement. In problem (1.1.1), the essential boundary condition is

$$u(0) = 0$$

whereas the natural boundary condition follows on integration by parts as

$$u'(1) = 3$$

The boundary conditions must be compatible with the differential equation in the sense that the operators involved satisfy a Green's formula. For instance, in the case of problem (1.1.1), we have

$$\int_0^1 (-u'' + u)v \, dx = \int_0^1 (u'v' + uv) \, dx - u'v \Big|_0^1 \qquad (1.1.3)$$

whenever u and v are sufficiently regular.

The Spaces of Admissible Functions: In a variational problem such as (1.1.1), the function u is said to belong to a class of *trial* functions $\tilde{H}$, whereas v is a member of a class of *test* functions H. Collectively, u and v are called admissible functions, and in this case, the classes of trial and test functions coincide.

The space H of test functions contains those functions that satisfy the homogeneous essential boundary conditions and are smooth enough for the integrals appearing in the variational problem to be well defined. Clearly, if we are given problem (1.1.1), the solution u and the test functions v need only be smooth enough that the squares of their first derivatives be integrable. Thus, in the case of problem (1.1.1), we can take

$$\tilde{H} = H = \left\{ v = v(x) \,\middle|\, \int_0^1 [(v')^2 + v^2] \, dx < \infty, \quad v(0) = 0 \right\} \qquad (1.1.4)$$

Solutions of the classical problem (1.1.2) must be more regular. However, if the solution of (1.1.1) is sufficiently regular, then, as noted earlier, it is also the solution of (1.1.2).

Galerkin Approximations: A Galerkin approximation of (1.1.1) is obtained by posing the problem on a finite-dimensional subspace H^h of the space H of admissible functions; that is, we seek $u_h \in H^h$ such that

$$\int_0^1 (u_h' v_h' + u_h v_h) \, dx = \int_0^1 f v_h \, dx + 3 v_h(1) \qquad \forall v_h \in H^h \qquad (1.1.5)$$

Here h is a real parameter related to the dimension of H^h. If $\phi_i, i = 1, 2, \ldots,$ N denotes a set of linearly independent basis functions spanning H^h, the approximate solution will be of the form

$$u_h(x) = \sum_{j=1}^N \alpha_j \phi_j(x), \qquad x \in [0, 1] \qquad (1.1.6)$$

The coefficients α_j are determined on substituting the expansion (1.1.6) in the problem statement (1.1.5) and solving the resulting linear algebraic system. Note that we identify a particular subspace H^h by simply specifying the basis functions ϕ_i.

Finite Element Approximations:* The finite element method provides a systematic method for constructing appropriate functions ϕ_i. The domain [0, 1] is first partitioned into subintervals, the finite elements. Over each element Ω_e, the solution is approximated as a linear combination of shape functions ψ_i^e, which are typically polynomials. The construction of the global basis functions ϕ_i can be viewed as a process of patching together element shape functions at the nodes. Hence, the global basis functions ϕ_i are nonzero only on the element or elements containing the node at x_i; that is, the basis functions have *local support*.

If we substitute the expansion (1.1.6) in (1.1.5) and select v_h to be the basis functions ϕ_i that span H^h, the resulting system is

$$\sum_{j=1}^N \left[\int_0^1 (\phi_i' \phi_j' + \phi_i \phi_j) \, dx \right] \alpha_j = \int_0^1 f \phi_i \, dx + 3 \phi_i(1) \qquad \forall \phi_i \in H^h \qquad (1.1.7)$$

Although the problem is stated in a variational form involving integrals over the entire domain, the fact that the function ϕ_i is nonzero only on those elements containing nodal point x_i in the finite element mesh makes it pos-

---

* For additional details, review the complete descriptions of the method outlined in Volume I, Chapters 2 and 4. We also furnish more particulars later in this chapter.

sible to generate the Galerkin approximation (1.1.7) by summing local contributions from each element in the mesh. This, of course, makes the method easy to implement on digital computers.

Other Remarks: Generally, the space H of admissible functions is endowed with a norm, often called the energy norm of the problem, which provides a natural measure of the quality of the approximation. In particular, we demand that the error in the energy norm converge to zero if the mesh is appropriately *refined* (i.e., the number of elements in the discretization of Ω increases while the maximum size of an element in successive meshes decreases). The quality of our approximation, as measured, for example, by the rate at which the method converges with respect to the energy norm, depends on several key factors:

1. The regularity (smoothness) of the solution;
2. The highest order of derivatives appearing in the definition of the energy;
3. The largest degree of complete polynomials in x contained in the space spanned by the shape functions ψ_i^e.

The procedure that is outlined above describes the finite element method only in a very broad sense. There are numerous variants of several of the steps mentioned, and each involves detailed considerations that affect the success or failure of the method.

First, the variational "principle" (1.1.1) is only one of many equivalent variational formulations that one could construct for the same boundary-value problem. For instance, if u and v are sufficiently regular, we might integrate the left side of (1.1.1) by parts and, with an appropriate change in the definition of the admissible functions H, obtain a different (but equivalent) statement of the problem. Another integration by parts would produce still another formulation. We could treat u' as a second dependent variable σ and augment (1.1.1) with the condition $\sigma = u'$, or we could treat the values of u or its derivatives at the element endpoints as additional unknowns. Each of these variational formulations provides the basis for different finite element methods for a single given problem, and each may have advantages or disadvantages compared with the standard finite element procedure. We will consider several of these alternative methods in Chapters 3 and 4.

Second, the selection of the basis functions ϕ_i (or, equivalently, the shape functions ψ_i^e) is by no means a trivial issue. The choice of the element to be used in a specific application involves considerations of the regularity of the solution, the form of the variational statement used, the rate of convergence desired, the ease of programming, and the accuracy one wishes to attain, weighed against the price one can pay in computational time and effort.

There are also many other purely computational issues of importance, such as the generation of the mesh, data management procedures in implementing the method on a computer, integration of the stiffness coefficients and the load vectors, element assembly, implementation of boundary conditions, the selection of numerical techniques to solve the resulting system of linear equations, the evaluation of the accuracy and quality of the solution, the display and synthesis of the output, and so on. We consider some of these computational aspects in Volume III of this series.

In the present volume, our principal concern is with formulative aspects of the method itself and of its principal variants. Beyond the standard finite element method outlined in the preceding section, what are the major alternatives? What are their advantages and disadvantages? What is the qualitative behavior of the approximate solution? How can the method be applied to other types of boundary-value problems such as eigenvalue problems? What choices of the shape functions ψ_i^e are available? What are their properties, and what criteria can we use to select appropriate elements for a problem of interest? These are the questions we address in the chapters that follow.

1.2 REGULARITY: THE SPACES $C^m(\Omega)$ AND $H^m(\Omega)$

If there is a single property of solutions of boundary-value problems that is of overriding importance in studying finite element methods, it is *regularity*, that is, the "degree of smoothness or lack of it" of the solution and its "companions" in the space of admissible functions. In our discussion in the preceding section, we have already encountered this term several times in our lists of important considerations. Much of the remainder of this chapter is devoted to the concept of regularity and to the establishment of some minimal mathematical techniques and properties that will allow us to use the concept intelligently in studying virtually all finite element methods.

The description of a given function as being "regular" or "irregular" is too vague and qualitative to be of much value. We need to *quantify*, in some sense, the notion of the "degree of regularity" of a function. There is a natural and elegant way of doing this. It is universally used and we have already had a hint of it in Volume I: *We identify the continuity class $C^m(\Omega)$ or the Sobolev class $H^m(\Omega)$ to which the function to be approximated belongs.*

1.2.1 The Space $C^m(\Omega)$

Suppose that Ω is a bounded region in $\mathbb{R}^3$, the three-dimensional Euclidean space, and that $u = u(x, y, z)$ is a given real-valued function of position in Ω. Then *u is said to be of class C^m on Ω* [or to belong to $C^m(\Omega)$, or " $\in C^m(\Omega)$," or, simply, to be a C^m-function] *if u and all of its partial derivatives of order less than or equal to m are continuous at every point* (x, y, z) *in* Ω, where m is a nonnegative integer. We often use the following type of notation as a definition of $C^m(\Omega)$:

$$C^m(\Omega) = \left\{ v = v(x, y, z), (x, y, z) \in \Omega \,\middle|\, v, \frac{\partial v}{\partial x}, \frac{\partial v}{\partial y}, \frac{\partial v}{\partial z}, \frac{\partial^2 v}{\partial x^2}, \dots, \right.$$

$$\left. \frac{\partial^m v}{\partial x^m}, \frac{\partial^m v}{\partial x^{m-1} \partial y}, \dots, \frac{\partial^m v}{\partial z^m} \text{ are continuous in } \Omega \subset \mathbb{R}^3 \right\} \tag{1.2.1}$$

which is read "$C^m(\Omega)$ is a set of functions, a typical member v of which has the property that $v, \partial v/\partial x, \dots, \partial^m v/\partial z^m$ are continuous in a region Ω contained in three-dimensional Euclidean space $\mathbb{R}^3$." A more compact notation would be

$$C^m(\Omega) = \left\{ v \,\middle|\, \frac{\partial^r v}{\partial x^i \, \partial y^j \, \partial z^k} \text{ is continuous in } \Omega \subset \mathbb{R}^3 \,;\, i, j, k \geq 0, \right.$$

$$\left. i + j + k = r, \quad r \leq m \right\} \tag{1.2.2}$$

The class $C^m(\Omega)$ is, in fact, a *linear space* of functions; in other words, if $u \in C^m(\Omega)$ and $v \in C^m(\Omega)$, then $\alpha u + \beta v \in C^m(\Omega)$ for any real scalars α and β.

Consider, for example, the function u defined on the one-dimensional domain $[0, a] = \{x \in \mathbb{R} \mid 0 \leq x \leq a\}$ in Fig. 1.1. Let $v(x) = u'(x)$ and $w(x) = v'(x) = u''(x)$. Clearly, w suffers a jump discontinuity at the point x_0. Hence, u cannot belong to $C^2([0, a])$. However, v is continuous, so u is a C^1-function. We write $u \in C^1([0, a])$, $v \in C^0([0, a])$. Similarly, the function f in Fig. 1.2 is a C^0-function but not a C^1-function because $\partial f/\partial x$ has a discontinuity along the line $x = x_0$. If u is infinitely differentiable on Ω, we write $u \in C^\infty(\Omega)$.

We sometimes wish to include boundary conditions in our definitions of classes of functions. A typical example is the space $C_0^m(\Omega)$, which is defined as the subspace of $C^m(\Omega)$ consisting of those C^m-functions u such that u and its normal derivatives $\partial u/\partial n, \partial^2 u/\partial n^2, \dots, \partial^{m-1} u/\partial n^{m-1}$ vanish on the boundary $\partial \Omega$ of Ω, n being the direction of the unit normal to $\partial \Omega$.

1.2.2 L^2-Functions

Now we are well aware that in the finite element approximation of variational boundary-value problems, we often deal with classes of functions with derivatives defined in only a "square-integrable" sense. These are more abstract classes than the C^m-functions. To see this, let f and g be two functions

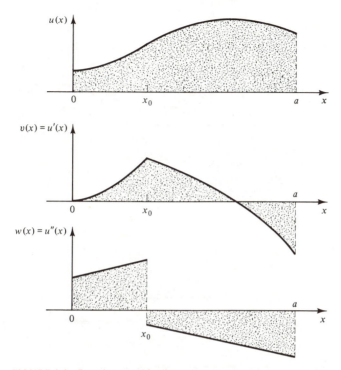

FIGURE 1.1 *Function u with discontinuous second derivative but continuous first derivative.*

that differ at a finite number or countably infinite number of points. Even though these functions differ at these points ("on a set of measure zero"), the integrals of f^2 and g^2 are identical numerical values. For instance, if

$$f(x) = x \quad \text{and} \quad g(x) = \begin{cases} x & \text{for } x \text{ irrational} \\ \sqrt{3} & \text{for } x \text{ rational} \end{cases} \qquad (1.2.3)$$

and $0 \le x \le 1$, then*

$$\int_0^1 f^2 \, dx = \int_0^1 g^2 \, dx \qquad (1.2.4)$$

even though $f(x) \ne g(x)$ at every x rational. Of particular importance here is the observation that functions such as w in Fig. 1.1 having simple jump discontinuities, although not C^0-functions, are still square integrable. Functions such as f in Fig. 1.2 that have square-integrable first derivatives occur frequently in the finite element formulation of boundary-value problems.

---

* Throughout this volume, all integrals are to be interpreted in the Lebesgue sense.

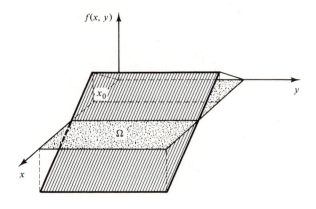

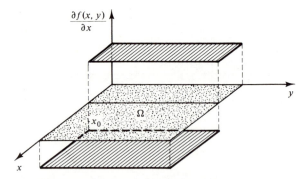

FIGURE 1.2 *C^0-function f defined on a two-dimensional domain Ω.*

We classify such functions by introducing a set, denoted $L^2(\Omega)$ (read "L-two omega"), which consists of "equivalence classes" $[f]$ of functions defined on Ω which are square-integrable over Ω. In other words, a function f in $L^2(\Omega)$ will have the property that $\int_{\Omega} f^2 \, dx < \infty$. Two functions f and g in Ω are *equivalent* (belong to the same equivalence class $[f]$ or $[g]$) if their values identically coincide at all points in Ω except possibly on sets of points of zero measure (such as on a countable set of points as mentioned earlier). We do not (and often *cannot*) distinguish between two equivalent functions f and g in $L^2(\Omega)$ since the distance between them (in an "L^2-sense") is zero: $(\int_{\Omega} |f - g|^2 \, dx)^{1/2} = 0$. Thus, if a given function $f \in L^2(\Omega)$, we know that $\int_{\Omega} f^2 \, dx < +\infty$, and that f is a representative of a class of functions $[f]$ which differ from f pointwise only on sets of zero measure and which have the property that any $g \in [f]$ will be such that $\int_{\Omega} g^2 \, dx = \int_{\Omega} f^2 \, dx$ (and, indeed, $\int_{\Omega} |f - g|^2 \, dx = 0$).

Why are we interested in square-integrable functions? The answer is simple: In most of the finite element methods we consider, we are concerned with approximations of variational boundary-value problems of the type: find u such that

$$\int_\Omega (\nabla u \cdot \nabla v + uv)\, dx = \int_\Omega fv\, dx \qquad (1.2.5)$$

for all v in some class of admissible functions H. This is equivalent, in some sense, to the problem: find u satisfying

$$-\Delta u + u = f \quad \text{in} \quad \Omega$$

and subject to boundary conditions determined by H and the functional in (1.2.5). If the solution u is also to be an "admissible function," (1.2.5). should also hold for the choice $v = u$. This means that u (and, in fact, all members of this class of admissible functions) must be such that

$$\int_\Omega (\nabla u \cdot \nabla u + u^2)\, dx < \infty \qquad (1.2.6)$$

In other words, u (and, in this case, its first partial derivatives in $\nabla u \cdot \nabla u = u_x^2 + u_y^2 + u_z^2$) must be square-integrable (or, technically speaking, in an equivalence class of square-integrable functions).

1.2.3 The Space $H^m(\Omega)$

If the reader is convinced of the importance of the class $L^2(\Omega)$ of functions in finite element methods, he should have little difficulty in accepting this class as an alternative and more general means for measuring the smoothness of functions than the C^m-classes discussed previously. For example, the function

$$v(x) = \begin{cases} x, & 0 \le x \le 1, \quad x \text{ irrational} \\ \sqrt{2}, & 0 \le x \le 1, \quad x \text{ rational} \end{cases}$$

is a member of $L^2(0, 1)$, but it does not belong to any class of C^m-functions, $m \ge 0$. Similarly, the continuous function

$$v(x) = \begin{cases} x, & 0 \le x \le 1 \\ 2 - x, & 1 \le x \le 2 \end{cases}$$

is not differentiable at $x = 1$, but its derivative $v'(x)$ is a well-defined member of $L^2(\Omega)$.

These observations lead us to a natural generalization of the C^m-classes as a device for quantifying the smoothness or regularity of functions—the

Sobolev classes $H^m(\Omega)$. A function v is in the class $H^m(\Omega)$ if v and all of its partial derivatives* of order less than or equal to m (m being a nonnegative integer) are members of the class $L^2(\Omega)$. Compactly, we write for $\Omega \subset \mathbb{R}^2$,

$$H^m(\Omega) = \left\{ v \,\middle|\, v, \frac{\partial v}{\partial x}, \frac{\partial v}{\partial y}, \ldots, \frac{\partial^m v}{\partial z^m} \in L^2(\Omega) \right\} \tag{1.2.7}$$

The class $H^m(\Omega)$ is a linear space of functions: If $u \in H^m(\Omega)$ and $v \in H^m(\Omega)$, then $\alpha u + \beta v \in H^m(\Omega)$ for any real scalars α and β. In fact, $H^m(\Omega)$ is much more than a linear space. We shall note below (but not use) the fact that $H^m(\Omega)$ is also a Hilbert space.

The function u in Fig. 1.1 provides a simple example. Clearly, $u \notin C^2([0, a])$ since u'' is discontinuous at $x = x_0$. However, u'' is certainly square-integrable: $\int_0^a (u'')^2 \, dx$ is merely the area under the curve $y = [u''(x)]^2$. Thus,

$$u \in H^2(0, a)$$

Similarly, $v' \in H^1(0, a)$ and $w = u'' \in H^0(0, a)$. Thus, it makes sense to use the notation

$$H^0(0, a) = L^2(0, a)$$

The function u satisfying (1.2.6) is clearly in $H^1(\Omega)$. As a further example, the function u shown in Fig. 1.3 is a member of $H^3(0, 1)$ since its third derivative is square-integrable, but its fourth derivative w (represented by Dirac deltas) is not square-integrable $\left(\int_0^1 w^2 \, dx \text{ is undefined} \right)$. Since the "antiderivative" of w, $\int_0^x w \, dx$, is the step function u''', which *is* in $L^2(0, 1)$, we often use the notation $w \in H^{-1}(0, 1)$. We discuss these "negative" Sobolev spaces in more detail later.

We often include boundary conditions in our definition of Sobolev spaces just as in the C^m-spaces earlier. For instance, we define the space $H_0^m(\Omega)$ as the space of H^m-functions with the property that†

$$u = 0, \quad \frac{\partial u}{\partial n} = 0, \quad \frac{\partial^2 u}{\partial n^2} = 0, \ldots, \quad \frac{\partial^{m-1} u}{\partial n^{m-1}} = 0$$

on the boundary $\partial \Omega$ of Ω, n being the direction of the unit outward normal

---

* Strictly speaking, these derivatives should be interpreted in a "weak" or "generalized" sense and, as noted earlier, the members of $H^m(\Omega)$ are equivalence classes of functions. See Chapter 1 of Volume IV.

† These normal derivatives, of course, must also be interpreted in some appropriate weak sense. This is made precise by the so-called "trace theorem" for Sobolev spaces. See Chapter 1, Volume IV, for more details.

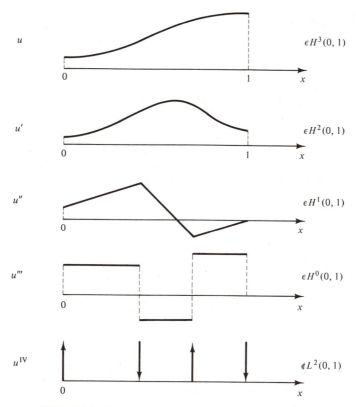

u $\epsilon H^3(0, 1)$

u' $\epsilon H^2(0, 1)$

u'' $\epsilon H^1(0, 1)$

u''' $\epsilon H^0(0, 1)$

u^{IV} $\notin L^2(0, 1)$

FIGURE 1.3 *Regularity of a function u and its derivatives.*

to $\partial\Omega$. Clearly, $H_0^m(\Omega)$ is a subspace of $H^m(\Omega)$. Occasionally, we also employ spaces of the type

$$H = \{v \in H^1(0, 1) \mid v = 0 \quad \text{at} \quad x = 0\}$$

Then v is in $H^1(0, 1)$ but vanishes at $x = 0$ and not necessarily at $x = 1$.

With these classes of functions now explicitly defined, we are able to make a concrete statement of various variational boundary-value problems. For instance, consider as an example the following problem: find $u \in H_0^m(0, 1)$ such that

$$\int_0^1 \frac{d^m u}{dx^m} \frac{d^m v}{dx^m} \, dx = \int_0^1 fv \, dx \qquad \forall v \in H_0^m(0, 1) \tag{1.2.8}$$

It is clear that the space H of admissible functions on $\Omega = (0, 1)$ is precisely the Sobolev space $H_0^m(\Omega)$. The test functions v are also in this class, and the boundary conditions are automatically incorporated in the definition of the problem. Indeed, $u \in H_0^m(\Omega)$ means that

$$u(x) = 0, \quad \frac{du(x)}{dx} = 0, \quad \frac{d^2u(x)}{dx^2} = 0, \quad \ldots, \quad \frac{d^{m-1}u(x)}{dx^{m-1}} = 0$$

at $x = 0$ and $x = 1$.

Similarly, a variational statement of the problem

$$-u'' + u = f, \quad 0 < x < 1$$
$$u(0) = 0, \quad u(1) = 1 \qquad (1.2.9)$$

is then: find $u \in H^1(0, 1)$ satisfying the prescribed end conditions and such that

$$\int_0^1 (u'v' + uv) \, dx = \int_0^1 fv \, dx \qquad \text{for all } v \in H_0^1(0, 1) \qquad (1.2.10)$$

whereas a variational statement of

$$-u'' + u = 0, \quad 0 < x < 1$$
$$u'(0) = 3, \quad u(1) = 0 \qquad (1.2.11)$$

is: find $u \in \tilde{H}$ such that

$$\int_0^1 (u'v' + uv) \, dx = -3v(0) \qquad \text{for all } v \in \tilde{H} \qquad (1.2.12)$$

where

$$\tilde{H} = \{v \in H^1(0, 1) \,|\, v(1) = 0\} \qquad (1.2.13)$$

1.2.4 Major Properties of $H^m(\Omega)$

The spaces $H^m(\Omega)$ have a rich structure that allows us to list some of the properties of finite element methods outlined in Section 1.1 with mathematical precision. We record their most important properties:

Linearity: $H^m(\Omega)$ is a linear space (a fact already pointed out in the preceding section). This means that, algebraically, the functions in $H^m(\Omega)$ can be treated like vectors. If u and v are in $H^m(\Omega)$, their "vector" sum $u + v$, their scalar multiples αu and βv, and, therefore, their linear combinations $\alpha u + \beta v$ are also in $H^m(\Omega)$.

Orthogonality: An inner product can be defined on $H^m(\Omega)$; for example, if

$$H^m(0, 1) = \left\{v = v(x), x \in (0, 1) \,|\, v, \frac{dv}{dx}, \frac{d^2v}{dx^2}, \ldots, \frac{d^mv}{dx^m} \in L^2(0, 1)\right\}$$

then the scalar

$$(u, v)_m \equiv \int_0^1 \left(\frac{d^m u}{dx^m} \frac{d^m v}{dx^m} + \frac{d^{m-1} u}{dx^{m-1}} \frac{d^{m-1} v}{dx^{m-1}} + \cdots + \frac{du}{dx} \frac{dv}{dx} + uv \right) dx \qquad (1.2.14)$$

defines an inner product on $H^m(0, 1)$. This means that, for any u, v, w in $H^m(0, 1)$, and any real numbers α and β,

$$\left. \begin{aligned}
(\alpha u + \beta v, w)_m &= \alpha(u, w)_m + \beta(v, w)_m \\
(u, v)_m &= (v, u)_m \\
(u, u)_m &\geq 0, \qquad \text{and} \\
(u, u)_m &= 0 \qquad \text{if and only if} \\
& \qquad\qquad u = 0
\end{aligned} \right\} \qquad (1.2.15)$$

Thus, we can speak of *orthogonality* of functions u and v in $H^m(0, 1)$:

$(u, v)_m = 0$ implies that u and v are "orthogonal in $H^m(\Omega)$."

Magnitude: $H^m(\Omega)$ is a *normed space*, the norm of a function $u \in H^m(\Omega)$ being defined as the nonnegative real number $\| u \|_m$ given by

$$\| u \|_m = \sqrt{(u, u)_m} \qquad (1.2.16)$$

The norm is a measure of the *magnitude* of the function and has the properties that for any $u, v \in H^m(\Omega)$, and any real number α,

$$\left. \begin{aligned}
\| u + v \|_m &\leq \| u \|_m + \| v \|_m \\
\| \alpha u \|_m &= | \alpha | \| u \|_m \\
\| u \|_m &\geq 0, \qquad \text{and} \\
\| u \|_m &= 0 \qquad \text{if and only if} \\
& \qquad\qquad u = 0
\end{aligned} \right\} \qquad (1.2.17)$$

For the space $H^m(0, 1)$,

$$\| u \|_m = \left\{ \int_0^1 \left[\left(\frac{d^m u}{dx^m} \right)^2 + \left(\frac{d^{m-1} u}{dx^{m-1}} \right)^2 + \cdots + \left(\frac{du}{dx} \right)^2 + u^2 \right] dx \right\}^{1/2} \qquad (1.2.18)$$

We note that the existence of an inner product on $H^m(\Omega)$ adds "geometric structure" to these spaces. For example, in the same way that $\mathbf{a} \cdot \mathbf{b} = | \mathbf{a} | | \mathbf{b} | \cos \theta$ for two vectors in $\mathbb{R}^3$ oriented at angle θ relative to one another, we can write

$$(u, v)_m = \| u \|_m \| v \|_m \cos \theta$$

where θ is now an abstract "angle between the functions u and v in $H^m(\Omega)$." In particular, we will always have the Schwarz inequality,

$$|(u, v)_m| \leq \|u\|_m \|v\|_m \qquad (1.2.19)$$

for all $u, v \in H^m(\Omega)$.

Distance: The *distance between functions* in $H^m(\Omega)$ is defined by the H^m-norm of their difference. Thus, the distance between u and v is $\|u - v\|_m$.

Duality: Suppose that **u** is a displacement vector in a three-dimensional Euclidean space. If $(\mathbf{i}_1, \mathbf{i}_2, \mathbf{i}_3)$ are mutually orthogonal unit vectors, then **u** has the form

$$\mathbf{u} = u_1 \mathbf{i}_1 + u_2 \mathbf{i}_2 + u_3 \mathbf{i}_3$$

Let **f** be a constant force vector,

$$\mathbf{f} = f_1 \mathbf{i}_1 + f_2 \mathbf{i}_2 + f_3 \mathbf{i}_3$$

Then the work done by **f** on **u** is the real number

$$\mathbf{f} \cdot \mathbf{u} = f_1 u_1 + f_2 u_2 + f_3 u_3 \qquad (1.2.20)$$

If **u** is an element of the three-dimensional space **H**, the operation (1.2.20) is said to define a *linear functional* on **H**; it depends linearly on **f**, linearly on **u**, and its value is a real number. If displacements belong to **H**, forces belong to a similar space **H′** called the *dual* of **H**; **H′** is defined as the space of all (continuous) linear functionals on **H**.

Analogously, the space of all continuous linear functionals on $H^m(\Omega)$ is called the dual of $H^m(\Omega)$. While the functions in $H^m(\Omega)$ may be very smooth, the elements of $H^m(\Omega)'$ may be very irregular; indeed, they may not even be functions! For example, if $v \in H_0^1(-1, 1)$, the Dirac delta, δ, produces a continuous linear functional on $H_0^1(-1, 1)$ by the rule

$$\delta(v) = v(0)$$

Hence, $\delta \in (H_0^1(-1, 1))'$. Since the antiderivative of δ is a step function, which is clearly in $L^2(-1, 1)$, δ belongs to a class whose (-1)-derivative (i.e., its integral) is an L^2-function. We express this symbolically by writing $\delta \in H^{-1}(-1, 1)$.

Carrying these observations a step further, we define the *negative spaces* $H^{-m}(\Omega)$ $(m \geq 0)$ as duals of the spaces $H_0^m(\Omega)$:

$$H^{-m}(\Omega) = (H_0^m(\Omega))' \qquad (1.2.21)$$

Additional properties of these spaces are covered in the exercises that follow.

EXERCISES

1.2.1 Specify the smoothness of the following functions u by indicating the classes $C^m(\Omega)$ and $H^m(\Omega)$ to which they belong:

(a) $u''(x) = \begin{cases} 0, & 0 < x < 1 \\ x - 1, & 1 \leq x < 2 \\ x^2 - 3, & 2 \leq x < 3 \end{cases}$

(b) $u(x) = 1/\sqrt{x}, \quad 0 \leq x \leq 1$

(c) $u'(x) = \begin{cases} x - \frac{1}{4}, & 0 \leq x \leq 1 \\ \frac{1}{4} - x, & 1 \leq x \leq 2 \end{cases}$

(d) $\dfrac{\partial u(x, y)}{\partial y} = \begin{cases} xy, & 0 \leq x \leq 1, \quad 0 \leq y \leq 1 \\ x(2 - y), & 0 \leq x \leq 1, \quad 1 \leq y \leq 2 \\ (2 - x)y, & 1 \leq x \leq 2, \quad 0 \leq y \leq 1 \\ (2 - x)(2 - y), & 1 \leq x \leq 2, \quad 1 \leq y \leq 2 \end{cases}$

$\dfrac{\partial u(x, y)}{\partial x} = 3x^2, \quad 0 \leq x \leq 2, \quad 0 \leq y \leq 2$

(e) $u(x, y) = \begin{cases} xy, & 0 \leq x \leq 1, \quad 0 \leq y < 1 \\ x(2 - y), & 0 \leq x \leq 1, \quad 1 < y \leq 2 \\ x, & 0 \leq x < \frac{1}{2}, \quad y = 1 \\ 3\pi, & x = \frac{1}{2}, \quad y = 1 \\ x, & \frac{1}{2} < x \leq 1, \quad y = 1 \end{cases}$

(f) $u''(x) = \begin{cases} x^2, & 0 < x < 1, \quad x \text{ irrational} \\ 100, & 0 < x < 1, \quad x \text{ rational} \end{cases}$

1.2.2 (a) Show that the functions

$$f(x) = \begin{cases} x, & 0 \leq x \leq 1 \\ 2 - x, & 1 \leq x \leq 2 \end{cases} \qquad g(x) = \sin \pi x, \quad 0 \leq x \leq 2$$

are orthogonal in $H^0(0, 2)$. Are they orthogonal in $H^1(0, 2)$?

(b) What is the distance between f and g in $H^0(0, 2)$? In $H^1(0, 2)$? Is it possible to compute the distance between these functions in $H^2(0, 2)$? Why?

1.2.3 Given $f(x) = x^3$, $g(x) = 1 + x + \frac{1}{4}x^4$, $0 \leq x \leq 1$:

(a) Compute the magnitude of f and g in $H^m(0, 1)$ for $m = 0, 1, 2$. What conclusions do you reach concerning the magnitudes of these functions in Sobolev spaces? Can you generalize these observations?

(b) Compute the angle between f and g in $H^m(0, 1)$ for $m = 0, 1, 2$. Verify that Schwarz's inequality holds.

(c) If $h(x) = x^5/5$, $0 \leq x \leq 1$, verify the triangle inequality $\|f - g\|_m \leq \|f - h\|_m + \|h - g\|_m$ for $m = 0, 1, 2$.

1.2.4 Let $v \in H_0^1(-1, 1)$ and recall that the Dirac δ is in $(H_0^1(-1, 1))' = H^{-1}(-1, 1)$, where

$$\delta(v) = v(0)$$

The "dipole" functional δ' is defined by

$$\delta'(v) = -v'(0)$$

for sufficiently smooth v.

(a) Give arguments that support the assertion that $\delta' \in H^{-2}(-1, 1)$.

(b) Using the ideas underlying part (a), give an example of an element of $H^{-3}(-1, 1)$; then give an example for $H^{-m}(-1, 1)$.

(c) Discuss the interpretation of a function u such that $d^3u/dx^3 \in H^{-1}(-1, 1)$. In what space does u lie?

1.2.5 If $u \in H_0^1(-1, 1)$, show that $(-u'' + u) \in H^{-1}(-1, 1)$. In particular, if we denote $Au \equiv -u'' + u$, show that $(Au, Av)_{-1} = (u, v)_1$ is an inner product on $H^{-1}(-1, 1)$.

1.3 THE FINITE ELEMENT METHOD IN $H^m(\Omega)$

Let us reexamine the properties of finite element methods listed in Section 1.1, now taking advantage of the powerful concepts presented in the preceding section.

Weak Formulations: Boundary-value problems to be solved by the finite element method are set in a weak or variational form typically of the following type:* find a real-valued function $u = u(x_1, x_2, \ldots, x_N) = u(\mathbf{x})$ in a class H of admissible functions such that

$$B(u, v) = \int_\Omega fv \, dx + \int_{\partial\Omega_2} gMv \, ds \qquad \forall v \in H \qquad (1.3.1)$$

where $B(\cdot, \cdot)$ is a bilinear form representing an integral of the type

$$B(u, v) = \int_\Omega \left(a_{mm}^{11} \frac{\partial^m u}{\partial x_1^m} \frac{\partial^m v}{\partial x_1^m} + a_{mm-1}^{11} \frac{\partial^m u}{\partial x_1^m} \frac{\partial^{m-1} v}{\partial x_1^{m-1}} + \cdots \right.$$
$$\left. + a_{mm}^{NN} \frac{\partial^m u}{\partial x_N^m} \frac{\partial^m v}{\partial x_N^m} + \cdots + a_{00}^{NN} uv \right) dx \qquad (1.3.2)$$

and M is a linear partial-differential operator of order $m - 1$. In (1.3.1) and (1.3.2), Ω is an open bounded domain in N-dimensional Euclidean space

---

* We confine our attention here to regular linear elliptic problems.

$\mathbb{R}^N$ (typically, $N = 1, 2$, or 3), $dx = dx_1\, dx_2 \cdots dx_N$ denotes an element of volume, $(x_1, x_2, \ldots, x_N)$ being the set of Cartesian coordinates of points $\mathbf{x}$ in $\mathbb{R}^N$, ds is an element of the boundary $\partial\Omega$ of Ω, and f and g are given data.

The quantity $B(u, v)$ is sometimes referred to as the *virtual work* of the problem, in view of its correspondence to the virtual work in classical mechanics. In (1.3.2), the quantities a_{kl}^{ij} are given functions of position $\mathbf{x}$ in Ω. In the general class of problems considered here, $B(u, v)$ contains all possible combinations of products of derivatives of u with respect to $x_1, x_2, \ldots,$ x_N of order m and less, with derivatives of v of order m and less. Thus, products of derivatives of order m appear in the statement of the problem and, for sufficiently regular coefficients a_{kl}^{ij}, we must have

$$H \subset H^m(\Omega)$$

As will be seen below, in this case the boundary operator M will involve linear combinations of partial differential operators of order $m - 1$ and less, evaluated on the boundary. We give some concrete examples later.

In (1.3.1), it is understood that the boundary $\partial\Omega$ of Ω is naturally divided into two parts, $\partial\Omega_1$ and $\partial\Omega_2$ ($\partial\Omega = \overline{\partial\Omega_1} \cup \overline{\partial\Omega_2}$), on which two distinct types of boundary conditions are respectively applied. On $\partial\Omega_1$, we assume that homogeneous *essential* boundary conditions are imposed of the form

$$Mu = 0 \quad \text{on} \quad \partial\Omega_1 \tag{1.3.3}$$

On $\partial\Omega_2$, *natural* boundary conditions of the type

$$Nu = g \quad \text{on} \quad \partial\Omega_2 \tag{1.3.4}$$

are to hold. For sufficiently smooth u and v, the boundary operators M and N are related to one another and to $B(\cdot, \cdot)$ through a Green's formula of the type

$$B(u, v) = \int_\Omega (Au)v\, dx + \oint_{\partial\Omega} Nu\, Mv\, ds \tag{1.3.5}$$

where A is a linear elliptic partial-differential operator of order $2m$. Then, *formally*, problem (1.3.1) corresponds to the "classical" boundary-value problem

$$\left.\begin{array}{l} Au = f \quad \text{in} \quad \Omega \\ Mu = 0 \quad \text{on} \quad \partial\Omega_1 \\ Nu = g \quad \text{on} \quad \partial\Omega_2 \end{array}\right\} \tag{1.3.6}$$

Clearly (see Exercise 1.3.1), any solution of (1.3.6) is also a solution of the given variational problem (1.3.1); moreover, any sufficiently smooth solution of (1.3.1) will satisfy (1.3.6). However, we again emphasize that the formulation (1.3.1) is considerably more general than (1.3.6), since its solution need only be such that $\partial^m u/\partial x_i^m \, \partial^m v/\partial x_i^m$, $i = 1, 2, \ldots, N$, are integrable, whereas the solution of (1.3.6) satisfies $Au = f$ at every point in Ω.

Essential and Natural Boundary Conditions: We are concerned primarily* with weak formulations of boundary-value problems in which the same smoothness requirements are imposed on the test function v in (1.3.1) as the trial function u; hence, *both $\partial^m u/\partial x^m$ and $\partial^m v/\partial x^m$ appear in the statement of this problem.* The operator A defining formally the classical boundary-value problem (1.3.6), is of order $2m$. For example, if Ω is the one-dimensional domain, consisting of the open unit interval, $\Omega = (0, 1) = \{x \in \mathbb{R} \mid 0 < x < 1\}$ and we are given

$$B(u, v) = \int_0^1 a \frac{d^m u}{dx^m} \frac{d^m v}{dx^m} dx \qquad (1.3.7)$$

where $a = a(x) > 0$ is a given smooth function, then

$$Au(x) = (-1)^m \frac{d^m}{dx^m} \left[a(x) \frac{d^m u(x)}{dx^m} \right] \qquad (1.3.8)$$

This can be seen by integrating $B(u, v)$ by parts m times to obtain the Green's formula,

$$\int_0^1 (-1)^m \frac{d^m}{dx^m} \left(a \frac{d^m u}{dx^m} \right) v \, dx = \sum_{r=0}^{m-1} (-1)^{m+r} \left[\frac{d^r v}{dx^r} \frac{d^{m-1-r}}{dx^{m-1-r}} \left(a \frac{d^m u}{dx^m} \right) \right] \Big|_0^1 \\ + \int_0^1 a \frac{d^m u}{dx^m} \frac{d^m v}{dx^m} dx \qquad (1.3.9)$$

The boundary operators M and N must be chosen so that $-Nu \, Mv$ produces the sum on the right-hand side of (1.3.9).

One possible variational boundary-value problem for this type of operator is: find a function u among a class H of admissible functions such that

$$\int_0^1 a \frac{d^m u}{dx^m} \frac{d^m v}{dx^m} dx = \int_0^1 fv \, dx + \sum_{r=0}^{m-1} (-1)^{m+r+1} g_r \frac{d^r v(1)}{dx^r} \quad \forall v \in H \quad (1.3.10)$$

where the g_r, $r = 0, 1, \ldots, m - 1$, are given constants. Clearly, an appropriate space H is then

$$H = \left\{ v \in H^m(0, 1) \Big| \frac{d^r v}{dx^r}(0) = 0, \quad r = 0, 1, 2, \ldots, m - 1 \right\} \quad (1.3.11)$$

---

* Some exceptions are considered in Chapters 3 and 4.

Formally, (1.3.10) represents the boundary-value problem

$$(-1)^m \frac{d^m}{dx^m}\left[a(x)\frac{d^m u(x)}{dx^m}\right] = f(x) \quad \text{in} \quad (0,1)$$

$$\frac{d^r u(0)}{dx^r} = 0, \qquad r = 0, 1, \ldots, m-1 \qquad (1.3.12)$$

$$\frac{d^{m-1-r}}{dx^{m-1-r}}\left[a(1)\frac{d^m u(1)}{dx^m}\right] = g_r, \qquad r = 0, 1, \ldots, m-1$$

At this point, three fundamental observations can be made:

1. As pointed out in (1.3.11), the space H of admissible functions must be contained in $H^m(0,1)$ since the variational problem (1.3.10) makes sense for functions u and v just smooth enough for $(d^m u/dx^m)^2$ and $(d^m v/dx^m)^2$ to be integrable on $\Omega = (0,1)$.

2. A total of m boundary conditions are applied at each end of the interval. Since functions in $H^m(0,1)$ have derivatives of order m which are only L^2-functions and, therefore, cannot necessarily be defined point by point, *it may be impossible to impose conditions on the mth derivatives of our admissible functions*. Thus, only conditions on derivatives of order $0, 1, 2, \ldots, m-1$ make sense in $H^m(0,1)$.

3. The remaining boundary conditions, those involving derivatives of order $m, m+1, \ldots, 2m-1$ of u, enter problem (1.3.10) through the actual form of the statement of the problem; the data g_r in these conditions appear in the right-hand side of (1.3.10).

Thus, boundary conditions in a linear elliptic boundary-value problem of order $2m$ fall into two categories:

1. *Essential boundary conditions* involve conditions on derivatives of order $0, 1, \ldots, m-1$, and are used to define the space H of admissible functions [as a subspace of $H^m(\Omega)$].

2. *Natural boundary conditions* involve conditions on derivatives of order $m, m+1, \ldots, 2m-1$ and enter in the statement of the variational boundary-value problem.

The ideas described here for problem (1.3.10) carry over without alteration to problems of any order in any number of dimensions.

The Space of Admissible Functions: The structure of the space H of admissible functions is now very clear. It must contain those functions that satisfy the homogenous essential boundary conditions and are smooth enough for the variational problem (1.3.1) to make sense. In particular, we can define H as the class of functions

$$H = \{v = v(\mathbf{x}) \,|\, B(v, v) < \infty \quad \text{and} \quad Mv = 0 \quad \text{on} \quad \partial\Omega_1\} \quad (1.3.13)$$

and since $B(u, v)$ contains products of derivatives of order m,

$$H = \{v \in H^m(\Omega) \,|\, Mv = 0 \quad \text{on} \quad \partial\Omega_1\} \quad (1.3.14)$$

where M cannot involve derivatives of order m or higher: for example,

$$Mv = \left\{v, \frac{\partial v}{\partial n}, \ldots, \frac{\partial^{m-1} v}{\partial n^{m-1}}\right\} \quad \text{on} \quad \partial\Omega_1 \quad (1.3.15)$$

Before continuing with our discussion of the underlying concepts of conventional finite element methods, we shall present a few examples illustrating the ideas covered thus far.

Example 1.3.1

Consider a two-dimensional domain $\Omega \subset \mathbb{R}^2$ and recall that sufficiently smooth functions defined on Ω satisfy a Green's formula of the type

$$\int_\Omega (-\Delta u) v \, dx = \int_\Omega \nabla u \cdot \nabla v \, dx - \int_{\partial\Omega} \frac{\partial u}{\partial n} v \, ds \quad (1.3.16)$$

where Δ is the Laplacian operator,

$$\Delta u = \text{div}\,(\text{grad}\,u) = \nabla \cdot \nabla u = \frac{\partial^2 u}{\partial x_1^2} + \frac{\partial^2 u}{\partial x_2^2} \quad (1.3.17)$$

and

$$\nabla u \cdot \nabla v = \frac{\partial u}{\partial x_1}\frac{\partial v}{\partial x_1} + \frac{\partial u}{\partial x_2}\frac{\partial v}{\partial x_2} \quad (1.3.18)$$

We have denoted $dx = dx_1\, dx_2$, and $\partial u/\partial n$ is the derivative of u on $\partial\Omega$ in the direction of a unit vector $\mathbf{n}$ exterior and normal to the boundary. From (1.3.5), we have

$$\left.\begin{array}{c} A = -\Delta = -\text{div}\,(\text{grad}), \\[6pt] Nu = \dfrac{\partial u}{\partial n}, \qquad Mv = v \quad (\text{on } \partial\Omega) \\[6pt] B(u, v) = \displaystyle\int_\Omega \nabla u \cdot \nabla v \, dx \end{array}\right\} \quad (1.3.19)$$

and the variational problem assumes the form: find $u \in H$ satisfying

$$\int_\Omega \nabla u \cdot \nabla v \, dx = \int_\Omega fv \, dx + \int_{\partial\Omega_2} gv \, ds \qquad \forall v \in H \quad (1.3.20)$$

In this case, the space H of admissible functions is

$$H = \{v \in H^1(\Omega)|v = 0 \quad \text{on} \quad \partial\Omega_1\} \tag{1.3.21}$$

where "$v = 0$ on $\partial\Omega_1$" means that v vanishes on $\partial\Omega_1$ everywhere except possibly at discrete points where nonzero v can contribute nothing to the boundary integral in (1.3.20).*

Sufficiently smooth solutions of (1.3.20) will also satisfy the classical boundary-value problem

$$\left.\begin{array}{rcl} -\Delta u &=& f \quad \text{in} \quad \Omega \\[4pt] u &=& 0 \quad \text{on} \quad \partial\Omega_1 \\[4pt] \dfrac{\partial u}{\partial n} &=& g \quad \text{on} \quad \partial\Omega_2 \end{array}\right\} \tag{1.3.22}$$

Conversely, any solution of (1.3.22) will automatically be a solution of (1.3.20).

Example 1.3.2

Consider a one-dimensional problem in which the domain is the unit interval

$$\Omega = (0,1) = \{x \in \mathbb{R}|0 < x < 1\}$$

and let u and v be sufficiently smooth functions defined on Ω. The integration-by-parts formula,

$$\int_0^1 u^{(\text{iv})}v \, dx = \left.(u'''v - u''v')\right|_0^1 + \int_0^1 u''v'' \, dx \tag{1.3.23}$$

[wherein $u^{(\text{iv})} = d^4u/dx^4$, $u''' = d^3u/dx^3$, etc.] serves as a Green's formula and, in this case, a comparison of (1.3.23) with (1.3.5) gives

$$\left.\begin{array}{c} A = \dfrac{d^4}{dx^4}, \qquad B(u,v) = \displaystyle\int_0^1 u''v'' \, dx \\[10pt] -NuMv\Big|_0^1 = u'''(1)v(1) - u'''(0)v(0) \\[6pt] -u''(1)v'(1) + u''(0)v'(0) \end{array}\right\} \tag{1.3.24}$$

The essential boundary conditions involve conditions on the value of the function and its first derivative, whereas natural boundary con-

---

* Similar interpretations of boundary conditions apply to all examples considered in this volume. Note also that we assume throughout this chapter that the boundary $\partial\Omega$ is sufficiently smooth. Complications due to irregular boundaries (e.g., cracks and reentrant corners) are taken up in Chapter 5.

ditions involve u'' and u'''. A typical boundary-value problem in this class is: find $u \in H$ satisfying

$$\int_0^1 u''v''\, dx = \int_0^1 fv\, dx + p_0 v(0) - m_0 v'(0) \qquad \forall v \in H \qquad (1.3.25)$$

where m_0 and p_0 are given constants, and

$$H = \{v \in H^2(0, 1)| v(1) = 0, \quad v'(1) = 0\} \qquad (1.3.26)$$

The corresponding classical boundary-value problem is

$$\left. \begin{aligned} u^{(\text{iv})}(x) &= f(x), & x \in (0, 1) \\ u(1) &= u'(1) = 0 \\ u''(0) &= m_0, & u'''(0) = p_0 \end{aligned} \right\} \qquad (1.3.27)$$

Example 1.3.3

For $\Omega \subset \mathbb{R}^2$, $dx = dx_1\, dx_2$, consider the biharmonic operator,

$$\Delta^2 u = \Delta(\Delta u) = \frac{\partial^4 u}{\partial x_1^4} + 2\frac{\partial^4 u}{\partial x_1^2\, \partial x_2^2} + \frac{\partial^4 u}{\partial x_2^4} \qquad (1.3.28)$$

The following Green's formula holds for sufficiently smooth u and v:

$$\int_\Omega (\Delta^2 u) v\, dx = \int_\Omega \Delta u\, \Delta v\, dx + \oint_{\partial\Omega} \left(\frac{\partial \Delta u}{\partial n} v - \Delta u \frac{\partial v}{\partial n} \right) ds \qquad (1.3.29)$$

This formula assumes the form (1.3.5) if, for example, we take

$$\left. \begin{aligned} Au &= \Delta^2 u \\ Nu &= \left\{ -\frac{\partial \Delta u}{\partial n}, +\Delta u \right\} \quad \text{on} \quad \partial\Omega \\ Mv &= \left\{ v, \frac{\partial v}{\partial n} \right\}^T \quad \text{on} \quad \partial\Omega, \quad \text{and} \\ B(u, v) &= \int_\Omega \Delta u\, \Delta v\, dx \end{aligned} \right\} \qquad (1.3.30)$$

The variational boundary-value problem of finding u in H such that

$$\int_\Omega \Delta u\, \Delta v\, dx = \int_\Omega fv\, dx \qquad \forall v \in H \qquad (1.3.31)$$

with

$$H = H_0^2(\Omega) = \left\{ v \in H^2(\Omega)| v = 0, \; \frac{\partial v}{\partial n} = 0 \; \text{ on } \; \partial\Omega \right\} \qquad (1.3.32)$$

then corresponds to the classical fourth-order problem

$$\Delta^2 u = f \quad \text{in} \quad \Omega$$

$$u = 0, \qquad \frac{\partial u}{\partial n} = 0 \quad \text{on} \quad \partial\Omega \tag{1.3.33}$$

We now return to our review of finite element concepts.

Galerkin Approximations: A Galerkin approximation of (1.3.1) is obtained by posing the variational problem on a finite-dimensional subspace H^h of the space of admissible functions. Specifically, we seek u_h in H^h such that*

$$B(u_h, v_h) = \int_\Omega f v_h \, dx + \int_{\partial\Omega_2} g M v_h \, ds \qquad \forall v_h \in H^h \tag{1.3.34}$$

The superscript h here is a real parameter that decreases as the dimension of the subspace H^h increases; for finite element approximations, h will correspond to a mesh parameter, as will be explained below.

The Galerkin approximation u_h of the solution u is of the form

$$u_h(\mathbf{x}) = \sum_{j=1}^N \alpha_j \phi_j(\mathbf{x}) \tag{1.3.35}$$

where N is the dimension of H^h, α_j are unknown constants, and ϕ_j are linearly independent basis functions spanning H^h. The essential conditions on $\partial\Omega_1$ are satisfied by u_h because $H^h \subset H$. Thus, problem (1.3.34) is equivalent to the linear system

$$\sum_{j=1}^N K_{ij}\alpha_j = F_i, \qquad i = 1, 2, \ldots, N \tag{1.3.36}$$

where K_{ij} and F_i are entries of the stiffness matrix and load vector, respectively:

$$\left. \begin{aligned} K_{ij} &= B(\phi_i, \phi_j) \\ F_i &= \int_\Omega f \phi_i \, dx + \int_{\partial\Omega_2} g M \phi_i \, ds \\ & \qquad 1 \le i, j \le N \end{aligned} \right\} \tag{1.3.37}$$

The Finite Element Method: The finite element method provides a general and systematic technique for constructing the basis functions ϕ_i:

---

* Throughout this text we shall assume, for convenience, that the domain approximation Ω_h coincides with Ω and that essential boundary conditions are represented exactly, unless stated otherwise.

The domain Ω is replaced by a collection Ω_h of simple domains Ω_e, the finite elements

$$\Omega_h \simeq \Omega, \qquad \bar{\Omega}_h = \bigcup_{e=1}^{E} \bar{\Omega}_e \qquad (1.3.38)$$

The functions ϕ_i are frequently chosen to be piecewise polynomials in $x_1, x_2,$ $\ldots, x_N$ and are constructed in a special way: Over each finite element Ω_e, local *shape functions* ψ_i^e, $i = 1, 2, \ldots, N_e$, are constructed so that the restriction of the approximate solution u_h to $\bar{\Omega}_e$ is of the form

$$\left. u_h \right|_{\bar{\Omega}_e} (\mathbf{x}) = u_h^e(\mathbf{x}) = \sum_{j=1}^{N_e} \alpha_j^e \psi_j^e(\mathbf{x})$$

$$(\mathbf{x} \in \bar{\Omega}_e), \qquad e = 1, 2, \ldots, E \qquad (1.3.39)$$

The *global* basis functions ϕ_i can then be regarded as being generated by patching together these *local* shape functions in the following way: The shape functions ψ_i^e and possibly their derivatives of a specified order are designed to assume a value of unity or zero at nodal points $\mathbf{x}_i^e$ within each element: for example,

$$\left. \begin{array}{ll} \psi_i^e(\mathbf{x}_j^e) = \delta_{ij}; & 1 \leq i, j \leq N_0^e \\[2mm] \dfrac{\partial \psi_i^e}{\partial x_k}(\mathbf{x}_j^e) = \delta_{ij}; & N_0^e + 1 \leq i, j \leq N_1^e \\[2mm] \dfrac{\partial^2 \psi_i^e}{\partial x_k \partial x_l}(\mathbf{x}_j^e) = \delta_{ij}; & N_1^e + 1 \leq i, j \leq N_2^e \end{array} \right\} \qquad (1.3.40)$$

with $N_0^e + N_1^e + N_2^e = N_e$, the number of degrees of freedom for the element, and $1 \leq k, l \leq N$. Then, at a nodal point $\mathbf{x}_j \in \Omega_h$, the basis functions ϕ_i will also have the property that their value or the values of their first or second derivatives will be unity or zero.

Suppose, for example, that nodal point i in the finite element mesh Ω_h is shared by $\hat{E}$ finite elements. Then the local shape functions corresponding to each of these elements are combined in such a way that a global basis function ϕ_i is produced for node i, which has three fundamental properties: (1) the proper interelement continuity is attained so that $\phi_i \in H$; (2) ϕ_i or its appropriate derivatives assume a value of unity or zero at node i in accordance with (1.3.40); and (3) ϕ_i is nonzero only over the particular patch of $\hat{E}$ elements meeting at node i and is zero elsewhere in the mesh.

Continuing our analysis, a key issue is that, upon generating the global basis functions ϕ_i, we obtain a basis for a space H^h such that

$$H^h \subset H$$

Collections of finite elements which have this property are called *conforming elements*. Suppose that $H \subset H^m(\Omega)$. To determine sufficient conditions for an element to be conforming, we note that the functions v_h are generally infinitely regular (e.g., polynomials) on the interior of all elements. Thus, their regularity over all of Ω depends on how many derivatives are continuous at interelement boundaries. If $v_h \in C^0(\Omega)$ but there is a jump in the derivative of v_h normal to an interelement boundary, $\partial^2 v_h/\partial n^2$ will be undefined. Then $v_h \in H^1(\Omega)$ but $v_h \notin H^2(\Omega)$. Generalizing, if v_h is such that its derivatives of order $m - 1$ are continuous in Ω [i.e., $v_h \in C^{m-1}(\Omega)$] then $v_h \in H^m(\Omega)$. We call such conforming elements in $H^m(\Omega)$, "C^{m-1} elements."

The above properties facilitate the development of the stiffness and load matrices by making it possible to compute contributions to the entries of these matrices from each finite element, and then to sum the contributions over the entire connected mesh. Let us consider the case in which $m = 1$ and $N_0^e = N_e$ ($N_1^e = N_2^e = 0$) in (1.3.40) so that $\psi_i^e(\mathbf{x}_j^e) = \delta_{ij}$, $1 \le i, j \le N_e$. Then

$$u_h^e(\mathbf{x}) = \sum_{i=1}^{N_e} u_i^e \psi_i^e(\mathbf{x}), \qquad u_i^e = u_h(\mathbf{x}_i^e)$$

By reformulating the variational problem (1.3.1) locally over each element Ω_e with $v_h = \psi_i^e$, we obtain the variational problem

$$B_e(u_h^e, \psi_i^e) = \int_{\Omega_e} f\psi_i^e \, dx + \int_{\partial\Omega_e} NuM\psi_i^e \, ds, \qquad i = 1, 2, \ldots, N_e \qquad (1.3.41)$$

or

$$\sum_{j=1}^{N_e} k_{ij}^e u_j^e = f_i^e - \sigma_i^e, \qquad\qquad i = 1, 2, \ldots, N_e \qquad (1.3.42)$$

where $B_e(\cdot, \cdot)$ is the bilinear form obtained by integrating over only Ω_e rather than Ω,

$$\left.\begin{aligned} k_{ij}^e &= B_e(\psi_i^e, \psi_j^e) \\ f_i^e &= \int_{\Omega_e} f\psi_i^e \, dx + \int_{\partial\Omega_{2e}} gM\psi_i^e \, ds \\ \sigma_i^e &= \int_{\partial\Omega - \partial\Omega_{2e}} - NuM\psi_i^e \, ds \end{aligned}\right\} \qquad (1.3.43)$$

Here k_{ij}^e and f_i^e are the entries in the element stiffness and load matrices for element Ω_e, $\partial\Omega_{2e} = \partial\Omega_2 \cap \partial\Omega_e$, and σ_i^e are entries in a *flux vector* representing the contributions to the natural conditions on the boundary of the element due to Nu, u being the exact solution.

By renumbering and expanding the ranges of the indices of the quantities in (1.3.42) so as to correspond to the numbering scheme used to label nodes

$\mathbf{x}_i$ in the connected mesh of finite elements, (1.3.42) is rewritten in the form

$$\sum_{j=1}^{N} K_{ij}^e u_j = F_i^e - \Sigma_i^e, \qquad i = 1, 2, \dots, N \qquad (1.3.44)$$

Then the final system of equations (1.3.36) is obtained by summing the contributions to $\mathbf{K}$ and $\mathbf{F}$ of each element,*

$$\sum_{e=1}^{E} K_{ij}^e = K_{ij}, \qquad \sum_{e=1}^{E} F_i^e = F_i, \qquad \sum_{e=1}^{E} (\Sigma_i^e) = 0 \qquad (1.3.45)$$

To summarize, we have thus represented the Galerkin approximation u_h in (1.3.35) by using basis functions ϕ_i built up from functions defined locally over the elements. Because of this special construction, the matrices in the final system of linear equations can be obtained by summing the contributions of each element, taken one element at a time, as in (1.3.43). As is well known, these properties make the method particularly convenient for implementation using digital computers. Having formulated (1.3.36) (*which includes essential boundary conditions*), we solve the resulting system and use (1.3.35) to obtain the final finite element approximation to our problem.

Errors: Let u be the solution of (1.3.1) and u_h, its Galerkin approximation, the solution of (1.3.34). The *error* in this approximation is defined as the function

$$e = u - u_h \qquad (1.3.46)$$

Assuming that $\Omega_h = \Omega$ and setting $v = v_h$ in (1.3.1) (which is possible because H^h is a subspace of H) and subtracting (1.3.34) from (1.3.1), we find that e satisfies the *orthogonality condition*

$$B(e, v_h) = 0 \qquad \forall v_h \in H^h \qquad (1.3.47)$$

The *strain energy* of the problem is

$$U(v) = \tfrac{1}{2} B(v, v) \qquad (1.3.48)$$

Recall that $H^h \subset H^m(\Omega)$. Then, under reasonable conditions on the coefficients a_{kl}^{ij} in the definition of $B(u, v)$, one can show that the virtual work satisfies inequalities of the type

$$B(u, v) \le M_0 \, \| u \|_m \| v \|_m, \qquad B(v, v) \ge \alpha \, \| v \|_m^2 \qquad (1.3.49)$$

---

* The quantities Σ_i^e may not sum to zero for certain data f, and in such cases $\sum_{e=1}^{E} \Sigma_i^e$ will lead to contributions to the stiffness and load matrices. These ideas are discussed in some detail in Chapter 4 of Volume I.

for arbitrary $u, v \in H$, where M_0 and α are positive constants. Thus, in such cases, the energy is equivalent to the H^m-norm on Ω in the sense that

$$\sqrt{\alpha}\,\|u\|_m \le \sqrt{2U(u)} \le \sqrt{M_0}\,\|u\|_m \qquad (1.3.50)$$

for any $u \in H$.

Assuming that conditions (1.3.49) hold, we note that

$$\|e\|_m^2 \le \frac{1}{\alpha} B(e, e)$$

$$= \frac{1}{\alpha} B(e, u - u_h + v_h - v_h)$$

$$= \frac{1}{\alpha} B(e, u - v_h)$$

$$\le \frac{M_0}{\alpha} \|e\|_m \|u - v_h\|_m$$

where v_h is an arbitrary test function in H^h, and, in the third step, we have used the orthogonality condition (1.3.47).

Thus, for any $v_h \in H^h$,

$$\|e\|_m \le \frac{M_0}{\alpha} \|u - v_h\|_m \qquad (1.3.51)$$

This is an important result. It establishes that *the magnitude of the approximation error, measured in the H^m-norm, is bounded above by the distance in $H^m(\Omega)$ of the exact solution u from any member v_h in the finite element subspace H^h*. This reduces the problem of establishing the convergence of the finite element solution to one of estimating the distance $\|u - v_h\|_m$, $v_h \in H^h$ (in terms of some appropriate measure of the quality of the approximation). This may accordingly be phrased as a problem of *interpolation* since, in particular, the finite element interpolant $\tilde{u}_h$ of u is a member of the finite element subspace H^h.

To determine a measure of the size of $\|u - \tilde{u}_h\|_m$ for various types of finite element approximations, we note that when we *refine* the finite element mesh (i.e., when we increase the number of elements), we generally expect the error to become smaller in some sense. In any acceptable finite element scheme, the error should approach zero as the size of the elements approaches zero. The finite element method is then *convergent*, and we can expect an improvement in our results if we continue to use more and more elements in our discretization of the problem.

To make the notion of convergence more precise, we need to introduce a parameter that will provide an index for our sequence of error functions

$\{e\}$ obtained by successively refining the mesh. A standard mesh parameter is h, the maximum diameter of any element in the mesh:

$$h = \max_e h_e, \qquad h_e = \text{diameter of element } \Omega_e$$
$$1 \leq e \leq E, \qquad E = E(h) = \text{number of elements in the mesh} \tag{1.3.52}$$

We must be careful not to distort the elements too much when we refine a mesh, else the local error may become infinitely large even though h approaches zero. We avoid such difficulties by always keeping the smallest interior angle in an element larger than some fixed angle θ or by always requiring that, for some constant σ_0, $\rho_e \geq \sigma_0 h_e$, where ρ_e is the diameter of the largest sphere that can be inscribed in element Ω_e. Thus, for a two-dimensional problem, a sequence of refinements such as that shown in Fig. 1.4 would be acceptable.

Now let us examine the behavior of the error. Suppose that the following conditions prevail:

1. Conforming (C^{m-1}) elements are used [i.e., $H^h \subset H^m(\Omega)$].

2. Within each element, the local shape functions are constructed using functions which contain complete polynomials in $\mathbf{x}$ of degree k.

3. In addition to being a member of $H^m(\Omega)$, the regularity of the *exact* solution u of our problem is such that

$$u \in H^r(\Omega), \qquad r > m$$

[i.e., $u \in H \cap H^r(\Omega)$].

Then it can be shown (see Chapter 2 of Volume IV) that as h tends to zero, there exists a function $\tilde{u}_h \in H^h$ (called an interpolant of u) such that

$$\| u - \tilde{u}_h \|_m \leq C h^\mu \| u \|_r \tag{1.3.53}$$

where

$$\mu = \min (k + 1 - m, r - m) \tag{1.3.54}$$

and C is a constant.

Using this interpolation error estimate on the right in (1.3.51), the approximation error satisfies

$$\| e \|_m \leq \frac{M_0 C}{\alpha} h^\mu \| u \|_r \tag{1.3.55}$$

Thus, the rate at which our finite element method converges is given by the exponent μ in (1.3.55). If u is very regular, so that $r - m$ is very large, we can always increase the rate of convergence by increasing the degree k of the

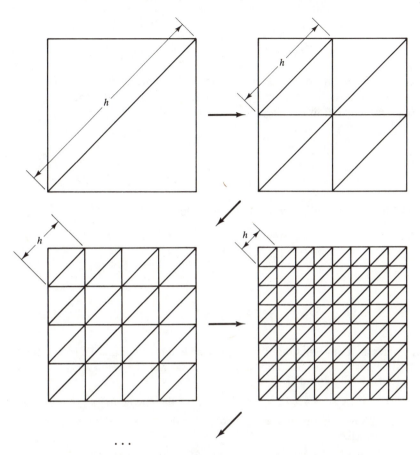

FIGURE 1.4 *Successive refinements of a finite element mesh for which $h \longrightarrow 0$.*

complete polynomials used in designing our elements. Of course, as k increases, so does the number of degrees of freedom (and hence the number of unknowns) in our model. Thus, we must generally reach some compromise between speed of convergence (large k) and computational effort (small k) to fit our needs. Note that if the actual solution is so irregular that $r - m < k + 1 - m$, the rate of convergence is completely unaffected by k; no matter how large the degree of polynomial k used, the asymptotic rate of convergence is still $r - m$.

Accuracy is lost in differentiation and it can be shown that the error estimates in "lower norms" are of the form

$$\|e\|_s \leq Ch^\nu \|u\|_r, \qquad 0 \leq s \leq m$$
$$\nu = \min(k + 1 - s, r - s) \tag{1.3.56}$$

Thus, if $k = 1, m = 1, r = 3$, we have

$$\| e \|_1 \leq Ch \| u \|_3 \qquad \text{but} \qquad \| e \|_0 \leq Ch^2 \| u \|_3$$

Finally, we mention that it is sometimes desirable to measure the rate of convergence in other norms. Of particular interest are the so-called "maximum" or "infinity" norms of e and its derivatives,*

$$\| e \|_\infty = \max_{\mathbf{x} \in \Omega} | e(\mathbf{x}) |$$

$$\| e \|_{m, \infty} = \max_{\mathbf{x} \in \Omega} \max_{0 \leq j \leq m} \left| \frac{\partial^j e(\mathbf{x})}{\partial x^r \, \partial y^s \, \partial z^t} \right|, \qquad (r + s + t = j) \qquad (1.3.57)$$

for which estimates of the following type hold:

$$\| e \|_\infty \leq C \begin{cases} h^2 | \log h |, & k = 1 \\ h^{k+1}, & k \geq 2 \end{cases} \| u \|_{k+1, \infty} \qquad (1.3.58)$$

EXERCISES

1.3.1 Show that any solution of (1.3.6) is also a solution of (1.3.1). In addition, show that if the solution u and the data f and g of (1.3.1) are very smooth, u is also a solution of (1.3.6).

1.3.2 (a) Derive the Green's formula (1.3.9) for the operator A given in (1.3.8).

(b) Show that any sufficiently smooth solution of (1.3.10) is also a solution of (1.3.12).

(c) Identify the boundary operators M and N (for essential and natural boundary conditions, respectively) in problem (1.3.10) [or, equivalently, problem (1.3.12)].

(d) Discuss modifications in the formulation necessary to handle the following boundary conditions:

$$\frac{d^{m-1}u(0)}{dx^{m-1}} = 3$$

$$\frac{d^{m-2}u(0)}{dx^{m-2}} = 4$$

$$\frac{d^{m-1-r}}{dx^{m-1-r}} \left[a(0) \frac{d^m u}{dx^m} (0) \right] = g_r + u(0) + \frac{d^2 u(0)}{dx^2} \qquad r = 0, 1, \ldots, m - 3$$

$$\frac{d^s u(1)}{dx^s} = 0, \qquad s = 0, 1, \ldots, m - 1 \quad (m > 2)$$

---

* Technically, we compute these norms over sets in Ω of nonzero measure. Thus, we may ignore isolated points where e may be large.

1.3.3 Consider the operator A of order $2m$ given by

$$Au(x) = a_{2m}(x)\frac{d^{2m}u(x)}{dx^{2m}} + a_{2m-1}(x)\frac{d^{2m-1}u(x)}{dx^{2m-1}}$$
$$+ \cdots + a_0(x)u(x), \qquad x \in (0, 1)$$

where $a_0, a_1, \ldots, a_{2m}$ are smooth (e.g., infinitely differentiable) functions.

(a) Show that a Green's formula for this operator is

$$\int_0^1 (Au)v \, dx = B(u, v) + \sum_{i=0}^{2m} \sum_{r=0}^{s_i} (-1)^r \frac{d^r a_{2m-i}v}{dx^r} \frac{d^{2m-1-i-r}u}{dx^{2m-1-i-r}}\bigg|_0^1$$

where

$$B(u, v) = \int_0^1 \left[\sum_{i=0}^{2m} (-1)^{1+s_i} \frac{d^{1+s_i}a_{2m-i}v}{dx^{1+s_i}} \frac{d^{2m-1-i-s_i}u}{dx^{2m-1-i-s_i}}\right] dx$$

and $m - 1 - i \le s_i \le m - 1$, $i = 0, 1, \ldots, 2m$ (i.e., the lower-order terms have been integrated unevenly s_i times so that no derivatives of order higher than m appear in the resulting integrand).

(b) Give examples of the operators M and N.

(c) Construct an example of variational boundary-value problem using the choices selected in part (b); what is the space H of admissible functions for this choice?

1.3.4 In problem (1.3.20), suppose that $\partial\Omega = \partial\Omega_2$; that is, the natural boundary condition, $\partial u/\partial n = g$ is applied on all of $\partial\Omega$. Show that the compatibility condition on the data f and g

$$\int_\Omega f \, dx + \oint_{\partial\Omega} g \, ds = 0$$

must hold.

1.3.5 Consider problem (1.3.25). As alternative boundary conditions, suppose that we have

$$u''(0) = m_0, \quad u'''(0) = p_0, \quad u''(1) = m_1, \quad u'''(1) = p_1$$

(a) Develop a variational statement of this boundary-value problem and identify the appropriate space H of admissible functions.

(b) Give necessary (compatibility) conditions on the data f, m_0, m_1, p_0, and p_1 for there to exist a solution to this problem.

1.3.6 Specify the spaces (classes of admissible functions) C^m, H^m, H_0^m, and so on, appropriate for the following problems:

(a) $\displaystyle\int_0^1 (u''v'' + xu'v' + uv)dx = \int_0^1 \sin x \, v \, dx$

for all appropriate v;

$$u(0) = 0, \quad u'(0) = 0, \quad u(1) = 0, \quad u'(1) = 0$$

(b) $\int_\Omega \Delta u \, \Delta v \, dx = \int_\Omega xyv \, dx$ for all appropriate v;

$$u = 0 \quad \text{and} \quad \frac{\partial u}{\partial n} = 0 \quad \text{on} \quad \partial \Omega \qquad (\Omega \subset \mathbb{R}^2)$$

(c) $\dfrac{d^2u}{dx^2} + \dfrac{du}{dx} + u = f, \qquad 0 < x < 1$

$$f(x) = \begin{cases} 1, & 0 < x < \frac{1}{2} \\ 0, & \frac{1}{2} \le x < 1 \end{cases}$$

$$u(0) = 2, \quad u'(0) = 3, \quad u(1) = 4, \quad u'(1) = 0$$

1.3.7 Consider the boundary-value problem

$$\frac{d^4u}{dx^4} = f, \qquad -1 < x < 1$$

$$u(-1) = u'(-1) = u(1) = u'(1) = 0$$

where f is a C^∞-function. By multiplying $u^{iv} - f$ by a test function v, a weaker form of the problem is obtained:

(a) By (formally) integrating by parts, develop weak statements of this problem for each of the following choices of trial and test functions.

Trial Functions*	Test Functions
$H^4(-1, 1)$	$H^0(-1, 1)$
$H^3(-1, 1)$	$H^1(-1, 1)$
$H^2(-1, 1)$	$H^2(-1, 1)$
$H^1(-1, 1)$	$H^3(-1, 1)$
$H^0(-1, 1)$	$H^4(-1, 1)$
$H^{-1}(-1, 1)$	$H^5(-1, 1)$

* It is understood that the trial functions are also in $H_0^2(-1, 1)$.

(b) Discuss the minimal continuity requirements of polynomial finite element approximations of both the trial functions and the test functions in each of the case studies in part (a).

(c) Discuss in detail how polynomial finite element approximations are constructed for the case in which trial functions are in $H^3(-1, 1) \cap H_0^2(-1, 1)$ and test functions are in $H_0^1(-1, 1)$.

(d) Discuss the "collocation" case: trial functions in $H^5(-1, 1) \cap H_0^2(-1, 1)$, test functions in $H^{-1}(-1, 1)$.

(e)   What advantages, from the point of view of finite element methods, can one identify for the case in which both trial and test functions are in the same space $H_0^2(-1, 1)$?

[*REMARK:*   Some of these ideas are explored further in Chapter 4.]

1.3.8   Consider the boundary-value problem

$$\frac{d^6u}{dx^6} - \frac{d^4u}{dx^4} + u = f, \qquad 0 < x < 1$$

$$u(0) = u'(0) = u''(0) = 0, \quad u(1) = u'(1) = u''(1) = 0$$

(a)   Describe a polynomial finite element approximation of this problem in which conforming elements are used.

(b)   If the exact solution u is in $H^6(0, 1) \cap H_0^3(0, 1)$, what is the rate of convergence of the error in the H^3-norm? In H^s, $s = 2, 1, 0$?

(c)   Suppose that $f(x)$ is replaced by a Dirac delta $f(x) = \delta(x - \frac{1}{2})$ and that the solution u is in $H^5(0, 1)$. Give the error estimates in the H^s-norms, $s = 0, 1, 2, 3$.

1.3.9   Suppose that C^0-piecewise-linear elements are used to solve the problem $-u'' + u = x$, $0 < x < 1$, $u(0) = u(1) = 0$. Give an estimate of the error in $H^1(0, 1)$ and in $H^0(0, 1)$. Repeat this exercise for C^0-quadratics, C^0-cubics, and C^1-cubics.

2

FINITE ELEMENT INTERPOLATION

2.1 INTRODUCTION

Quite apart from, but basic to its use as a method of *approximation* of solutions to boundary-value problems, the finite element method can also be viewed as a method of *interpolation*. For instance, given a smooth function $u = u(x, y)$ defined on a region Ω in the (x, y)-plane, we may consider the problem of constructing a finite element representation $u_h = u_h(x, y)$ of u which coincides with u (and, possibly, has various partial derivatives which coincide with those of u) at prescribed nodal points in a finite element discretization of Ω.

Although our principal objective in this volume is to consider approximation methods based on finite element concepts, a consideration of the interpolation question allows us to study a variety of general properties of finite elements without serious regard to the actual problems to which these elements are to be applied. Our objective in this chapter is to consider examples of *finite element families* (i.e., different classes of finite elements) that may be used in applications. We give criteria for the selection of appropriate elements for interpolation or approximation and we provide a catalogue of some of the more common types of finite elements for one-, two-, and three-dimensional problems.

2.2 ELEMENT SELECTION AND INTERPOLATION ESTIMATES

The choice of which type of element one uses in a specific application depends on several factors; among them are:

1. The regularity of the function and domain being approximated (or interpolated); that is, the smoothness of the function and domain

2. The accuracy we wish to attain for a given mesh

3. The price we are willing to pay in terms of computational time and effort for our approximations

The last of these factors involves a number of considerations: the selection of the number of degrees of freedom for the elements; the choice of a numerical scheme to solve the system of equations obtained from the finite element model; efficient coding of various algorithms needed to generate the finite element mesh, manage data, and solve the equations; peculiarities of the available computing machinery; the resources (money) available for the calculation; and so on. We take up some of these purely computational aspects of finite elements in Volume III of this series. Our main concern here is with factors 1 and 2 above: regularity and accuracy. The major criteria governed by these considerations were discussed in Chapter 1:

1. For variational problems set in the space $H^m(\Omega)$, conforming finite element approximations belong to $C^{m-1}(\Omega)$.

2. Let $u \in H^r(\Omega)$ for r sufficiently large. Suppose that the shape functions ψ_j contain complete polynomials of degree $\leq k$. Then the error in the interpolant $\tilde{u}_h$ of u in the $H^s(\Omega)$-norm satisfies* the following estimate as $h \to 0$:

$$\| u - \tilde{u}_h \|_s \leq C h^{k+1-s} \| u \|_r, \qquad 0 \leq s \leq m \qquad (2.2.1)$$

provided that $r \geq k + 1 > m$, where C is a constant independent of u and h.

It is customary to refer to an estimate such as (2.2.1) as specifying the "error in $H^s(\Omega)$ as $O(h^{k+1-s})$," where $O(\cdot)$ is called the order symbol: $O(h^{k+1-s})$ is read "order h^{k+1-s}" and $k + 1 - s$ is termed the asymptotic rate of convergence. The integer k appearing in the asymptotic rate of convergence in

---

* Estimates such as this are derived in Chapter 2 of Volume IV.

(2.2.1) denotes the degree of the *complete* polynomial of greatest order appearing in the definition of the element shape functions. Thus, if

$$v_h(x, y) = a_0 + a_1 x + a_2 y + a_3 xy + a_4 x^2 + a_5 y^3$$
$$+ a_6 x^3 y + a_7 xy^3 + a_8 x^2 y^2$$

then $k = 1$; while quadratic, cubic, and quartic terms are present, only enough terms to define all polynomials in x and y of degree ≤ 1 appear. The local approximation

$$v_h(x) = a_1 x + a_2 x^2 + a_3 x^3$$

is unacceptable, since no constant term is present, whereas if

$$v_h(x) = a_0 + a_1 x^3$$

$k = 0$; and if

$$v_h(x) = a_0 + a_1 x + a_2 x^2 + a_3 x^5 + a_4 x^6$$

$k = 2$; and so on.

Results analogous to (2.2.1) hold for functions in the $C^m(\Omega)$-spaces. If $\Omega \subset \mathbb{R}^2$, and

$$\| u \|_{C^m(\Omega)} = \max_{(x,y) \in \Omega} \max_{0 \leq \alpha \leq m} \left| \frac{\partial^\alpha u(x, y)}{\partial x^{\alpha_1} \partial y^{\alpha_2}} \right| \tag{2.2.2}$$

with $\alpha_1 + \alpha_2 = \alpha > 0$, then estimates of the type

$$\| u - \tilde{u}_h \|_{C^m(\Omega)} \leq C h^{k+1-m} \| u \|_{C^{k+1}(\Omega)} \tag{2.2.3}$$

hold, provided that u has bounded derivatives of order $k + 1$ in Ω.

In virtually all of the finite element methods we consider in the remainder of this volume, we confine ourselves to second-order boundary-value problems (such as Poisson's equation, $-\Delta u = f$) or to fourth-order problems (such as biharmonic problems, $\Delta^2 u = f$). For second-order problems, $2m = 2$; hence only C^0-elements are needed. For fourth-order problems, $2m = 4$; hence C^1-elements are needed. For these reasons, most of the examples of finite element families given in the remaining sections of this chapter involve C^0- or C^1-finite elements.

We see that there are two properties which must be considered simultaneously in studying finite element interpolation: (1) *local* accuracy as determined by the degree of the polynomial approximation on an element, and (2) *global* smoothness dictated by the continuity of the interpolant and its derivatives across the element interface. In describing finite element interpolation, we shall develop the theory of local polynomial interpolation on an element Ω_e in one and then higher dimensions and shall assume that the interpolated function u is smooth.

EXERCISES

2.2.1 Let u'' be continuous on the interval $\bar{\Omega} = [0, 1]$ and $\tilde{u}_h$ be a piecewise-linear finite element interpolant of u on a uniform mesh of size h. Let $E = u - \tilde{u}_h$ denote the interpolation error. Expand $E(x)$ in a Taylor series on an arbitrary element Ω_e and about the point where $|E|$ is a maximum in Ω_e. Use this expansion to obtain the pointwise error bound

$$\max_{0 \le x \le 1} |u - \tilde{u}_h| \le \tfrac{1}{8} h^2 \max_{0 \le x \le 1} |u''(x)|$$

[*REMARK*: The constant $\tfrac{1}{8}$ is optimal for an arbitrary point in Ω.]

2.2.2 By again observing $E'(x) = 0$ at the point in Ω_e where $|E|$ is a maximum and using the fact that $f(x) = f(a) + \int_a^x f'(z)\, dz$, deduce the pointwise bound on the first derivative

$$\max_{0 \le x \le 1} |u' - \tilde{u}_h'| \le h \max_{0 \le x \le 1} |u''(x)|$$

[*REMARK*: The constant here is not optimal; $\tfrac{1}{2}$ is the optimal value.]

2.2.3 Use the results of Exercises 2.2.1 and 2.2.2 to show that the interpolation error in the L^2- and H^1-norms is $O(h^2)$ and $O(h)$, respectively.

2.2.4 Note that the result of Exercise 2.2.3 is based on the assumption that u'' is continuous, and this is much stronger than the requirements indicated in our general theoretical estimates in (2.2.1). To obtain the improved result, expand the error $E(x)$ in a sine series on Ω_e to verify that

$$\int_{\Omega_e} (E')^2 \, dz \le \frac{h^2}{\pi^2} \int_{\Omega_e} (u'')^2 \, dz$$

and

$$\int_{\Omega_e} E^2 \, dz \le \frac{h^4}{\pi^4} \int_{\Omega_e} (u'')^2 \, dz$$

2.2.5 Let $u \in C^3[0, 1]$ and denote by $\tilde{u}_h$ a finite element interpolant of u, defined by piecewise C^0-quadratics on a uniform mesh. Using Taylor's formula, derive an estimate of the type (2.2.3) for the interpolation error in the

$C^0(\Omega)$-norm under the assumption that u is interpolated by $\tilde{u}_h$ by setting the values of $\tilde{u}_h$ equal to those of u at nodal points in $[0, 1]$.

[**HINT**: See a similar result for $k = 1$ in Chapter 2, Volume I.]

2.3 LAGRANGE C^0-ELEMENTS

We begin our study of element families with the most common class of elements—Lagrange C^0-elements. As noted in the preceding section, with second-order problems $(2m = 2)$ and for conforming elements we require C^0 global continuity. Lagrange-type families are, as the expression suggests, based on applying the classical Lagrange polynomial interpolation theory locally on each element. We were introduced to the fundamental ideas in Volume I and here shall develop them in a systematic manner to construct C^0-families of Lagrange elements in one, two, and three dimensions.

2.3.1 One-Dimensional Interpolation

We begin by considering a sufficiently smooth function u of x defined on the interval $\bar{\Omega} = [a, b]$. We introduce a partition $a = x_0 < x_1 < \cdots < x_N = b$ of $\bar{\Omega}$ into N elements with typical element $\bar{\Omega}_e = [x_{i-1}, x_i]$. We wish to develop continuous piecewise-polynomial interpolants of u using Lagrange polynomial interpolation on each element.

The simplest example is piecewise-linear interpolation: Locally on each element $\bar{\Omega}_e$, the approximation is linear. Since, specifically, the end values u_{i-1} and u_i of u are interpolated, global continuity of the interpolant is assured. If the local linear interpolant had involved an interior point or points, the approximation would have discontinuities at interface nodes between elements and the stipulation that it be globally C^0 would be violated.

Let u_h be the piecewise-linear Lagrange interpolant of u.* The restriction of u_h to a typical element $\bar{\Omega}_e$ is of the form

$$u_h\bigg|_{\Omega_e}(x) = u_h^e(x) = u_1^e\psi_1^e(x) + u_2^e\psi_2^e(x) \qquad (2.3.1)$$

where ψ_1^e and ψ_2^e are the local linear shape functions corresponding to the left endpoint $(x = x_{i-1})$ and the right endpoint $(x = x_i)$, respectively, of Ω_e; u_1^e and u_2^e are the values of u_h^e at these interpolation points. Clearly, by

---

* Throughout the remainder of this chapter, for notational convenience we use u_h to denote the interpolant of u. Elsewhere in the volume we reserve u_h for the finite element solution of the boundary-value problem.

virtue of this interpolation property, the shape functions must satisfy

$$\psi_1^e(x_{i-1}) = 1, \qquad \psi_1^e(x_i) = 0$$
$$\psi_2^e(x_{i-1}) = 0, \qquad \psi_2^e(x_i) = 1 \tag{2.3.2}$$

Suppose that there are M elements in the mesh and $N + 1$ nodes. For piecewise-linear elements, $M = N$. Globally, the piecewise-linear interpolant u_h has the form

$$u_h(x) = \sum_{j=0}^{N} u_j \phi_j(x), \qquad x \in [a, b] \tag{2.3.3}$$

where u_j are the interpolated nodal values of u and ϕ_j are the piecewise-linear global basis functions constructed from shape functions on elements adjacent to node j. In view of (2.3.2), we must have

$$\phi_j(x_i) = \delta_{ij}, \qquad i, j = 0, 1, \ldots, N \tag{2.3.4}$$

We refer to the element $\bar{\Omega}_e$ and the patch $\bar{\Omega}_{e-1} \cup \bar{\Omega}_e$ as the *support* of the local shape functions and global basis functions, respectively. By the *degree* k of an element, we mean the degree of the complete polynomial defining the approximation on an element. By simply including additional interpolation points in the interior of each element, polynomial interpolations of higher degree can be employed locally. Provided that the end nodes are interpolation points, the global approximation remains C^0. Examples of element polynomials and the corresponding global piecewise polynomials are given in Fig. 2.1. Notice that the element and global basis functions for nodes interior to an element coincide and have support on this single element alone.

Within any element the approximation is constructed using classical Lagrange interpolation theory. For convenience of subsequent element derivations, we introduce a transformation to a master element $\hat{\Omega}$. Let $x \in \bar{\Omega}_e$ transform to $\xi \in \hat{\Omega} = [-1, 1]$, according to

$$\xi = \frac{2x - (x_{i-1} + x_i)}{x_i - x_{i-1}} \tag{2.3.5}$$

with Jacobian $|J| = \xi_x = 2/(x_i - x_{i-1}) = 2/h_e$, where $h_e = x_i - x_{i-1}$ is the element length. Under this linear transformation, $u_e(x)$ becomes $\hat{u}_e(\xi)$ and the element Lagrange interpolant $\hat{U}$ of degree k may be expressed as

$$\hat{U}(\xi) = \sum_{j=1}^{k+1} \hat{u}_j^e \hat{\psi}_j(\xi), \qquad \xi \in \hat{\Omega} \tag{2.3.6}$$

where $\hat{\psi}_j(\xi)$ are the Lagrange interpolation polynomials on the master element $\hat{\Omega}$ and $\hat{u}_j^e = u_j^e$ are interpolated nodal values at points ξ_j, including the

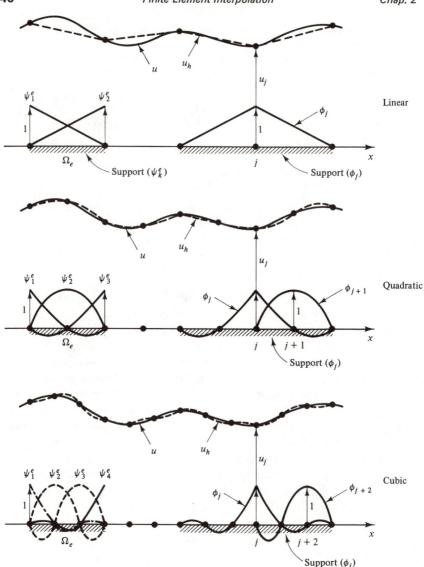

FIGURE 2.1 *Representative local element interpolation functions and global basis functions for C^0-piecewise linears, quadratics, and cubics.*

endpoints $\xi_1 = -1$ and $\xi_{k+1} = +1$. The local interpolation error is $\hat{E}(\xi) = \hat{u}_e(\xi) - \hat{U}(\xi)$ and at the interpolation points $\hat{E}(\xi_i) = 0$, $i = 1, 2, \ldots, k + 1$.

Since $\hat{U}(\xi_i) = \hat{u}_e(\xi_i)$, $i = 1, 2, \ldots, k + 1$, the Lagrange polynomials must satisfy $\hat{\psi}_j(\xi_i) = \delta_{ij}$ for $i, j = 1, 2, \ldots, k + 1$ (where $\delta_{ij} = 1$ if $j = i$ and $\delta_{ij} = 0$ if $j \neq i$). Factoring the Lagrange polynomials at the interpolation points and normalizing, we obtain the general Lagrange polynomial of degree k on the master element:

$$\hat{\psi}_j(\xi) = \frac{(\xi - \xi_1) \cdots (\xi - \xi_{j-1})(\xi - \xi_{j+1}) \cdots (\xi - \xi_k)(\xi - \xi_{k+1})}{(\xi_j - \xi_1) \cdots (\xi_j - \xi_{j-1})(\xi_j - \xi_{j+1}) \cdots (\xi_j - \xi_k)(\xi_j - \xi_{k+1})}$$

$$(2.3.7)$$

Introducing the product polynomial

$$p_{k+1}(\xi) = \prod_{i=1}^{k+1} (\xi - \xi_i) \tag{2.3.8}$$

we can write (2.3.7) more compactly as

$$\hat{\psi}_j(\xi) = \frac{p_{k+1}(\xi)}{(\xi - \xi_j)p'_{k+1}(\xi_j)}, \qquad \text{where } p'_{k+1}(\xi_j) = \frac{dp_{k+1}}{d\xi} \text{ at } \xi = \xi_j \quad (2.3.9)$$

The interpolation error $\hat{E}_e(\xi) = \hat{u}_e(\xi) - \hat{U}(\xi)$ is (Exercise 2.3.3)

$$\hat{E}_e(\xi) = \frac{p_{k+1}(\xi)}{(k+1)!} \hat{u}_e^{(k+1)}(\eta), \qquad \eta \in [-1, 1] \tag{2.3.10}$$

Mapping back to the actual element Ω_e in the discretization, we have

$$\hat{u}_e^{(k+1)}(\eta) = \left(\frac{h_e}{2}\right)^{k+1} \frac{d^{k+1}}{dx^{k+1}} u_e(x^*), \qquad x^* \in [x_{i-1}, x_i] \tag{2.3.11}$$

so that the local pointwise interpolation error is $O(h_e^{k+1})$ on element $\bar{\Omega}_e = [x_{i-1}, x_i]$.

Formulas (2.3.6) and (2.3.7) define a family of Lagrange elements in one dimension. The inverse of the linear map (2.3.5) can be used to determine the element shape functions $\psi_j^e(x)$ and the local approximation $u_h^e(x)$ on Ω_e. Since the values at the ends of each element are interpolated, by "patching together" these element shape functions in the manner described earlier for piecewise linears in (2.3.3), we can construct C^0-global basis functions that are polynomials of degree k on the open interior of each element.

EXERCISES

2.3.1 Derive and sketch the element and global basis functions for piecewise-quartic polynomial interpolation in one dimension. If the mesh has M elements, what is the dimension of the global approximation space?

2.3.2 Let $u(x) = x^2$ and interpolate u using piecewise-linear elements on a uniform mesh of size $h = x_i - x_{i-1}$. Bound the interpolation error E on interval $[x_{i-1}, x_i]$ to show that the pointwise error is $O(h^2)$.

2.3.3   Let $F(z) = E(z) - E(\xi)p_{k+1}(z)/p_{k+1}(\xi)$ and apply Rolle's theorem $k + 1$ times to derive the Lagrange interpolation error estimate (2.3.10).

2.3.4   Formulate a least-squares method for approximating $u(x)$ and $u'(x)$ using piecewise-linear finite elements. Comment on the continuity of the approximation and derivatives.

2.3.2   Two and Three Dimensions

The extension of the foregoing ideas to two and three dimensions is quite straightforward. Again, polynomial interpolation may be employed locally on each individual element. Global continuity is now to hold across the interface lines or surfaces between adjacent elements. There are some subtle distinctions between the two standard classes of elements employed: (1) the simplex elements typified by the triangle and tetrahedron, and (2) the tensor-product elements characterized by the quadrilateral and its analogue in three dimensions. These distinctions stem from considerations of the completeness of the local element polynomial approximation and the manner in which the element shape functions are generated.

Simplex Elements: Consider a triangulation of a two-dimensional domain Ω such that the interiors of no two triangular elements overlap ($\Omega_e \cap \Omega_f$ is empty if $e \neq f$). Also, let us assume that any two elements sharing a side have that entire side in common; that is, triangles may meet only at vertices or along sides defined by common vertices.

We begin again with the linear case. A linear polynomial on a triangle may be expressed in the form $u_h^e(x, y) = a_1 + a_2 x + a_3 y$. On selecting any three distinct interpolation points that are not collinear within the element, the coefficients a_i may be expressed in terms of interpolated values u_i at these points to define a linear interpolant over the triangle. Now let $\overline{pq}$ be the interface between two adjacent triangles with vertices p and q on that side. The global approximation will be continuous across side $\overline{pq}$ only if the linear approximations on adjacent elements agree along this line. The same argument holds for elements adjacent to the remaining two sides of each triangle. This implies that the interpolation nodes must be located at triangle vertices. Thus, the simplest C^0 piecewise-polynomial interpolant corresponds to linear interpolation on a triangle with interpolation points specified as the vertices (Fig. 2.2).

Other choices of interpolation points, such as the midside points, are admissible in finite element interpolation. Adjacent linear interpolants then agree only at the midpoint along a side and the global approximation is not C^0. Recall that whenever the number of continuous derivatives across element interfaces is less than that sufficient for the variational problem, we term the elements *nonconforming*. For example, the triangular element with three

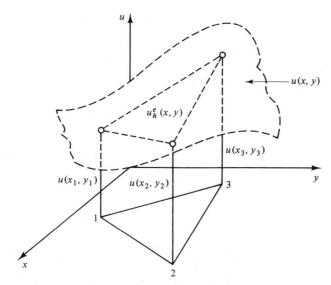

FIGURE 2.2 Linear interpolation on a triangular element.

nodes, one located at the midpoint of each side, is nonconforming for the variational statement of Laplace's equation which requires that the derivatives of the finite element approximation be in $L^2(\Omega)$. We see more of nonconforming elements in Chapter 5 in connection with both second-order problems and also fourth-order problems such as plate-bending and viscous flow problems, where they are more frequently encountered.

As in the one-dimensional example, we may associate with the interpolation points linear shape functions ψ_i^e for element Ω_e with local nodal numbers $i = 1, 2, 3$. The interpolation formula on Ω_e is

$$u_h^e(x, y) = \sum_{j=1}^{3} u_j^e \psi_j^e(x, y) \tag{2.3.12}$$

and $\psi_j^e(x_i, y_i) = \delta_{ij}$ as before. Each local basis function ψ_j^e assumes unit value at node j and is zero at the other two nodes. The shape functions ψ_j^e are easily derived by selecting the coefficients a_i in $u_h^e(x, y) = a_1 + a_2 x + a_3 y$ so that u_h^e interpolates u at the vertices.

This elementary calculation yields the shape functions ψ_j^e as, in vector form,

$$\begin{bmatrix} \psi_1^e \\ \psi_2^e \\ \psi_3^e \end{bmatrix} = \frac{1}{2A_e} \begin{bmatrix} x_2 y_3 - x_3 y_2 & y_2 - y_3 & x_3 - x_2 \\ x_3 y_1 - x_1 y_3 & y_3 - y_1 & x_1 - x_3 \\ x_1 y_2 - x_2 y_1 & y_1 - y_2 & x_2 - x_1 \end{bmatrix} \begin{bmatrix} 1 \\ x \\ y \end{bmatrix} \tag{2.3.13}$$

where $A_e = \frac{1}{2}\{(x_2 y_3 - x_3 y_2) + (x_3 y_1 - x_1 y_3) + (x_1 y_2 - x_2 y_1)\}$ and $|A_e|$ is the triangle area.

It is more convenient to pose the interpolation problem and its extension to higher-degree interpolation on a simpler master triangle. To this end, we first make the important observation that the coordinate functions x and y are themselves linear functions on an element. Interpolating these coordinate functions, we have exactly the relations

$$x = \sum_{j=1}^{3} x_j^e \psi_j^e(x, y), \qquad y = \sum_{j=1}^{3} y_j^e \psi_j^e(x, y) \qquad (2.3.14)$$

Since $\psi_1^e(x, y) + \psi_2^e(x, y) + \psi_3^e(x, y) = 1$ for (x, y) in $\bar{\Omega}_e$, then $\psi_1^e(x, y) = 1 - \psi_2^e(x, y) - \psi_3^e(x, y)$ and

$$\begin{aligned}
x &= x_1^e + (x_2^e - x_1^e)\psi_2^e + (x_3^e - x_1^e)\psi_3^e \\
y &= y_1^e + (y_2^e - y_1^e)\psi_2^e + (y_3^e - y_1^e)\psi_3^e
\end{aligned} \qquad (2.3.15)$$

defines a mapping between the (x, y)-coordinates and new ψ_2^e, ψ_3^e reference coordinates.

The linear transformation between the coordinate functions (x, y) and interpolation functions $\psi_1^e, \psi_2^e, \psi_3^e$ has a simple geometric interpretation. From (2.3.13), shape functions ψ_j^e are the ratios of areas of subtriangles formed by interior point (x, y) and the side opposite vertex j to the area of the triangular element. As (x, y) traverses a line parallel to a side, the area of the subtriangle on that side is constant. In particular, for (x, y) on side 1–2, ψ_3^e is zero and ψ_2^e ranges from 0 at node 1 to 1 at node 2. Similarly, for (x, y) on side 1–3, ψ_2^e is zero and ψ_3^e ranges from 0 at node 1 to 1 at node 3. Therefore, sides 1–2 and 1–3 represent skew coordinate axes for ψ_2^e and ψ_3^e and the linear transformations (2.3.13) and (2.3.15) map the general triangle in (x, y) to a right-isoceles triangle in (ψ_2^e, ψ_3^e), and vice versa. The actual element Ω_e and master element $\hat{\Omega}$ are shown in Fig. 2.3, where we now use ζ_2, ζ_3 for the new coordinates ψ_2^e, ψ_3^e to facilitate derivation of higher-degree C^0-Lagrange bases for triangular elements. Further details are given in Chapter 5 of Volume I.

On the right-isosceles master element $\hat{\Omega}$, under the transformation $(x, y) \rightarrow (\zeta_2, \zeta_3)$, $u_e(x, y)$ becomes $\hat{u}_e(\zeta_2, \zeta_3)$. The Lagrange polynomials are $\hat{\psi}_2(\zeta) = \zeta_2, \hat{\psi}_3(\zeta) = \zeta_3, \hat{\psi}_1(\zeta) = \zeta_1 = 1 - \zeta_2 - \zeta_3$ and correspond to the three sides of this reference triangle. The linear interpolant of $\hat{u}_e(\zeta)$ on $\hat{\Omega}$ is

$$\hat{U}(\zeta) = \sum_{j=1}^{3} \hat{u}_j^e \hat{\psi}_j(\zeta) = \hat{u}_1^e \zeta_1 + \hat{u}_2^e \zeta_2 + \hat{u}_3^e \zeta_3 \qquad (2.3.16)$$

where $\hat{u}_j^e = u_j^e$.

By similar arguments, the complete quadratic polynomial on the master element is of the form

$$\hat{U}(\zeta) = a_1 + a_2 \zeta_2 + a_3 \zeta_3 + a_4 \zeta_2^2 + a_5 \zeta_2 \zeta_3 + a_6 \zeta_3^2 \qquad (2.3.17)$$

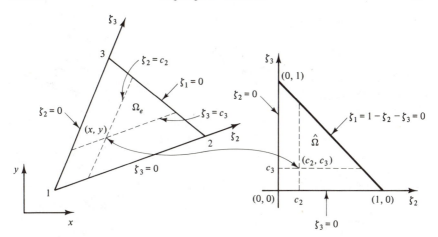

FIGURE 2.3 Linear mapping from Ω_e to master element $\hat{\Omega}$ defined by area coordinate transformation.

and requires six interpolation points to specify a_j uniquely. If we select the vertices 1, 2, 3 and midside nodes 4, 5, 6 as interpolation points, the quadratic Lagrange polynomial shape functions can be derived by inspection from the one-dimensional shape functions on each side as

$$\hat{\psi}_1(\zeta) = \zeta_1(2\zeta_1 - 1), \quad \hat{\psi}_2(\zeta) = \zeta_2(2\zeta_2 - 1), \quad \hat{\psi}_3(\zeta) = \zeta_3(2\zeta_3 - 1),$$
$$\hat{\psi}_4(\zeta) = 4\zeta_1\zeta_2, \qquad \hat{\psi}_5(\zeta) = 4\zeta_2\zeta_3, \qquad \hat{\psi}_6(\zeta) = 4\zeta_3\zeta_1 \tag{2.3.18}$$

where the coordinate functions ζ_i are defined by (2.3.13).

The piecewise-quadratic global approximation is continuous across each element interface since the three nodal values on each side define a unique quadratic on the interface between adjacent elements. Representative element shape functions and corresponding C^0-basis functions are shown in Fig. 2.4.

The Pascal Triangle: Blaise Pascal, in his studies of mathematical probability, observed that the classical arithmetic triangle could be used as a geometric pattern to solve problems in combinatorics. In particular, he noted that the binomial expansion is

$$(x + y)^k = x^k + \binom{k}{1} x^{k-1}y + \cdots + \binom{k}{k-1} xy^{k-1} + y^k,$$
$$\binom{m}{n} = \frac{m!}{n!(m-n)!} \tag{2.3.19}$$

and observed that its coefficients could be directly interpreted as row k of the arithmetic triangle. In view of (2.3.19), this new Pascal triangle provides a

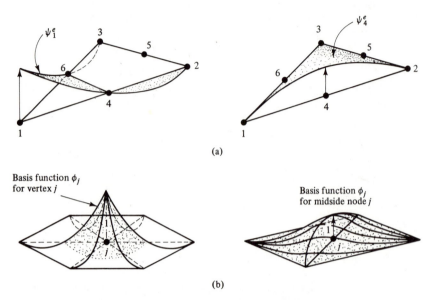

(a)

(b)

FIGURE 2.4 (a) Quadratic element shape functions ψ_j^e; (b) their corresponding global piecewise-quadratic basis functions ϕ_j.

simple pattern for characterizing complete polynomials on triangular elements and associating the requisite nodal interpolation points.

Assume that the Pascal triangle is aligned as in Fig. 2.5, so that the entries in each horizontal row k correspond to the functions in the binomial expansion of $(x + y)^k$ as marked. The first entry in row 0 is the constant function; in row 1 are the two linear functions; in row 2 the three quadratics, and so on. The collection of all entries from row 0 up to and including row k defines a basis for a complete polynomial expansion of degree k on an element. Summing the arithmetic progression $1, 2, \ldots, k + 1$ the complete polynomial of degree k requires $(k + 1)(k + 2)/2$ interpolation nodes. Of these $k + 1$ lie on each side of the element with $k(k + 1)/2 - 3(k + 2)$ in the interior.

By viewing the Pascal triangle in the figure as a hierarchic pattern of nested triangular elements, we can directly define complete C^0-Lagrange elements of any degree. For example, the innermost triangle is defined by rows 0 and 1. By locating the interpolation nodes at the vertices $1, x, y$, we define the C^0-linear triangle. Similarly, the triangle within rows 0–2, with nodes at the vertices and midsides again positioned according to the Pascal pattern, defines the C^0-quadratic triangle.

There is an analogous pattern for C^0-tetrahedral elements in three dimensions (Fig. 2.6). Simply note now that each "row" k is a plane that must correspond to the entries of the trinomial expansion $(x + y + z)^k$. The

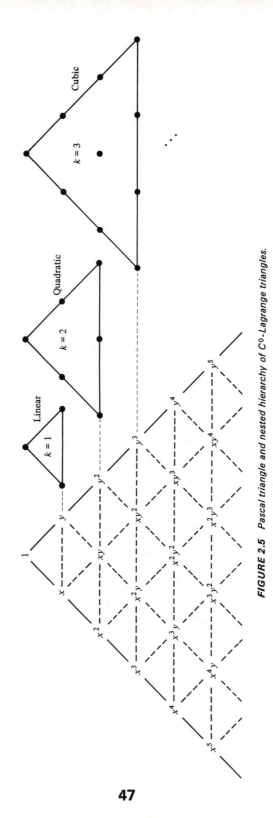

FIGURE 2.5 Pascal triangle and nested hierarchy of C^0-Lagrange triangles.

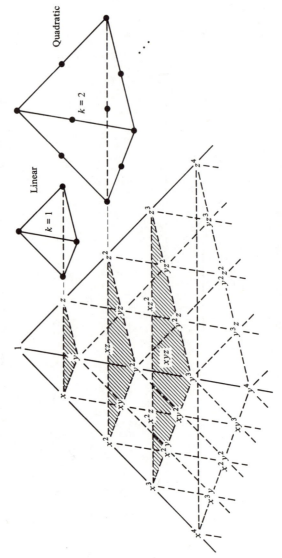

FIGURE 2.6 Pascal tetrahedron and nested C^0-Lagrange tetrahedra.

pattern is formed as a tetrahedron with generators x^k, y^k, and z^k. The shaded triangular "base" sections in the figure correspond to the trinomial expansions in the coordinate variables x, y, z to degrees $k = 1, 2, 3$, and 4. Again, the hierarchies of C^0-tetrahedra are easily identified. The simplest C^0-tetrahedron is defined by rows 0 and 1. The interpolation nodes are the vertices $1, x, y$, and z. In this case the local element expansion becomes

$$u_h^e(x, y, z) = a_1 + a_2 x + a_3 y + a_4 z \qquad (2.3.20)$$

Interpolating the nodal values $u_i^e = u_e(x_i, y_i, z_i)$, $i = 1, 2, 3, 4$, and solving for a_i, we determine the linear interpolant

$$u_h^e(x, y, z) = \sum_{j=1}^{4} u_j^e \psi_j^e(x, y, z) \qquad (2.3.21)$$

Here $\psi_j^e(x, y, z)$ are linear Lagrange polynomials on the tetrahedron and satisfy $\psi_j^e(x_i, y_i, z_i) = \delta_{ij}$. Since the three nodal values on any face define a unique linear function, the interpolant is continuous across the interface between adjacent elements.

Generalizing our notion of a patch, let *rspq* denote the vertices of a tetrahedral element. Consider those tetrahedra with common vertex at node *r* (*r* is the nodal center of the patch). The global function ϕ_r for node *r* is defined uniquely by the union of element shape functions at node *r* as described on each element by ψ_j^e in (2.3.21). Then ϕ_r is zero outside and on the exterior surface of the patch and is a linear function of the coordinates x, y, z in each of the tetrahedra constituting the patch.

Applying the reasoning used for triangular elements, we note that the coordinate functions x, y, z are themselves linear on a tetrahedron. Redefining the linear Lagrange functions ψ_j^e in (2.3.21) as coordinate variables $\zeta_j, j = 1, 2, 3, 4$,

$$x = \sum_{j=1}^{4} x_j \zeta_j, \quad y = \sum_{j=1}^{4} y_j \zeta_j, \quad z = \sum_{j=1}^{4} z_j \zeta_j \quad \text{with} \sum_{j=1}^{4} \zeta_j = 1 \qquad (2.3.22)$$

In fact, from (2.3.21) or by analogy with the triangle, the normalized coordinates ζ_i, $i = 1, 2, 3, 4$, are volume coordinates for the tetrahedron. Given a general point (x, y, z) inside the tetrahedron of volume V defined by (x_i, y_i, z_i), $i = 1, 2, 3, 4$, let the volume of the subtetrahedron formed by (x, y, z) and the triangular face opposite node *j* be v_j. Then $\zeta_j = v_j/V$ are the normalized coordinates. As (x, y, z) traverses a plane parallel to face *j*, the volume v_j is constant. Surfaces parallel to each of the faces thus constitute coordinate surfaces ζ_i, $i = 1, 2, 3, 4$. The master tetrahedron $\hat{\Omega}$ is the right tetrahedron with generators $\zeta_2, \zeta_3, \zeta_4$ and with $\zeta_1 = 1 - \zeta_2 - \zeta_3 - \zeta_4$.

The C^0-quadratic tetrahedron has 10 nodes, four at the vertices and one at the midside of each edge. The quadratic shape functions are easily constructed on $\hat{\Omega}$ and are:

$$\zeta_1(2\zeta_1 - 1), \quad \zeta_2(2\zeta_2 - 1), \quad \zeta_3(2\zeta_3 - 1), \quad \zeta_4(2\zeta_4 - 1)$$

at the vertices

and (2.3.23)

$$4\zeta_1\zeta_2, \quad 4\zeta_2\zeta_3, \quad 4\zeta_3\zeta_1, \quad 4\zeta_1\zeta_4, \quad 4\zeta_2\zeta_4, \quad 4\zeta_3\zeta_4 \quad \text{at respective}$$
edge nodes

The interval, triangle, and tetrahedron are the first three members ($N = 1, 2, 3$) of a general category termed N-simplex domains. An N-simplex is a closed region $\Omega_e \subset \mathbb{R}^N$ defined by $N + 1$ distinct points and such that these $(N + 1)$ points not lie in any $(N - 1)$-dimensional hyperplane.

Tensor-Product Elements: The triangle and tetrahedron are the polygon and polyhedral shapes with the minimum number of sides in two and three dimensions. From the Pascal triangle, we observe that they are optimal for defining complete C^0-polynomials in two and three dimensions. The theory of Lagrange interpolation, however, is most extensively developed for the interval and its most natural extension is to the square in two dimensions, cube in three dimensions, and N-hypercube in N dimensions. If $x \in [x_{i-1}, x_i] \equiv I_{x_i}$, $y \in [y_{j-1}, y_j] \equiv I_{y_j}$, and $z \in [z_{k-1}, z_k] \equiv I_{z_k}$ are intervals in each of the Cartesian coordinate directions, the corresponding rectangle and rectangular prism in two and three dimensions are the product domains

$$I_{x_i y_j} \equiv I_{x_i} \times I_{y_j} = [x_{i-1}, x_i] \times [y_{j-1}, y_j]$$
$$= \{(x, y) \mid x \in I_{x_i}, y \in I_{y_j}\}$$

(2.3.24)

and

$$I_{x_i y_j z_k} \equiv I_{x_i} \times I_{y_j} \times I_{z_k} = [x_{i-1}, x_i] \times [y_{j-1}, y_j] \times [z_{k-1}, z_k]$$
$$= \{(x, y, z) \mid x \in I_{x_i}, y \in I_{y_j}, z \in I_{z_k}\}$$

(2.3.25)

Simple stretching transformations $\xi(x)$, $\eta(y)$, $\zeta(z)$ of the standard form $\xi = [2x - (x_{i-1} + x_i)]/(x_i - x_{i-1})$ map the rectangle and prism to the master square and cube elements $\hat{\Omega}$. In addition, as we shall see subsequently, simple bilinear and trilinear transformations map these master elements to the general quadrilateral and quadrilateral prism.

If the discretization consists of rectangles Ω_e, the interfaces between adjacent rectangles are lines $x = $ constant and $y = $ constant. In this case the bilinear interpolant of function values at the corners reduces to a linear function along any side and the finite element approximation is obviously continuous across the interfaces between adjacent elements. On a rectangle

Ω_e the bilinear function may be easily factored as a product of linear functions of x and y, respectively. That is,

$$u_h^e(x, y) = (b_1 + b_2 x)(c_1 + c_2 y) = a_1 + a_2 x + a_3 y + a_4 xy \qquad (2.3.26)$$

where $a_1, \ldots, c_2$ are constants. We refer to the product of monomial bases $(1, x)$ and $(1, y)$ as defined in (2.3.26) as a *tensor-product basis*, written

$$
\left.
\begin{aligned}
(1, x) \times (1, y) &\equiv (1, x, y, xy) \\[6pt]
\{x^i\} \times \{y^j\} &= \{x^i y^j\}, \qquad i = 0, 1; \quad j = 0, 1
\end{aligned}
\right\} \qquad (2.3.27)
$$

or

   A general tensor-product basis in x and y consists of all possible combinations $\{x^i y^j\}$, $i = 0, 1, \ldots, m$ and $j = 0, 1, \ldots, n$, where m and n are the degrees of the constituent monomial bases $\{x^i\}$, $i = 0, 1, \ldots, m$, and $\{y^j\}$, $j = 0, 1, \ldots, n$. A tensor-product formulation can be utilized to construct the element shape functions directly from those introduced for Lagrange interpolation in one dimension. For bilinear interpolation on a rectangle, instead of monomial basis functions $(1, x)$ and $(1, y)$ we may directly employ the linear Lagrange interpolation functions $(\psi_1^e(x), \psi_2^e(x))$ and $(\psi_1^e(y), \psi_2^e(y))$. The tensor-product basis is

$$
\left.
\begin{aligned}
\{\psi_1^e(x), \psi_2^e(x)\} &\times \{\psi_1^e(y), \psi_2^e(y)\} \\[6pt]
\{\psi_{ij}^e(x, y)\} &= \{\psi_i^e(x)\psi_j^e(y)\}, \qquad i = 1, 2; \quad j = 1, 2
\end{aligned}
\right\} \qquad (2.3.28)
$$

or

For each corner node let $k \equiv (i, j)$, where i and j are local nodal identifiers in the x and y directions on the element. The interpolation shape functions are simply products of one-dimensional Lagrange polynomials (Fig. 2.7). The bilinear Lagrange interpolation polynomial on Ω_e is then

$$u_h^e(x, y) = \sum_{k=1}^{4} u_k^e \psi_k^e(x, y) = \sum_{i,j=1}^{2} u_{ij}^e \psi_i^e(x)\psi_j^e(y) \qquad (2.3.29)$$

where $u_{ij}^e = u_e(x_i, y_j) \equiv u_k^e$, for node k at (x_i, y_j).

   Similarly, if quadratic interpolation functions are used in the x and y directions, the tensor-product basis is biquadratic. In general,

$$\{\psi_{ij}^e(x, y)\} = \{\psi_i^e(x)\psi_j^e(y)\}, \qquad i = 1, 2, \ldots, m; \quad j = 1, 2, \ldots, n \qquad (2.3.30)$$

defines a tensor-product basis of Lagrange type that is of degree $m - 1$ in x and $n - 1$ in y and the element interpolant is

$$u_h^e(x, y) = \sum_{k=1}^{mn} u_k^e \psi_k^e(x, y) = \sum_{i=1}^{m} \sum_{j=1}^{n} u_{ij}^e \psi_i^e(x)\psi_j^e(y) \qquad (2.3.31)$$

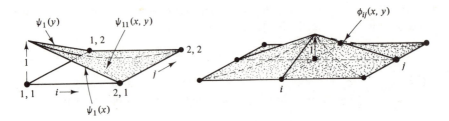

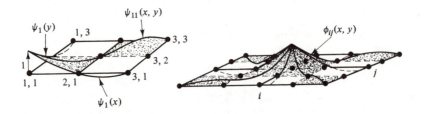

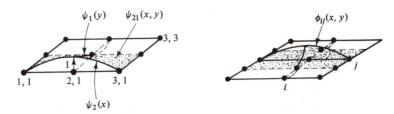

FIGURE 2.7 *Representative element and global basis functions for bilinear and biquadratic interpolation.*

From (2.3.30) there are mn basis functions associated with nodal points $l = (r, s)$ such that $\psi_k^e(\mathbf{x}_l) = \delta_{kl}$ or

$$\{\psi_{ij}^e(x_r, y_s)\} = \{\psi_i^e(x_r)\psi_j^e(y_s)\} = \delta_{ir}\,\delta_{js}$$

where δ_{pq} is the Kronecker delta.

The coordinates of the nodal points $l = (r, s)$ are obtained as the tensor product of the interpolation points x_r and y_s on the x and y intervals I_x and I_y defining $\bar{\Omega}_e = I_x \times I_y$.

The global basis functions ϕ_k are derived analogously as the tensor product of piecewise-polynomial basis functions in x and y or, alternatively, by patching together appropriate local shape functions on those elements adjacent to a given node. Along the interface between adjacent rectangular elements the interpolation polynomials on either side match, so the global basis is continuous.

The extension to three-dimensional domains discretized as rectangular prisms is immediate. On $\bar{\Omega}_e = I_{x_i} \times I_{y_j} \times I_{z_k}$, where I_{x_i}, I_{y_j}, and I_{z_k} are intervals in the coordinate directions, the tensor-product basis becomes

$$\{\psi_i^e(x)\} \times \{\psi_j^e(y)\} \times \{\psi_k^e(z)\} = \{\psi_i^e(x)\psi_j^e(y)\psi_k^e(z)\} = \{\psi_{ijk}^e(x, y, z)\}$$
$$\text{for } i = 1, 2, \ldots, l; \quad j = 1, 2, \ldots, m; \quad k = 1, 2, \ldots, n \tag{2.3.32}$$

The Lagrange polynomial interpolant of $u(x, y, z)$ on Ω_e is

$$u_h^e(x, y, z) = \sum_{i=1}^{l} \sum_{j=1}^{m} \sum_{k=1}^{n} u_{ijk}^e \psi_i^e(x) \psi_j^e(y) \psi_k^e(z) \tag{2.3.33}$$

where u_{ijk}^e are the interpolated nodal values. The element shape functions satisfy $\psi_{ijk}^e(x_r, y_s, z_t) = \delta_{ir} \delta_{js} \delta_{kt}$ at nodes (x_r, y_s, z_t). On an interface between two elements the approximation reduces to a function of two variables alone (e.g., on $z = $ constant u_h^e and ψ_{ijk}^e are functions of x and y alone) and the approximation is globally continuous.

Once again, it is convenient in describing tensor-product elements to appeal to a modified pattern of the Pascal triangle and tetrahedron. The generating lines from the apex are identified with the monomials x^i, y^j, z^k from which the tensor product bases $\{x^i y^j\}$ and $\{x^i y^j z^k\}$, respectively, for $i = 1, 2, \ldots, l; j = 1, 2, \ldots, m; k = 1, 2, \ldots, n$ are determined. Assuming that $l = m = n$, the relevant segments of the Pascal triangle and Pascal tetrahedron are the Pascal square and Pascal cube, respectively, shown in Figs. 2.8 and 2.9.

The nested hierarchies marked in the figures indicate the degree of the tensor-product basis, the functions defining the basis and the location of the nodes for the element interpolation functions. Note that the interpolation polynomials are complete only to the degree k of the generating monomials, although terms of degree up to $2k$ or $3k$ are included in the basis for two- or three-dimensional elements, respectively.

All of the preceding element interpolation results may be set on a master square or cube. The interpolant in the (x, y)-plane can then be constructed through a mapping from $\hat{\Omega}$ to Ω_e.* This is precisely the strategy employed in treating convex† quadrilateral elements in the (x, y)-plane. A polynomial basis on $\hat{\Omega}$ is transformed by a linear transformation to a basis on Ω_e. How-

---

* Henceforth, we shall not distinguish between the interior Ω_e of a finite element and its closure $\bar{\Omega}_e$ unless confusion is likely or the situation requires such a distinction.

† A domain is convex if the line joining any two points in the domain is entirely within the domain. For a quadrilateral this implies that no interior corner angle may exceed π.

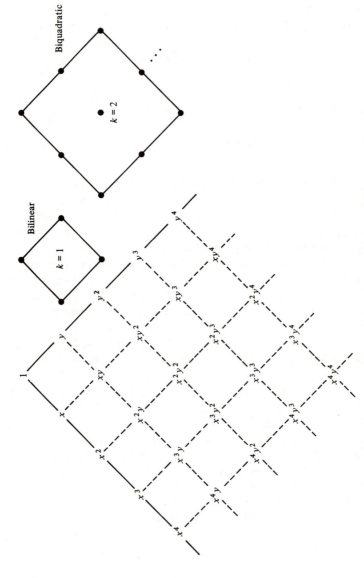

FIGURE 2.8 *Pascal square for tensor-product C^0-Lagrange interpolation. Nested squares define bilinears ($k = 1$), biquadratics ($k = 2$), and so on.*

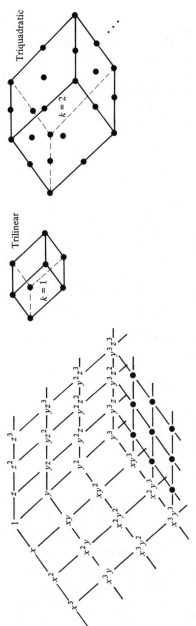

FIGURE 2.9 Pascal cube for tensor-product C^0-Lagrange interpolation in three dimensions.

ever, the map is not linear for quadrilaterals except in the special case of rectangular elements. We discuss the ramifications of this point in greater detail in Section 2.7.

The forward map from Ω_e to the master square $\hat{\Omega}$ is difficult to construct but the inverse map from $\hat{\Omega}$ to Ω_e which is essential to the finite element formulation is easily constructed. For convenience, let $\hat{\Omega} = \{(\xi, \eta) : \xi \in [0, 1], \eta \in [0, 1]\}$. Since the boundary lines $\xi = 0, 1$ and $\eta = 0, 1$ of $\hat{\Omega}$ map to the linear sides of Ω_e, the transformation must be linear on the boundaries. Considering the four sides in turn, we can write immediately the bilinear mapping for a given element $\bar{\Omega}_e$ with corners (x_i, y_i), $i = 1, 2, 3, 4$.

$$x(\xi, \eta) = x_1 + (x_2 - x_1)\xi + (x_3 - x_1)\eta + (x_4 - x_3 - x_2 + x_1)\xi\eta$$
$$y(\xi, \eta) = y_1 + (y_2 - y_1)\xi + (y_3 - y_1)\eta + (y_4 - y_3 - y_2 + y_1)\xi\eta$$

Regrouping terms, we obtain

$$x = \sum_{j=1}^{4} x_j \hat{\psi}_j(\xi, \eta), \qquad y = \sum_{j=1}^{4} y_j \hat{\psi}_j(\xi, \eta) \qquad (2.3.34)$$

where

$$\{\hat{\psi}_j(\xi, \eta)\} = \{(1 - \xi)(1 - \eta), \xi(1 - \eta), (1 - \xi)\eta, \xi\eta\}$$
$$= \{(1 - \xi, \xi) \times (1 - \eta, \eta)\}$$

is the underlying tensor-product basis on the master square $\hat{\Omega}$.

The Jacobian of the transformation (2.3.34) is

$$|\mathbf{J}| = x_\xi y_\eta - x_\eta y_\xi \qquad (2.3.35)$$

Writing $a_1 = x_2 - x_1, b_1 = x_3 - x_1, c_1 = x_4 - x_3 - x_2 + x_1$ and $a_2 = y_2 - y_1, b_2 = y_3 - y_1, c_2 = y_4 - y_3 - y_2 + y_1$ in (2.3.34) and evaluating $|\mathbf{J}|$, terms in $\xi\eta$ cancel to yield

$$|\mathbf{J}| = (a_1 b_2 - a_2 b_1) + (a_1 c_2 - a_2 c_1)\xi + (b_2 c_1 - b_1 c_2)\eta \qquad (2.3.36)$$

Since $|\mathbf{J}|$ is linear, the map will be invertible if $|\mathbf{J}| \neq 0$ in $\hat{\Omega}$. To examine this condition further, we note that at a corner $|\mathbf{J}|$ is the magnitude of the cross product of two vectors aligned along the sides forming the corner. The sign of this cross product is determined by $\sin \theta$, where θ is the interior angle at the corner. It follows that $\sin \theta$ and so $|\mathbf{J}|$ will be negative at a reentrant corner since $\theta > \pi$ here. Consequently, $|\mathbf{J}|$ will not change sign in $\hat{\Omega}$ provided that the quadrilateral Ω_e is convex (Fig. 2.10). Similarly, in order that the trilinear map from the master cube to a quadrilateral brick be invertible, the three-dimensional element must also be convex.

Finally, observe that the *sides* of the master square and the faces of the

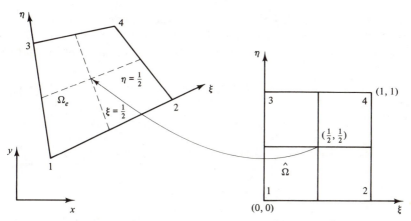

FIGURE 2.10 Bilinear map from $\hat{\Omega}$ to Ω_e is invertible if Ω_e is convex.

master cube map linearly to the corresponding sides and faces of Ω_e in $\mathbb{R}^2$ and $\mathbb{R}^3$. Hence, the degree of the polynomial interpolant on the interface between adjacent elements is unchanged from that on the master element, and this ensures that the approximation is continuous across an interface. Note that in two dimensions we can use triangles and quadrilaterals together and maintain C^0-continuity provided that the polynomial degrees match on interfaces between such elements. For example, we may use linear triangles with bilinear quadrilaterals, quadratic triangles with biquadratic quadrilaterals, and so on.

Transfinite Finite Elements: It is convenient to express symbolically the interpolation of a function $u = u(x)$ by Lagrange interpolation polynomials as the result of a *projection* Π_x of a smooth function u onto a class of piecewise polynomials to which its interpolant u_h belongs. Thus, it is meaningful to write $u_h(x) = \Pi_x u(x)$ for the Lagrange interpolant of u if we define

$$\Pi_x u(x) = \sum_{i=1}^{M+1} u(x_i) L_i^{(M)}(x) \tag{2.3.37}$$

where $L_i^{(M)}$ is the Lagrange interpolation polynomial of degree M for node i defined as in (2.3.7). Our local shape functions for rectangular elements based on tensor products of polynomials of degree k in x and m in y are then of the form

$$\psi_{ij}^e(x, y) = L_i^{(k)}(x) L_j^{(m)}(y) \tag{2.3.38}$$

and we have for element Ω_e,

$$u_h^e(x, y) = \Pi_x \Pi_y u(x, y) = \sum_{i=1}^{k+1} \sum_{j=1}^{m+1} u(x_i, y_j) \psi_{ij}^e(x, y) \tag{2.3.39}$$

as noted earlier.

There is an alternative method for constructing interpolants which may have some advantages over the use of tensor products in certain situations. This involves the so-called *Boolean sum* of projections defined for two dimensions by

$$\Pi_x \oplus \Pi_y = \Pi_x + \Pi_y - \Pi_x\Pi_y \tag{2.3.40}$$

For example,

$$\begin{aligned} \Pi_x \oplus \Pi_y u(x, y) = \sum_{i=1}^{k+1} u(x_i, y)L_i^{(k)}(x) + \sum_{j=1}^{m+1} u(x, y_j)L_j^{(m)}(y) \\ - \sum_{i=1}^{k+1}\sum_{j=1}^{m+1} u(x_i, y_j)L_i^{(k)}(x)L_j^{(m)}(y) \end{aligned} \tag{2.3.41}$$

The interesting property of such Boolean sums is that $\Pi_x \oplus \Pi_y u(x_i, y) = u(x_i, y)$ and $\Pi_x \oplus \Pi_y u(x, y_j) = u(x, y_j)$. Thus, the operation $\Pi_x \oplus \Pi_y$ interpolates $u(x, y)$ along the *lines* (not nodal points) $x = x_i, y = y_j$. These interpolants have been used by Gordon [1971] and Gordon and Hall [1973] to construct finite element interpolants of smooth functions. Because the interpolation is along lines rather than nodes, Gordon refers to these as *transfinite finite elements*.

One interesting property of transfinite finite element interpolation is its superior accuracy over ordinary tensor products. Of course, considerably more data are needed to apply these interpolation methods. As an indication of why this improved accuracy is obtained, let R_x and R_y denote the remainder projections,

$$R_x = I - \Pi_x, \qquad R_y = I - \Pi_y \tag{2.3.42}$$

I being the identity. Then R_x and R_y reflect the truncation error in the interpolants $\Pi_x u$ and $\Pi_y u$. Now

$$I - \Pi_x\Pi_y = R_x + R_y - R_xR_y = R_x \oplus R_y \tag{2.3.43}$$

whereas

$$I - \Pi_x \oplus \Pi_y = R_xR_y \tag{2.3.44}$$

Thus, in the usual tensor-product schemes the truncation error is a linear combination of the individual errors produced by Π_x and Π_y, while the Boolean sum leads to a product of R_x and R_y and, consequently, a higher-order error.

As an example of transfinite finite element interpolation, we mention the case in which $u = u(x, y)$ is given on the sides of the unit square $\Omega = [0, 1] \times$

[0, 1] and linear Lagrange interpolation is to be used in x and y. The Boolean sum then yields a "bilinearly blended" transfinite Lagrange interpolant of the form

$$
\begin{aligned}
\Pi_x \oplus \Pi_y u = {}& (1 - x)u(0, y) + xu(1, y) + (1 - y)u(x, 0) \\
& + yu(x, 1) - (1 - x)(1 - y)u(0, 0) \\
& - (1 - x)yu(0, 1) - x(1 - y)u(1, 0) \\
& - xyu(1, 1)
\end{aligned}
\tag{2.3.45}
$$

Finite elements based on formulas such as this have been used for generating smooth surfaces for the interpolation of discrete data for the design of automobile shapes.

EXERCISES

2.3.5 Derive the linear shape functions ψ_j^e, $j = 1, 2, 3$, given in equation (2.3.13). Sketch these local functions and a representative global basis function.

2.3.6 Sketch representative element and global functions for piecewise-linear interpolation on a triangulation of Ω, where u is specified only at the midside locations.

2.3.7 Derive cubic Lagrange polynomials for a triangle.

2.3.8 Formulate a least-squares approximation for $u(x, y)$ on Ω_h using piecewise-linear triangular elements. Leave your results in integral form but give both the global and element contributions.

2.3.9 Consider a master square and its bilinear map to a convex quadrilateral. Allow one corner of the quadrilateral to approach the opposite corner so that a reentrant corner is produced. Use geometric reasoning to conclude that quadrilaterals that are not convex should not be used in a finite element mesh.

2.3.10 Form the Jacobian for the trilinear map from a reference cube to a "quadrilateral brick" to determine the geometric conditions on the element for invertibility of the map.

[*HINT*: Motivated by the analysis for the quadrilateral in the text, use a similar argument employing the fact that the triple cross product of edge vectors determines the volume of the solid element. Verify your result using geometric reasoning as in Exercise 2.3.9.]

2.4 CONSTRAINED C⁰-ELEMENTS

2.4.1 Interior Constraints

Biquadratic approximation on a master square $\xi \in [-1, 1]$, $\eta \in [-1, 1]$ is defined by

$$\hat{U}(\xi, \eta) = \sum_{i,j=1}^{3} u_{ij}^e \hat{\psi}_i(\xi) \hat{\psi}_j(\eta) \tag{2.4.1}$$

where $\{\hat{\psi}_i(z)\} = \{\frac{1}{2}z(z-1), 1-z^2, \frac{1}{2}z(z+1)\}$ are the Lagrange polynomials for quadratic interpolation at $z = -1$, $z = 0$, and $z = 1$ in one dimension. There are nine nodal points (ξ_i, η_j), eight being on the boundary and one in the interior at $(0, 0)$. In the (x, y) discretization, the global basis function associated with the interior node is nonzero on the element interior alone, whereas the global basis functions for nodes on the element boundary are nonzero on adjacent elements as well. Since the basis function associated with the internal node is local to the element, it is particularly easy to modify the element interpolation formula locally so that the only interpolated values are on the boundary.

First we separate the shape function associated with the node at $(\xi_2, \eta_2) = (0, 0)$ in (2.4.1) according to

$$\hat{U}(\xi, \eta) = \sum_{\substack{i,j=1 \\ (i,j \neq 2)}}^{3} u_{ij}^e \hat{\psi}_i(\xi) \hat{\psi}_j(\eta) + u_{22}^e \hat{\psi}_2(\xi) \hat{\psi}_2(\eta) \tag{2.4.2}$$

Now we constrain u_{22}^e to be a linear combination of boundary values u_{ij}^e, $i, j \neq 2$, of the form

$$u_{22}^e = \sum_{\substack{i,j=1 \\ (i,j \neq 2)}}^{3} a_{ij} u_{ij}^e \tag{2.4.3}$$

where a_{ij} are constants defining the form of this constraint. Substituting this result into (2.4.2), we obtain the new expression

$$\bar{U}(\xi, \eta) = \sum_{\substack{i,j=1 \\ (i,j \neq 2)}}^{3} u_{ij}^e [\hat{\psi}_i(\xi) \hat{\psi}_j(\eta) + a_{ij} \hat{\psi}_2(\xi) \hat{\psi}_2(\eta)] \tag{2.4.4}$$

If we define the bracketed expression in (2.4.4) as $\bar{\psi}_{ij}(\xi, \eta)$ for each $i, j \neq 2$, $\bar{\psi}_{ij}$ represents a new shape function associated with the boundary node at (ξ_i, η_j). The element interpolant is now of the form

$$\bar{U}(\xi, \eta) = \sum_{\substack{i,j=1 \\ (i,j \neq 2)}}^{3} u_{ij}^e \bar{\psi}_{ij}(\xi, \eta) \tag{2.4.5}$$

For example, if u_{22}^e is constrained to be the arithmetic average of the four midside values then $a_{21} = a_{32} = a_{23} = a_{12} = \frac{1}{4}$ and all other $a_{ij} = 0$, whence

$$\bar{\psi}_{ij}(\xi, \eta) = \begin{cases} \hat{\psi}_i(\xi)\hat{\psi}_j(\eta), & \text{corner points } (i, j) \\ \hat{\psi}_i(\xi)\hat{\psi}_j(\eta) + \frac{1}{4}\hat{\psi}_2(\xi)\hat{\psi}_2(\eta), & \text{midside points } (i, j) \end{cases} \tag{2.4.6}$$

Similarly, we can set $a_{ij} = \frac{1}{2}$ to yield the form

$$\bar{\psi}_{ij}(\xi, \eta) = \hat{\psi}_i(\xi)\hat{\psi}_j(\eta) + \frac{1}{2}\hat{\psi}_2(\xi)\hat{\psi}_2(\eta)$$

for midside node (i, j) as shown in Fig. 2.11 for $i = 1, j = 2$.

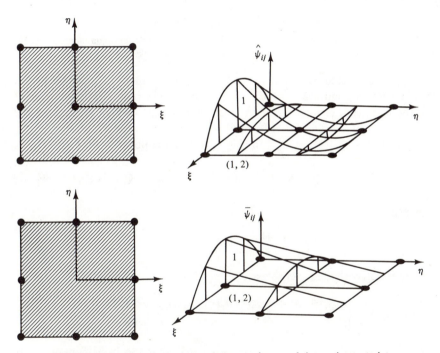

FIGURE 2.11 *Biquadratic interpolation and use of internal constraints on the element basis to "remove" an interior node.*

Other averaging strategies may be employed to constrain the behavior of the interpolation polynomial in the interior, thus removing the internal degree of freedom, yet retaining the quadratic polynomial degree on the boundary and also in the interior. For example, the constraint

$$u_{22} = \frac{1}{2}(u_{21} + u_{32} + u_{23} + u_{12}) - (u_{11} + u_{31} + u_{33} + u_{13})$$

yields the 8-node "serendipity" element and corresponds to suppressing the quartic term $\xi^2\eta^2$ in the biquadratic basis.

2.4.2 Interface Constraints: Transition Elements

The idea of constraining the polynomial form can be utilized to mix elements of different degree and maintain C^0-continuity by means of "transition" elements. For example, in Fig. 2.12 an eight-node transition element

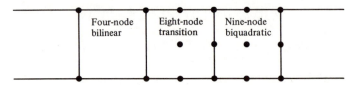

FIGURE 2.12 *Global C^0-interpolation using eight-node transition element between four-node bilinear and nine-node biquadratic elements.*

is to be constructed such that the polynomial interpolant is linear on its left boundary and quadratic on the right boundary. The biquadratic interpolant on the transition element is again given by (2.4.1), to which we append the constraint that on $\xi = -1$, $\hat{U}(-1, \eta)$ be linear.

Setting $u_{12}^e = \frac{1}{2}(u_{11}^e + u_{13}^e)$ in (2.4.1) to provide linearity on $\xi = -1$,

$$\hat{U}(\xi, \eta) = \sum_{\substack{i,j=1 \\ (i,j)\neq(1,2)}}^{3} u_{ij}^e \hat{\psi}_i(\xi)\hat{\psi}_j(\eta) + \frac{1}{2}(u_{11}^e + u_{13}^e)\hat{\psi}_1(\xi)\hat{\psi}_2(\eta) \qquad (2.4.7)$$

whence

$$\hat{U}(\xi, \eta) = \sum_{\substack{i,j=1 \\ (i,j)\neq(1,2)}}^{3} u_{ij}^e \bar{\psi}_{ij}(\xi, \eta) \qquad (2.4.8)$$

where the new shape functions are

$$\bar{\psi}_{ij}(\xi, \eta) = \begin{cases} \hat{\psi}_i(\xi)\hat{\psi}_j(\eta) + \frac{1}{2}\hat{\psi}_1(\xi)\hat{\psi}_2(\eta), & (i,j) = (1,1), (1,3) \\ \hat{\psi}_i(\xi)\hat{\psi}_j(\eta), & \text{otherwise} \end{cases} \qquad (2.4.9)$$

and $(i,j) \neq (1,2)$. On $\xi = -1$ we can verify that the basis functions are $\bar{\psi}_{11}(-1, \eta) = \frac{1}{2}(1 - \eta)$ and $\bar{\psi}_{13} = \frac{1}{2}(1 + \eta)$, as stipulated through the constraint.

The Lagrange families described above are C^0-elements on Ω. The global interpolation functions are continuous across interelement boundaries and hence the first derivatives are square-integrable. This enables us to apply these elements in finite element Galerkin formulations of second-order boundary-value problems in the usual way. We note in passing that interelement continuity might also be interpreted as a constraint on the global approximation. In this case we might seek to satisfy this continuity constraint in some weighted-average sense, perhaps within the variational formulation of the problem, as in the Lagrange multiplier and hybrid methods discussed in Chapter 3.

EXERCISES

2.4.1 Develop a class of rectangular transition elements to permit concurrent use of elements generated by polynomials of degree m and n, respectively.

2.4.2 Develop the interpolation functions for a "quadratic" triangle which is constrained so that the approximation on two sides is linear. Let $u = u(x, y)$ be a complete quadratic and find the local error estimate for this constrained interpolation formula.

2.4.3 Verify that the 8-node serendipity basis is obtained from the 9-node biquadratics using the constraint given in Section 2.4.1.

2.5 HERMITE FAMILIES OF ELEMENTS

When fourth-order problems, such as beam or plate-bending problems and certain viscous flow problems are considered, the variational functional involves second derivatives of the admissible trial and test functions. This implies that first and second derivatives should be square-integrable so that both function and normal derivative are to be continuous across an interface between elements for conformity. The need for smoother (C^1) global basis functions is also encountered in second-order problems when collocation finite element methods are utilized. With these situations in mind, we next examine Hermite families of elements in which the global basis functions are smooth C^1-piecewise polynomials.

In Hermite interpolation both function values and derivatives of various orders are interpolated. Since the approximation is smooth in the element interior, interpolation of derivatives is confined to nodes on the interelement boundaries to ensure that the derivative of the interpolant is continuous across the interface. Again, we limit consideration to piecewise-polynomial interpolation.

2.5.1 One-Dimensional Hermite Elements

We begin by examining Hermite interpolation in one dimension and consider the simple example of cubic interpolation on an element. Let $\bar{\Omega}_e = [x_{i-1}, x_i]$ be an arbitrary element in a finite element partition of the interval $[a, b]$. A C^1-cubic element is obtained by interpolating the function and derivatives at the ends x_{i-1} and x_i of each element.

As in the treatment of Lagrange elements, we develop the shape functions on a master element. The element $\bar{\Omega}_e$ is transformed linearly to $\hat{\Omega} = [-1, 1]$ by the map $\xi = [2x - (x_{i-1} + x_i)]/(x_i - x_{i-1})$. On $\hat{\Omega}$, interpolating $\hat{u}$ and $\hat{u}' \equiv d\hat{u}/d\xi$ at the end nodes $j = 1, 2$, the cubic Hermite expansion has the form

$$\hat{U}(\xi) = \sum_{j=1}^{2} \hat{u}_j \hat{\psi}_j^0(\xi) + \sum_{j=1}^{2} \hat{u}_j' \hat{\psi}_j^1(\xi) \tag{2.5.1}$$

where the Hermite basis functions $\{\hat{\psi}_j^0, \hat{\psi}_j^1\}$ satisfy the interpolation properties at end nodes $\xi_1 = -1, \xi_2 = 1$:

$$\hat{\psi}_j^0(\xi_k) = \delta_{jk}, \qquad \hat{\psi}_j^1(\xi_k) = 0$$
$$\frac{d\hat{\psi}_j^0}{d\xi}(\xi_k) = 0, \qquad \frac{d\hat{\psi}_j^1}{d\xi}(\xi_k) = \delta_{jk} \tag{2.5.2}$$

for local nodal indices $j = 1, 2$ and $k = 1, 2$. Using properties (2.5.2) we may construct the Hermite cubics directly as

$$\hat{\psi}_1^0(\xi) = \tfrac{1}{4}(\xi - 1)^2(\xi + 2), \qquad \hat{\psi}_2^0(\xi) = \tfrac{1}{4}(\xi + 1)^2(2 - \xi)$$
$$\hat{\psi}_1^1(\xi) = \tfrac{1}{4}(\xi - 1)^2(\xi + 1), \qquad \hat{\psi}_2^1(\xi) = \tfrac{1}{4}(\xi + 1)^2(\xi - 1) \tag{2.5.3}$$

By means of the linear map $\xi \longrightarrow x$ we can transform the shape functions on $\hat{\Omega}$ to the corresponding shape functions ψ_j^{0e} and ψ_j^{1e} on $\bar{\Omega}_e$ (Fig. 2.13).

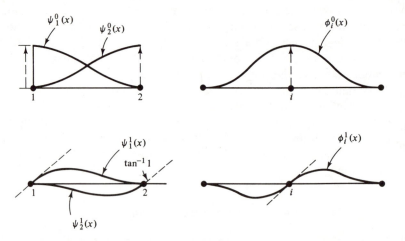

FIGURE 2.13 *Local and global basis functions for C^1-Hermite cubics in one dimension.*

The form of the approximation (2.5.1) on $\bar{\Omega}_e$ then becomes

$$u_h^e(x) = \sum_{j=1}^{2} u_j^e \psi_j^{0e}(x) + \frac{h_e}{2} \sum_{j=1}^{2} \left(\frac{du}{dx}\right)_j^e \psi_j^{1e}(x) \tag{2.5.4}$$

where we have used the chain rule to express $(d\hat{u}/d\xi)^e$ as $(du/dx)^e(h_e/2)$ with $h_e = x_i - x_{i-1}$, the element length.

In the more general case of a Hermite polynomial of degree $k = 2n - 1$ on $\hat{\Omega}$, we interpolate $\hat{u}_e$ and $\hat{u}_e' = (d\hat{u}/d\xi)^e$ at points $\xi_j, j = 1, 2, \ldots, n$. The resulting interpolation polynomial on $\hat{\Omega}$ is

$$\hat{U}(\xi) = \sum_{j=1}^{n} \hat{u}_j^e \hat{\psi}_j^0(\xi) + \sum_{j=1}^{n} (\hat{u}_j')^e \hat{\psi}_j^1(\xi) \tag{2.5.5}$$

The pointwise interpolation error is (Exercise 2.5.2)

$$\hat{E}_e(\xi) = \frac{p_n^2(\xi)}{(2n)!} \hat{u}_e^{(2n)}(\rho), \qquad p_n(\xi) = \prod_{i=1}^{n} (\xi - \xi_i) \tag{2.5.6}$$

where $\rho \in \hat{\Omega}$.

A more useful generalization for C^1 families of finite elements is *modified Hermite interpolation*. Derivative values are interpolated at select points, typically the end nodes, and function values interpolated at all nodes. Assume for notational convenience that we locally renumber the interpolation points so that the derivative is interpolated at the first $r \leq n$ points listed. In finite element applications the case $r = 2$ (the endpoints) is most meaningful. The modified formula is

$$\hat{U}(\xi) = \sum_{j=1}^{n} \hat{u}_j^e \hat{\psi}_j^0(\xi) + \sum_{j=1}^{r} (\hat{u}_j')^e \hat{\psi}_j^1(\xi) \tag{2.5.7}$$

and the error becomes

$$\hat{E}_e(\xi) = \frac{p_n(\xi)p_r(\xi)}{(n+r)!} \hat{u}_e^{(n+r)}(\rho^*), \qquad \rho^* \in \hat{\Omega} \tag{2.5.8}$$

where $p_n(\xi), p_r(\xi)$ are again the associated product polynomials. The modified Hermite family of shape functions for an element are:

$$\hat{\psi}_j^0(\xi) = \begin{cases} [1 - (\xi - \xi_j)\{l_{jn}'(\xi_j) + l_{jr}'(\xi_j)\}]l_{jn}(\xi)l_{jr}(\xi), & j = 1, 2, \ldots, r \\ l_{jn}(\xi) \left\{\dfrac{p_r(\xi)}{p_r(\xi_j)}\right\}, & j = r+1, \ldots, n \end{cases} \tag{2.5.9}$$

$$\hat{\psi}_j^1(\xi) = (\xi - \xi_j)l_{jr}(\xi)l_{jn}(\xi), \qquad j = 1, 2, \ldots, r$$

where the modified Lagrange polynomials $l_{jn}(\xi)$, $l_{jr}(\xi)$ are given by

$$l_{jn}(\xi) = \frac{p_n(\xi)}{(\xi - \xi_j)p_n'(\xi_j)}, \qquad j = 1, 2, \ldots, n$$

$$l_{jr}(\xi) = \frac{p_r(\xi)}{(\xi - \xi_j)p_r'(\xi_j)}, \qquad j = 1, 2, \ldots, r$$

In this manner we have constructed a family of C^1-elements of degree $n + r - 1$. It is straightforward to extend these results to C^q-conforming elements in one dimension by simply interpolating derivatives of order up to q at the interface nodes. This interpolation polynomial has the general form

$$\hat{U}(\xi) = \sum_{j=1}^{n} \hat{u}_j^e \hat{\psi}_j^0(\xi) + \sum_{j=1}^{r} \sum_{i=1}^{q} [\hat{u}_j^{(i)}]^e \hat{\psi}_j^i(\xi) \tag{2.5.10}$$

where $\hat{u}^{(i)} = d^i \hat{u}/d\xi^i$ and $\hat{\psi}_j^i(\xi)$ are the corresponding generalized Hermites. These element interpolation procedures in one dimension are very readily extended to tensor-product interpolation on discretizations of rectangles which we examine next.

2.5.2 Tensor-Product Hermites

Tensor-product Hermite elements may be generated in the same manner as the tensor-product Lagrange elements in Section 2.3. As an illustrative example we consider the tensor-product Hermite cubics (Fig. 2.14). The

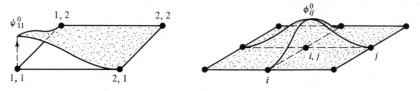

FIGURE 2.14 *Representative element shape function and global basis function for tensor-product Hermites.*

map from a rectangular element $\bar{\Omega}_e$ to the master square $\hat{\Omega}$ is simply the composition of two linear one-dimensional transformations $x \to \xi$ and $y \to \eta$. Hence, we can again develop the interpolant on $\hat{\Omega}$ and then map back to $\bar{\Omega}_e$ without any loss of generality.

Consider the Hermite cubics $\hat{\psi}_j^0(\xi)$, $\hat{\psi}_j^1(\xi)$ for $j = 1, 2$ on $\xi \in [-1, 1]$ defined in (2.5.3) and introduce $\hat{\psi}_j^0(\eta)$, $\hat{\psi}_j^1(\eta)$ in the orthogonal coordinate η. The tensor-product basis on $\hat{\Omega}$ consists of the 16 functions

$$\{\hat{\psi}_{kl}^{mn}(\xi, \eta)\} = \{\hat{\psi}_k^m(\xi)\} \times \{\hat{\psi}_l^n(\eta)\} \tag{2.5.11}$$

on $\hat{\Omega} = [-1, 1] \times [-1, 1]$ with $m = 0, 1$, and $n = 0, 1$, for nodes $k = 1, 2$ and $l = 1, 2$. The interpolant on $\hat{\Omega}$ is

$$\hat{U}(\xi, \eta) = \sum_{k,l=1}^{2} \sum_{m,n=0}^{1} \hat{u}_{kl}^{mn} \hat{\psi}_k^m(\xi) \hat{\psi}_l^n(\eta) \tag{2.5.12}$$

where $\hat{u}_{kl}^{00} = \hat{u}(\xi_k, \eta_l)$, $\hat{u}_{kl}^{10} = \partial \hat{u}/\partial \xi$ at (ξ_k, η_l), and so on.

Using the maps $\xi \to x$ and $\eta \to y$, we obtain the corresponding tensor-product shape functions on the rectangle $\bar{\Omega}_e$. The derivatives in the ξ and η directions in (2.5.12) transform by the chain rule to partial derivatives in x and y. There are now four element shape functions associated with each node to interpolate the corner values of u, u_x, u_y, and u_{xy}. The element has 16 degrees of freedom. The form of the element interpolant (2.5.12) transforms to

$$u_h^e(x, y) = \sum_{k,l=1}^{2} \sum_{m,n=0}^{1} u_{kl}^{mn} \psi_k^m(x) \psi_l^n(y) \tag{2.5.13}$$

where $\{u_{kl}^{mn}\} = \{u(x_k, y_l), \partial u(x_k, y_l)/\partial x \ldots\}$ and we have suppressed the superscript e on the element shape functions for notational simplicity.

This tensor-product basis ensures C^1-continuity, as is evident from the form of the global basis functions at an interface between two elements in the discretization. Equivalently, differentiating (2.5.13) with respect to x and evaluating $u_x(x, y)$ on side x_k yields a cubic function $u_x(y)$ on this side that is uniquely determined by u_x and $(u_x)_y$ at the endpoints of that side.

By using the modified Hermite polynomials in (2.5.7), a family of C^1-rectangular elements of increasing degree is readily constructed. Recall that the derivatives are interpolated at the ends of the generating intervals, but additional function values and derivatives may be interpolated at interior points of an interval. For an interval with n nodes, the one-dimensional shape functions interpolate the solution and derivatives at the end nodes and only solution values at the $n - 2$ interior nodes. The tensor-product element is obtained again by forming the tensor product of the one-dimensional shape functions. There are $n + 2$ shape functions in one dimension and $(n + 2)^2$ for the tensor-product basis. A corner node has basis functions associated with each of the interpolated values u, u_x, u_y, and u_{xy}; side nodes each have two basis functions (to interpolate either u, u_y, or u, u_x); and interior nodes have a single basis function to interpolate u that is local to the element. For example, the tensor product of Hermite quartics with nodes at the ends and midpoint of an interval produces a C^1-element with 25 shape functions, 16 being associated with the four corners, eight with the midsides, and one with the central node in the interior (Fig. 2.15).

We can deduce the form of the C^q-conforming tensor-product element using the Pascal square. If derivatives of order q are interpolated at each

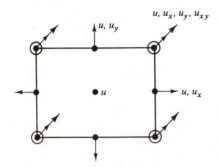

FIGURE 2.15 *Tensor-product element formed from modified Hermite quartics. The symbols at the nodes imply the degrees of freedom as marked:(•) $\equiv u$; (⊙) $\equiv u, u_x, u_y$; (↗) $\equiv u_{xy}$; (→) $\equiv u_x$; and (↑) $\equiv u_y$.*

endpoint of the base interval, the element basis is of degree $(2q + 1)$ and the tensor-product basis corresponds to a Pascal square of size $2q + 1 \times 2q + 1$. By block partitioning the square pattern to $(q + 1) \times (q + 1)$ blocks, we associate $(q + 1)^2$ shape functions with each of the corner nodes. The entries in the pattern identify the interpolated values at the corner nodes: For tensor-product cubics, the leading 2×2 block is associated with a corner node and contains the functions 1, x, y, and xy, implying that u, u_x, u_y, and u_{xy} are interpolated at each node; for the tensor-product quintic, the leading 3×3 block contains products of $x^i y^j$, $i, j = 0, 1, 2$, so that interpolated values at the corners are $\partial_x^i \partial_y^j u$, $i, j = 0, 1, 2$. Modified C^q-conforming rectangles can be generated in the same manner as the modified C^1-elements above.

Unfortunately, there is one serious shortcoming regarding these tensor-product Hermite elements—to retain C^1-continuity, the discretization must be restricted to consist of rectangles in two dimensions and rectangular prisms in three dimensions. This restriction is very prohibitive insofar as practical applications are concerned. One option to resolve the geometric restriction is to adopt nonconforming elements, but this may only be exercised with appropriate caution. Another alternative is to develop special simplex elements that are C^1.

2.5.3 Simplex Hermites

Hermite interpolation on simplex elements such as triangles is, in itself, a straightforward task. Ensuring that the normal derivative is continuous across the interfaces is more demanding. In fact, accomplishing this C^1-requirement economically—that is, with a low-degree interpolant involving very few interpolated nodal values—has led to some very inventive C^1-simplex elements in connection with plate-bending applications. Initially, let us ignore the C^1-requirement and examine Hermite interpolation polynomials on triangles.

As an illustrative example, consider a triangle in which u, u_x, and u_y are interpolated at each of the vertices. The complete cubic has 10 coefficients.

Interpolating u at the centroid in addition to the nine vertex values defines a unique cubic on the triangle (Fig. 2.16). The tangential and normal derivatives in directions (s, n) on a side of the triangle are $u_s = u_x x_s + u_y y_s$ and $u_n = u_x x_n + u_y y_n$. From u_x, u_y at the end nodes of the side, we may determine equivalently u_s, u_n as alternative variables at the vertices. The interpolation polynomial is, of course, unchanged.

The interpolation function on the triangle may be written

$$\sum_{i=1}^{3} [u_i \psi_i(x, y) + (u_x)_i \psi_i^{(x)}(x, y) + (u_y)_i \psi_i^{(y)}(x, y)] + u_4 \psi_4(x, y) \qquad (2.5.14)$$

where $i = 1, 2, 3, 4$ are local node numbers corresponding to the three vertices and centroid; the functions $\psi_i, \psi_i^{(x)}$, and $\psi_i^{(y)}$ are the cubic element shape functions associated with interpolating u, u_x, and u_y, respectively. We have again omitted the element superscript e from the expansion (2.5.14) to simplify notation.

The function u is cubic in the interior of the triangle and, using the tangential-normal (s, n) coordinate transformation, u is thus a cubic function of s along any given side. Since specifying (u, u_s) at the ends of the side defines a unique cubic in s on that side, the global interpolation function must be continuous across adjacent elements. The normal derivative $u_n = u_x x_n + u_y y_n$ is a quadratic function of s along each side. The two end values of u_n are insufficient to define a unique quadratic on the side, so the normal derivative is *not* continuous across interfaces between adjacent elements.

The Hermite cubic interpolant for the tetrahedron exhibits similar behavior. The complete cubic has 20 coefficients and we may interpolate u, u_x, u_y, u_z at each of the four vertices and u at the four midfaces to determine a Hermite cubic (Fig. 2.17). By directly relating u_x, u_y, u_z to skew-normal coordinates on any face, we verify directly that u is not C^1.

The global basis function associated with the centroid nodal point of the cubic triangle is local to the element. This degree of freedom can thus be

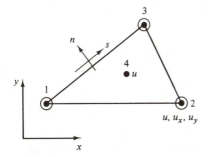

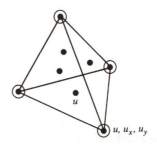

FIGURE 2.16 *Complete C^0-cubic Hermite triangle.*

FIGURE 2.17 *Complete C^0-cubic Hermite tetrahedron.*

removed by constraining the interpolant in the interior. The result is a C^0-Hermite triangle with interpolated values u, u_x, u_y at the vertices alone. As the cubic is now incomplete, interpolation accuracy in L^2 drops from $O(h^4)$ to $O(h^3)$. This procedure can be generalized to n-simplex Hermite elements by constraining the interpolation polynomials so that interpolation at the midface points is not required.

By merely using a polynomial expansion of sufficiently high degree, we can interpolate enough derivatives at the vertices and midsides of a triangle to make the normal derivative of the approximation continuous across the interfaces between elements. Let (s, n) be tangential-normal coordinates along the side of a triangle and assume that the, as yet unspecified, polynomial interpolant on the triangle is $u_h^e(x, y)$. The normal derivative is

$$\frac{\partial u_h^e}{\partial n} = u_x n_1 + u_y n_2 \qquad (2.5.15)$$

where n_1, and n_2 are components of the unit outward normal to the side.

On the side $x = x(s)$, $y = y(s)$, let $\partial u_h^e/\partial n = g(s)$. If g and dg/ds are specified at the end nodes and g at the midside node of the side, we can use these values to define a quartic along the side. That is, $\partial u/\partial n$ and $\partial^2 u/\partial s\, \partial n$ are required at the vertices and $\partial u/\partial n$ at midsides in this construction. Now we can infer from this the form of the desired interpolation polynomials determining u_h^e on Ω_e. Expanding $\partial^2 u/\partial s\, \partial n$ using (2.5.15) introduces nodal values u_{xx}, u_{xy}, and u_{yy} in addition to u_x and u_y from $\partial u/\partial n$. This implies that the nodal degrees of freedom are $u, u_x, u_y, u_{xx}, u_{xy}, u_{yy}$ at each vertex and $\partial u/\partial n$ at the three midside nodes (Fig. 2.18). From the Pascal triangle in Fig.

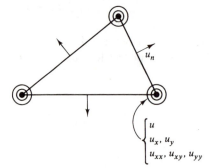

$$\begin{cases} u \\ u_x, u_y \\ u_{xx}, u_{xy}, u_{yy} \end{cases}$$

FIGURE 2.18 *Complete C^1-quintic triangle. Symbols indicate the degrees of freedom as marked.*

2.5, the corresponding polynomial is the complete quintic. Note that the leading 3×3 triangle involves the terms 1, x, y, x^2, xy, and y^2, which designate the interpolated derivatives at each vertex. The complete quintic has 21 terms, 18 of which are specified by interpolating these derivative values and the remaining 3 by the midside normal derivatives. The quintic along the edge between two triangles is matched exactly from either side, as u, u_s, and u_{ss} are known from the vertex degrees of freedom. By construction, the normal derivative $\partial u/\partial n$ is a unique quartic on a side, so the element is C^1.

This approach is perhaps the most direct formulation of a C^1-element in the spirit of the foregoing C^0-Lagrange and C^1-tensor-product interpolation theories. However, the element has a large number of degrees of freedom. By once again introducing constraints on the interpolant, we are able to reduce the degree of the interpolant and yet retain derivative continuity across the interfaces. The midside nodes are inconvenient for practical finite element computations and may be "removed" by simply constraining u_n to be cubic on the edges. The four nodal values of u_n and u_{ns} (from u_x, u_y, u_{xx}, u_{xy}, and u_{yy}) at the vertices will define the cubic on a given side. The number of degrees of freedom of this element has thus been reduced from 21 to 18, but the reduced shape functions are not complete quintic polynomials. Thus, this simplification in form has been achieved at a loss of accuracy in the L^2 norm for the interpolant from $O(h^6)$ to $O(h^5)$ for a triangulation of mesh size h.

The strategy employed to determine the C^1-triangle can be applied to determine C^q-triangulations. A major result is: For a C^q-triangle, the interpolated nodal variables must include *all* derivatives at the vertices of order up to and including $2q$ (Bramble and Zlamal [1970]). Summing the arithmetic progression in the Pascal triangle, a complete polynomial in x and y of degree k has $(k + 1)(k + 2)/2$ coefficients. For Hermite elements, the quintic triangle is the first "natural" triangle in which the expansion is complete and the global approximation C^1. More generally, complete polynomials of degree $k = 4q + 1$, $q = 0, 1, 2, \ldots$, have $(2q + 1)(4q + 3)$ coefficients. This implies that we can generate Hermite elements that produce global C^q-approximations by interpolating for $i, j = 0, 1, \ldots, q$:

$$\frac{\partial^{i+j} u}{\partial x^i\, \partial y^j}, \qquad 0 \leq i + j \leq 2q \qquad\qquad \text{at the vertices}$$

$$\frac{\partial^{i+j} u}{\partial x^i\, \partial y^j}, \qquad 0 \leq i + j \leq q - 2 \;\; (q > 1) \qquad \text{at the centroid} \tag{2.5.16}$$

and on each side the normal derivative of order r:

$$\frac{\partial^r u}{\partial n^r}(\mathbf{x}_{p,r}), \qquad p = 1, 2, \ldots, r; \quad r = 1, 2, \ldots, q$$

at equispaced interior points $\mathbf{x}_{p,r}$ on the side.

Setting $q = 0$, we recover the C^0-linear Lagrange triangle; $q = 1$ yields the C^1-quintic with derivatives of order up to two at the vertices, and normal first derivatives at the midsides as degrees of freedom; $q = 2$ generates a

C^2-triangle with derivatives of order up to four at the vertices, the function at the centroid, and derivatives of order up to two at the interior trisection points on each side as degrees of freedom.

2.5.4 Composite Hermite Elements

In an attempt to produce C^1-Hermite triangles of low degree, several composite elements have been devised. The strategy is to combine subtriangles and use interface constraints to obtain a composite element that is C^1-conforming. We consider briefly two prominent examples—a composite triangle and a quadrilateral.

To obtain a C^1-triangular element using only cubic polynomials, the parent triangle is partitioned to three subtriangles, as shown in Fig. 2.19.

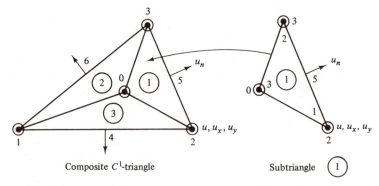

Composite C^1-triangle   Subtriangle ①

FIGURE 2.19 *Formation of composite C^1-triangle from cubic subtriangles. Along interior interfaces, constraints are imposed to require C^1-continuity.*

The point 0 is the centroid of the original triangle and the subtriangles are numbered to correspond to the opposite corner node. The 15 degrees of freedom are indicated on the composite element: these interpolated values are u, u_x, and u_y at the vertices and centroid, and u_n at the midside points (Clough and Tocher [1965]).

On each subelement cubic interpolation is utilized and subelements are then combined. However, the normal derivative u_n is not continuous across interfaces between subtriangles. By matching (not interpolating) u_n at midside nodes on an interface, the interpolant is constrained to be C^1 in the interior of the parent element. If the normal derivative u_n on the outer boundaries of this composite element is constrained to vary linearly along a side, the final element has only 12 degrees of freedom—u, u_x, and u_y at the vertices and at the centroid node. If four such triangles are combined as shown in Fig. 2.20, we obtain a composite quadrilateral. In solving boundary-value

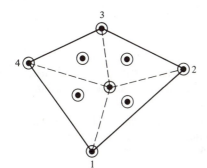

FIGURE 2.20 *Construction of C^1-composite quadrilateral from C^1-composite triangles of Fig. 2.19. Internal degrees of freedom are to be pre-eliminated during system solution in subsequent analysis.*

problems, the degrees of freedom associated with the internal nodes could be pre-eliminated by condensation in the element matrix routines during finite element computations. This would yield a quadrilateral stiffness matrix of reduced size (Clough and Felippa [1968]).

EXERCISES

2.5.1 Find the dimension of the approximation space for a mesh on $[0, 1]$ consisting of n piecewise-cubic Hermite elements and compare this result with that of the Lagrange cubics. For $u(x) = x^4$, determine bounds on the local error $E(x)$.

2.5.2 Following the same procedure as in Lagrange interpolation theory, set $F(z) = E(z) - E(\zeta)[p_n(z)/p_n(\zeta)]^2$, where ζ is not a node. By use of a generalization of Rolle's theorem, deduce the interpolation estimate (2.5.6) for Hermite interpolation.

2.5.3 Using the general interpolation estimates in Section 2.2, find the interpolation error in $H^m(\Omega)$, $m = 0, 1, 2, \ldots$, for Hermite elements with pointwise errors given by (2.5.6).

2.5.4 Sketch the shape functions for modified Hermite quartics in one dimension in which the function is interpolated at the midpoint of the element; sketch representative biquartic basis functions generated from this modified Hermite.

2.5.5 Verify that the C^1-continuity of tensor-product Hermites on the reference square $\hat{\Omega}$ is not retained on bilinear transformation to a general convex quadrilateral.

2.5.6 Show that the global basis derived from the cubic Hermite tetrahedron is only C^0 across element interfaces.

2.5.7 Find the degree of the highest complete polynomial basis contained in the composite C^1-triangle of Fig. 2.19.

2.6 *ISOPARAMETRIC ELEMENTS: CURVED BOUNDARIES*

Earlier, in studying C^0-Lagrange interpolation on triangles or quadrilaterals and their extension to three dimensions, a master element was employed on which interpolation and quadrature formulas are more easily devised. In isoparametric elements, the master element $\hat{\Omega}$ is related to the element Ω_e by means of a parametric map obtained using the master element shape functions. For example, a linear transformation

$$x = \sum_{j=1}^{3} x_j \hat{\psi}_j(\zeta), \qquad y = \sum_{j=1}^{3} y_j \hat{\psi}_j(\zeta) \tag{2.6.1}$$

maps the master triangle with unit sides on the ζ_2, ζ_3 axes to the triangle $\bar{\Omega}_e$ with vertices (x_i, y_i), $i = 1, 2, 3$. Recall that ζ_i, $i = 1, 2, 3$, are area coordinates with $\zeta_1 + \zeta_2 + \zeta_3 = 1$ and $\hat{\psi}_i(\zeta) = \zeta_i$, $i = 1, 2, 3$, are the linear element shape functions. Note that the map and its inverse are linear.

Similarly, the master square $\hat{\Omega} = [-1, 1] \times [-1, 1]$ is mapped bilinearly by

$$x = \sum_{i,j=1}^{2} x_{ij} \hat{\psi}_{ij}(\xi, \eta), \qquad y = \sum_{i,j=1}^{2} y_{ij} \hat{\psi}_{ij}(\xi, \eta) \tag{2.6.2}$$

to the quadrilateral with corner nodes (x_{ij}, y_{ij}), $i = 1, 2$ and $j = 1, 2$. The shape functions are the tensor-product basis functions $\hat{\psi}_i(\xi)\hat{\psi}_j(\eta)$ on the master element, where the index pairs (i, j), $i = 1, 2$ and $j = 1, 2$, define the corner nodes as in (2.3.29). Note that the map and inverse are no longer linear in general.

In defining the map we have restricted the choice of mapping functions to precisely those types that we employ in developing the element interpolant —namely, piecewise-polynomial shape functions. Accordingly, we can associate with the mapping exactly the same *parametric* form as that used to interpolate u both locally and globally—hence, the terminology "isoparametric" (Fig. 2.21). Each of the examples (2.6.1) and (2.6.2) above is a case in point. The coordinate variables x and y are recognized to be linear functions and interpolated. The interpolation polynomials are the same element shape functions $\hat{\psi}_i$ or $\hat{\psi}_{ij}$ that are used to interpolate $\hat{u}_e$ on the reference element.

The notion of using the same local interpolation basis for mapping the domain and for approximating u is a very natural one in view of the manner in which the domain is discretized. Furthermore, it is attractive from a practical standpoint as the mapping functions are available as part of the standard interpolation process and need not be regenerated. There are,

74

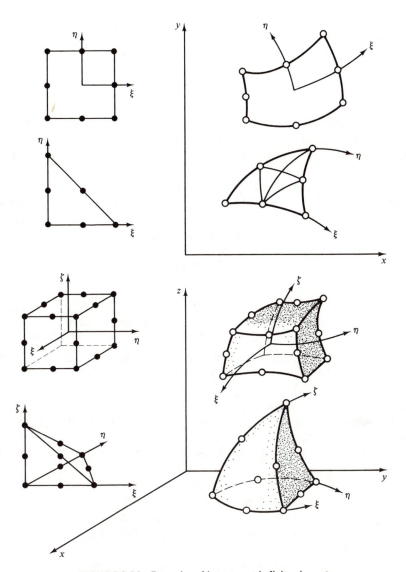

FIGURE 2.21 *Examples of isoparametric finite elements.*

however, some subtle shortcomings to this approach. These stem from the fact that a polynomial basis of degree k on $\hat{\Omega}$ is generally transformed to a basis that contains polynomials of lower degree on Ω_e, implying a possible loss of accuracy.

Formulas (2.6.1) and (2.6.2) permit immediate generalization to mappings from the master elements to triangles and quadrilaterals that have curved boundaries. The map to the general parametric triangle becomes

$$x = \sum_{j=1}^{n} x_j \hat{\psi}_j(\zeta), \qquad y = \sum_{j=1}^{n} y_j \hat{\psi}_j(\zeta) \qquad (2.6.3)$$

where $\{\hat{\psi}_j(\zeta)\}, j = 1, 2, \ldots, n$, are the C^0-Lagrange shape functions for polynomial interpolation on a triangle. Here $n = (k + 1)(k + 2)/2$ for complete polynomial interpolation of degree k on the element. The corresponding triangles in the (x, y)-plane have curvilinear boundaries that are polynomials of degree k.

For example, if the six quadratic shape functions are utilized, the coordinates x_i and y_i of the vertices and, say, the mid-arclength points on the sides must be provided in (2.6.3). The right-isosceles triangle $\hat{\Omega}$ is thereby mapped by (2.6.3) to a triangle with quadratic boundary curves passing through the specified end and midside positions.

The basis functions for the quadratic map are

$$\begin{aligned} \hat{\psi}_i(\zeta) &= \zeta_i(2\zeta_i - 1), &\quad \text{corner nodes } i = 1, 2, 3 \\ \hat{\psi}_m(\zeta) &= 4\zeta_i\zeta_j, &\quad \text{node } m \text{ midside } i - j \end{aligned} \qquad (2.6.4)$$

The map for tetrahedral elements with curved surfaces can be formed simply by including the relation $z = \sum z_j \hat{\psi}_j(\zeta)$ in (2.6.3), where ζ_i are volume coordinates. Obviously, for an N-simplex, each of the $N - 1$ coordinate variables is to be defined as in (2.6.3).

The same concern regarding C^0-continuity of the global interpolant of u resurfaces here when we construct maps from a master element to elements with curved boundaries. If the piecewise-polynomial global basis is not continuous, sides of adjacent curvilinear elements will not match exactly, resulting in "overlap" of elements and "openings" between elements. However, the results established earlier for finite element Lagrange interpolation apply to these piecewise-polynomial transformations. Consequently, continuity of the transformation holds across the polynomial interface between adjacent elements.

Similar comments may be inferred for quadrilateral elements with curved sides. The bilinear basis in (2.6.2) can be replaced by a general tensor-product of Lagrange polynomials of degree k,

$$x = \sum_{i,j=1}^{k+1} x_{ij}\hat{\psi}_i(\xi)\hat{\psi}_j(\eta) \quad \text{and} \quad y = \sum_{i,j=1}^{k+1} y_{ij}\hat{\psi}_i(\xi)\hat{\psi}_j(\eta) \qquad (2.6.5)$$

where the product basis functions $\hat{\psi}_i$ are the Lagrange polynomials on

$[-1, 1]$. From equation (2.3.9),

$$\hat{\psi}_i(\xi) = \frac{p_{k+1}(\xi)}{(\xi - \xi_i)p'_{k+1}(\xi_i)}, \qquad p_{k+1}(\xi) = \prod_{j=1}^{k+1} (\xi - \xi_j) \qquad (2.6.6)$$

where $p'_{k+1}(\xi_i) = dp_{k+1}/d\xi$ at $\xi = \xi_i$.

The biquadratic basis functions are derived from $\hat{\psi}_1(\xi) = \frac{1}{2}\xi(\xi - 1)$, $\hat{\psi}_2(\xi) = 1 - \xi^2$, $\hat{\psi}_3(\xi) = \frac{1}{2}\xi(1 + \xi)$ as tensor products of $\hat{\psi}_i(\xi)\hat{\psi}_j(\eta)$, $i, j = 1, 2, 3$. Using these nine basis functions in the coordinate transformation (2.6.5) with prescribed function values $\{x_{ij}\}$, $\{y_{ij}\}$, $i, j = 1, 2, 3$, yields a quadrilateral element that has each quadratically curved side through two endpoints and a side point. The remaining interior point is mapped biquadratically from the centroid $(0, 0)$ of the master square.

In the same vein as above, the "quadrilateral prism" in three dimensions that has curved polynomial boundary faces is obtained by a map of the form (2.6.5) for each of the coordinate functions x, y, and z. For a map defined on the N-dimensional hypercube to a "quadrilateral"-type element with curved boundary surfaces, each of the coordinate variables again transforms in this manner.

A brief remark on some generalizations of the terminology "parametric" is appropriate here. When interpolation functions of the same degree are chosen for the coordinate transformation as for the interpolation, the element is termed *isoparametric*. This would be the case if we used, for example, biquadratics for the map and also for interpolating u. If the map is defined by a lower-degree basis or a subset of that for the interpolation, the map is *subparametric*. Finally, if the interpolant of u is defined by a lower-degree basis or a subset of that for the mapping, the map is *superparametric*.

Note that under a nonlinear isoparametric transformation, the interpolation functions defining u in the domain Ω_e generally include functions that are not polynomials. This is of little concern in practical computations, as element calculations such as quadratures are performed directly on the master element. We need to verify, though, that the minimal-degree polynomials for approximability in the given variational boundary-value problem are preserved in the transformed bases. In structural analysis, this is equivalent to requiring that rigid body modes and constant strain states be represented in the approximate variational problem. For second-order equations in two dimensions this implies that 1, x, and y are represented in the transformed basis on $\hat{\Omega}$. (Recall Section 2.1.)

To verify this property for an isoparametric triangle, we prescribe at the nodes (x_i, y_i) point values u_i^e that correspond to a complete linear polynomial $u_e(x, y) = a + bx + cy$,

$$u_i^e = a + bx_i^e + cy_i^e, \qquad i = 1, 2, 3 \qquad (2.6.7)$$

The interpolating function for an element with n nodes is

$$\hat{U}(\zeta) = \sum_{j=1}^{n} u_j^e \hat{\psi}_j(\zeta) \tag{2.6.8}$$

and the isoparametric map is (2.6.3):

$$x = \sum_{j=1}^{n} x_j^e \hat{\psi}_j(\zeta), \qquad y = \sum_{j=1}^{n} y_j^e \hat{\psi}_j(\zeta) \tag{2.6.9}$$

Substituting the point values u_j^e of (2.6.7) in (2.6.8), we obtain

$$\begin{aligned}
\hat{U}(\zeta) &= \sum_{j=1}^{n} (a + bx_j^e + cy_j^e) \hat{\psi}_j(\zeta) \\
&= a \sum_{j=1}^{n} \hat{\psi}_j(\zeta) + b \sum_{j=1}^{n} x_j^e \hat{\psi}_j(\zeta) + c \sum_{j=1}^{n} y_j^e \hat{\psi}_j(\zeta)
\end{aligned} \tag{2.6.10}$$

Applying (2.6.9) and the property $\sum \hat{\psi}_j(\zeta) = 1$, we recover from $\hat{U}(\zeta)$

$$u_e(x, y) = a + bx + cy \tag{2.6.11}$$

Note that completeness of the basis to higher degree, however, may not be preserved and hence there may be a loss of accuracy.

The need for elements with curved sides or surfaces arises at the boundaries of the domain Ω. Usually only one, or perhaps two, sides of an element are to form part of the exterior boundary of the idealized domain. The remaining boundaries of these elements are interfaces between adjacent elements in the interior and may be straight. Accordingly, we wish to devise maps for which only one or two sides of $\bar{\Omega}_e$ are curved.

As an example, consider a triangle with one curved side (Fig. 2.22). The full quadratic map

$$x = \sum_{j=1}^{6} x_j \hat{\psi}_j(\zeta), \qquad y = \sum_{j=1}^{6} y_j \hat{\psi}_j(\zeta) \tag{2.6.12}$$

determines a triangle with quadratically curved sides.

To embed the constraint of linearity at nodes 4 and 6 on sides 1–2 and 1–3, we set $x_4 = (x_1 + x_2)/2, \ldots, y_6 = (y_1 + y_3)/2$. The map for x is then

$$x = \sum_{j \neq 4,6} x_j \hat{\psi}_j(\zeta) + \frac{(x_1 + x_2)}{2} \hat{\psi}_4(\zeta) + \frac{(x_1 + x_3)}{2} \hat{\psi}_6(\zeta)$$

so that

$$x = x_1 \bar{\psi}_1(\zeta) + x_2 \bar{\psi}_2(\zeta) + x_3 \bar{\psi}_3(\zeta) + x_5 \bar{\psi}_5(\zeta) \tag{2.6.13}$$

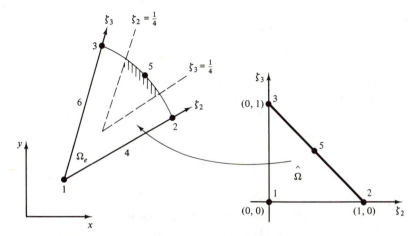

FIGURE 2.22 *Parametric map from master element $\hat{\Omega}$ to triangle Ω_e which has one quadratic side. The shaded region indicates the permissible zone for positions of node 5 which will allow an invertible transformation.*

where $\bar{\psi}_1 = \hat{\psi}_1 + \frac{1}{2}\hat{\psi}_4 + \frac{1}{2}\hat{\psi}_6$, $\bar{\psi}_2 = \hat{\psi}_2 + \frac{1}{2}\hat{\psi}_4$, $\bar{\psi}_3 = \hat{\psi}_3 + \frac{1}{2}\hat{\psi}_6$, and $\bar{\psi}_5 = \hat{\psi}_5$, are the transformation functions, including constraints.

Substituting the expression (2.6.4) for $\hat{\psi}_j(\zeta), j = 1, 2, \ldots, 6$, in $\bar{\psi}_j(\zeta)$ and simplifying yields

$$\left.\begin{array}{l} \bar{\psi}_1(\zeta) = 1 - \zeta_2 - \zeta_3 \\ \bar{\psi}_2(\zeta) = \zeta_2(1 - 2\zeta_3) \\ \bar{\psi}_3(\zeta) = \zeta_3(1 - 2\zeta_2) \\ \bar{\psi}_5(\zeta) = 4\zeta_2\zeta_3 \end{array}\right\} \qquad (2.6.14)$$

On sides 1–2 and 1–3, where $\zeta_3 = 0$ and $\zeta_2 = 0$, respectively, the functions $\bar{\psi}_j(\zeta)$ in (2.6.14) are linear or zero, which confirms the linearity of the map. On side 2–3, $\zeta_1 = 0$, $\zeta_2 + \zeta_3 = 1$, and the side is quadratic. The same form of transformation (2.6.13)–(2.6.14) holds for the coordinate variable y.

In the foregoing treatment, there is no guarantee that the coordinate map is invertible. Certainly, we anticipate some degeneracy, at least in the limit as node 5 in the curved side is allowed to approach the corner nodes 2 and 3. Let us examine in further detail invertibility of this map. Since node 5 is the only node whose position we wish to vary, let us assume, without loss of generality that $(x_1, y_1) = (0, 0)$, $(x_2, y_2) = (1, 0)$, and $(x_3, y_3) = (0, 1)$ so that $\bar{\Omega}_e$ in Fig. 2.22 is a right triangle with unit sides 1–4–2, 1–6–3, and curved side 2–5–3. Thus, the straight sides map undistorted onto the master element sides $\zeta_3 = 0$ and $\zeta_2 = 0$ and the bilinear map (2.6.13)–(2.6.14) becomes

$$x = \zeta_2(1 - 2\zeta_3) + 4x_5\zeta_2\zeta_3, \qquad y = \zeta_3(1 - 2\zeta_2) + 4y_5\zeta_2\zeta_3 \qquad (2.6.15)$$

The Jacobian $|\mathbf{J}| = |\partial(x, y)/\partial(\zeta_2, \zeta_3)|$ is

$$|\mathbf{J}| = \begin{vmatrix} (1 - 2\zeta_3) + 4x_5\zeta_3 & -2\zeta_2 + 4x_5\zeta_2 \\ -2\zeta_3 + 4y_5\zeta_3 & (1 - 2\zeta_2) + 4y_5\zeta_2 \end{vmatrix}$$
$$= [1 + (4x_5 - 2)\zeta_3][1 + (4y_5 - 2)\zeta_2] - (4x_5 - 2)(4y_5 - 2)\zeta_2\zeta_3$$

On evaluation, bilinear terms cancel and $|\mathbf{J}|$ is linear:

$$|\mathbf{J}| = 1 + (4x_5 - 2)\zeta_3 + (4y_5 - 2)\zeta_2 \qquad (2.6.16)$$

The map (2.6.15) is invertible if $|\mathbf{J}| \neq 0$ in $\hat{\Omega}$. Since $|\mathbf{J}|$ is linear and $|\mathbf{J}| = 1$ at $(\zeta_2, \zeta_3) = (0, 0)$, then $|\mathbf{J}|$ must be positive at the remaining vertices. Setting $(\zeta_2, \zeta_3) = (0, 1)$ and $(1, 0)$, we have $|\mathbf{J}| > 0$ here if

$$4x_5 - 2 > -1 \quad \text{and} \quad 4y_5 - 2 > -1$$

or
$$\left. \begin{array}{c} \\ \\ x_5 > \tfrac{1}{4} \quad \text{and} \quad y_5 > \tfrac{1}{4} \end{array} \right\} \qquad (2.6.17)$$

Thus node 5 must be placed in the "middle half" of the side, and this implies that in general node 5 must lie in the shaded "quadrant" marked in Fig. 2.22. Notice that this result holds even if side 2–5–3 is straight with node 5 positioned arbitrarily on that side.

EXERCISES

2.6.1 Write down the transformations between Ω_e and $\hat{\Omega}$ for a sector element defined in polar coordinates (r, θ). Describe tensor-product interpolation for this special element. Comment on the nature of the shape functions on Ω_e in Cartesian coordinates.

2.6.2 Consider the triangle with vertices (x_i, y_i), $i = 1, 2, 3$, and with side 2–3 a quadratic arc. Using the procedure (2.6.14)–(2.6.17), establish restrictions on the location of side nodes for this element for the more general form of Ω_e in Fig. 2.22.

2.6.3 Determine the restriction on midside node location for biquadratic transformation to a quadrilateral with curved sides.

2.6.4 Show that the constant 1 is in the approximation space for the four-node isoparametric quadrilateral by using the isoparametric map.

2.7 RATIONAL ELEMENTS AND BASIS COMPLETION

2.7.1 Rational Bases

In this subsection we consider the use of rational polynomials as a basis for finite-element interpolation following Wachspress (Wachspress [1975]; Wachspress and McLeod [1979]). Such rational functions may be utilized to develop general polygonal elements with straight or curved sides. Consider the quadrilateral in Fig. 2.23a with nodes and sides numbered as indicated: side i is defined by nodes i and $i + 1$, where $i = 1, 2, 3, 4$ with cyclic permutation. An interpolant is now sought directly on the quadrilateral in the form

$$u_h^e(x, y) = \sum_{j=1}^{4} u_j^e r_j(x, y) \tag{2.7.1}$$

where the shape functions now are rational functions; that is, $\psi_j^e(x, y) \equiv r_j(x, y) = p(x, y)/q(x, y)$, where p and q are polynomials. In order that the rational functions satisfy the usual interpolation property, we require that $r_j(x_i, y_i) = \delta_{ij}$; each rational function r_i takes on a unit value at a designated node i and is to be zero on all sides except those adjacent to node i.

Let $f_i(x, y) = 0$ represent the equation of side i. Then the product function $p_i(x, y) = f_{i+1}(x, y)f_{i+2}(x, y)$ is zero on the sides not adjacent to node i. For example, $f_2(x, y)f_3(x, y)$ is zero on sides 2 and 3 opposite node 1 in the figure. If this quadratic is divided by the linear function f_{AB} defined by joining extrapolated sides as shown in Fig. 2.23a and scaled with $C_1 = f_{AB}(x_1, y_1)/\{f_2(x_1, y_1)f_3(x_1, y_1)\}$, we obtain r_1.

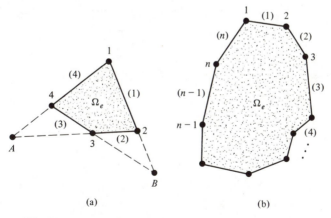

(a)   (b)

FIGURE 2.23 (a) Quadrilateral and (b) polygon for rational basis construction.

Proceeding in similar fashion with the remaining shape functions, we have

$$r_i(x, y) = C_i \frac{f_{i+1}(x, y) f_{i+2}(x, y)}{f_{AB}(x, y)} \qquad (2.7.2)$$

This approach can be generalized directly to the polygon in Fig. 2.23b by again specifying the equations to the sides j as

$$f_j(x, y) = 0, \qquad j = 1, 2, \ldots, n \qquad (2.7.3)$$

and defining the rational functions

$$r_i(x, y) = C_i \frac{f_{i+1}(x, y) f_{i+2}(x, y) \cdots f_{i-2}(x, y)}{g_1(x, y) g_2(x, y) \cdots g_{n-2}(x, y)} \qquad (2.7.4)$$

Here

$$C_i = \left[\frac{f_{i+1}(x_i, y_i) \cdots f_{i-2}(x_i, y_i)}{g_1(x_i, y_i) \cdots g_{n-2}(x_i, y_i)} \right]^{-1}$$

and $g_s(x, y)$ are constructed such that f_{i+1}/g_s and f_{i+2}/g_s are constants on sides i and $i - 1$, respectively.

By including more nodal points in each straight side, higher-order formulas can be immediately developed. Moreover, if the equations $f_i(x, y) = 0$ define curves rather than straight lines, rational function interpolation can be utilized in the same way as above. Let us examine rational function interpolation for the previous example of a "right-isosceles triangle" with one quadratic side. Number the nodes as shown in Fig. 2.24 and extrapolate the sides to intersection points 5 and 6. Let f_1, f_2, f_4, f_5, and $f_6 = 0$ be the equations of straight segments 3–1–6, 2–1–5, 4–6, 4–5, and 5–6. Let $q(x, y)$ be the quadratic side 2–4–3.

The shape functions are then defined as in the quadrilateral example as

$$r_1 = C_1 \frac{q}{f_6}, \quad r_2 = C_2 \frac{f_1 f_5}{f_6}, \quad r_3 = C_3 \frac{f_2 f_4}{f_6}, \quad r_4 = C_4 \frac{f_1 f_2}{f_6} \qquad (2.7.5)$$

where $C_1 = f_6(x_1, y_1)/q(x_1, y_1)$, $C_2 = f_6(x_1, y_1)/f_1(x_1, y_1) f_5(x_1, y_1)$, $\ldots$.

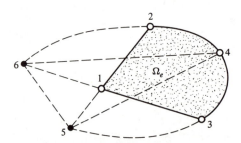

FIGURE 2.24 *Right isosceles triangle with one quadratically curved side 3–4–2.*

Our earlier interpolation estimates were based on polynomial interpolation. We saw that the interpolation error in L^2 is $O(h^{k+1})$ for a complete polynomial basis of degree k provided that the function u is sufficiently regular. In the present instance the basis is constructed of rational functions, and to ascertain interpolation estimates we need to indicate the degree of the complete polynomial basis contained in a given rational basis for an element. We refer to this degree as the "degree of the approximation" and say that a rational approximation is "degree k" if the complete polynomial of greatest degree that can be represented in the basis has degree k.

Each element of this type in a given discretization has a node at each of its vertices, nodes on each side sufficient to satisfy interelement continuity of the global approximation, and additional interior nodes to complete the degree of approximation required by the variational problem under consideration. To achieve the desired degree-k accuracy using rational shape functions, we need $(k - 1)(k - 2)/2$ interior nodes and require that these interior nodes not all lie on a curve of degree less than $k - 2$. For example, the complete linear polynomial basis is contained in the rational approximation space for the quadrilateral of Fig. 2.23a if there is simply a node at each vertex as shown in the figure. If the basis is to contain the complete quadratic ($k = 2$), we require nodes at the vertices, one further node on each side, and no interior nodes. The complete cubic ($k = 3$) requires nodes at the vertices, two more nodes on each side, and one interior node.

2.7.2 Basis Completion

A more intriguing point arises if we compare the degree of the rational function approximation with those of more standard isoparametric quadrilaterals. First recall that in the case of straight-sided simplex elements such as triangles, the master element map to the right isosceles master triangle is linear (affine). The Jacobian is then constant. Moreover, a polynomial basis on the master element $\hat{\Omega}$ is mapped to a polynomial basis *of the same degree* on the actual element Ω_e in the (x, y)-plane. This property does not hold if the map is not linear, as is the case when the element Ω_e has a curved side. With tensor-product elements such as quadrilaterals, the degree of the basis is usually not preserved under the map from $\hat{\Omega}$, even for elements with straight sides, since the mapping is not linear. On the other hand, the rational approximation is constructed in the (x, y)-plane to be of degree k. That is, the rational basis contains the complete polynomials of degree k. There is no intermediate map and thus no uncertainty regarding the degree of the complete polynomial exactly attainable.

As an example let us consider the standard four-node quadrilateral. Provided there are no reentrant corners in the quadrilateral Ω_e, the isoparametric bilinear map from Ω is invertible (the Jacobian is positive at all points in the element). The isoparametrically transformed basis is

$$\psi_{ij}(x, y) = \hat{\psi}_{ij}[\xi(x, y), \eta(x, y)], \qquad i, j = 1, 2 \qquad (2.7.6)$$

and is not bilinear in x and y.

The bilinearity is unimportant, however, as approximability only requires that the linear basis be included. We can verify that this is true here by simply noting that the map is actually defined by

$$x = \sum_{i,j=1}^{2} x_{ij}\hat{\psi}_i(\xi)\hat{\psi}_j(\eta)$$

$$y = \sum_{i,j=1}^{2} y_{ij}\hat{\psi}_i(\xi)\hat{\psi}_j(\eta) \qquad (2.7.7)$$

where x_{ij} and y_{ij} are the coordinates of node (i, j). Thus, the linear functions x and y are automatically contained in the transformed basis. Finally, the constant 1 can be shown to be a linear combination of these shape functions on further use of the transformation.

The situation becomes more interesting when biquadratics are used on the master square with a bilinear map to the straight-sided quadrilateral. This is the standard subparametric nine-node quadrilateral (subparametric since the map is only bilinear). By constraining the degree of freedom at the interior node, as discussed earlier, we obtain the eight-node quadrilateral. The basis functions on the master square contain the complete quadratic polynomial basis in master coordinates ξ and η. Under the bilinear map, however, this property is not retained. Since the isoparametric map is again defined by an expansion similar to (2.7.7), the linear functions x and y are present. The constant 1 is also included, so the linears are complete. Analysis similar to that for the constant reveals that the basis in the (x, y)-plane does *not* contain the complete quadratic. This suggests that the approximability property is weakened by the map.

Ciarlet and Raviart [1972a,b] have developed an interpolation theory for finite elements generated by affine maps and subsequently have extended their analysis to more general nonaffine elements. Their analysis rests on the assumption that the curved element Ω_e is obtained by a small distortion of an element with straight sides (e.g., a rectangular element). That is, the map is "almost affine." In the limit as the mesh size is reduced, the distortions diminish and the elements approach the corresponding affine elements. Hence, the optimal approximability predicted for affine maps is attained. In the present case, the L^2-interpolation error estimate is thus $O(h^3)$, even though the basis is not a complete quadratic, provided that the distortion of the element is small.

On the other hand, an eight-node quadrilateral element can also be developed using rational basis functions. The quadrilateral element is shown in Fig. 2.25. Using the construction established earlier, we may define representative basis functions for the corner nodes and midside nodes, respectively, of the form

$$r_1(x, y) = C_1 \frac{f_{23}(x, y) f_{34}(x, y) f_{58}(x, y)}{f_{AB}(x, y)} \tag{2.7.8}$$

and

$$r_5(x, y) = C_5 \frac{f_{23}(x, y) f_{34}(x, y) f_{41}(x, y)}{f_{AB}(x, y)} \tag{2.7.9}$$

where $f_{ij}(x, y) = 0$ is the equation to the line joining nodes i and j. This rational basis is of degree 2 and quadratic approximability on Ω_e is assured (Wachspress and McLeod [1979]).

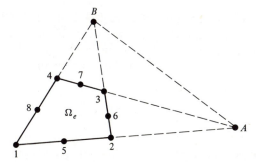

FIGURE 2.25 *Construction of degree 2 rational basis for a quadrilateral.*

The reduction in degree introduced by the isoparametric map has motivated research on methods for augmenting the transformed basis in the (x, y)-plane to restore completeness to a specified degree. This has led to "higher-order transformation methods." The basic idea is to add additional basis functions in the (x, y)-coordinates that will complete the polynomial basis to the required degree. Here we illustrate the construction by developing a complete quadratic basis for a triangle with one curved side (McLeod and Murphy [1979]).

The quadratic Lagrange basis on the master triangle in Fig. 2.26 is

$$\hat{\psi}_1(\zeta) = \zeta_1(2\zeta_1 - 1), \quad \hat{\psi}_2(\zeta) = \zeta_2(2\zeta_2 - 1), \quad \hat{\psi}_3(\zeta) = \zeta_3(2\zeta_3 - 1),$$
$$\hat{\psi}_4(\zeta) = 4\zeta_1\zeta_2, \quad \hat{\psi}_5(\zeta) = 4\zeta_2\zeta_3, \quad \hat{\psi}_6(\zeta) = 4\zeta_3\zeta_1 \tag{2.7.10}$$

where ζ_i, $i = 1, 2, 3$, are area coordinates. The isoparametric map to the

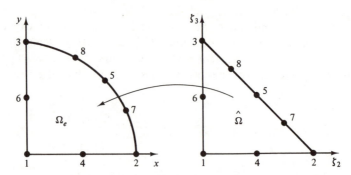

FIGURE 2.26 *Isoparametric (six-node) map from $\hat{\Omega}$ to Ω_e. Additional nodes 7 and 8 are required for completion of basis in (x, y) to degree 2.*

curved element Ω_e shown in the figure is formally defined in the usual way by

$$x = \sum_{j=1}^{6} x_j \hat{\psi}_j(\zeta), \qquad y = \sum_{j=1}^{6} y_j \hat{\psi}_j(\zeta) \qquad (2.7.11)$$

In the present case the linearity of sides 1–4–2 and 1–6–3 leads to a corresponding simplification of the transformation. If, in addition, we confine our attention to the particular orientation in the figure in which the sides of Ω_e are parallel to the x and y axes and of unit length, (2.7.11) simplifies to

$$x = \zeta_2[1 + 2(2x_5 - 1)\zeta_3]$$
$$y = \zeta_3[1 + 2(2y_5 - 1)\zeta_2] \qquad (2.7.12)$$

This transformation maps the master triangle $\hat{\Omega}$ onto the triangle with quadratically curved side 2–5–3. The additional nodes 7 and 8 marked on the curved side will be used to complete the basis for approximation but do not enter in defining the map. If the side 2–5–3 in the (x, y)-plane were straight, the polynomial basis would simply have the same form as (2.7.10) on $\hat{\Omega}$. (The map would reduce to the identity.) Now consider node 5 in its actual position on the quadratic arc 2–5–3 and let $q_i(x, y)$ be the quadratic Lagrange polynomials that satisfy

$$q_i(x_j, y_j) = \delta_{ij}, \qquad i, j = 1, 2, \ldots, 6 \qquad (2.7.13)$$

Next, let $(f_{ij})_k$ denote the linear polynomial which is zero at (x_i, y_i), (x_j, y_j) and has unit value at (x_k, y_k). We use the additional nodes 7 and 8 to add additional shape functions to enrich the basis. The new basis becomes

$$\left.\begin{aligned}
\psi_7(x, y, \zeta_2, \zeta_3) &= \tfrac{16}{3}\zeta_2\zeta_3(f_{85})_7 \\
\psi_8(x, y, \zeta_2, \zeta_3) &= \tfrac{16}{3}\zeta_2\zeta_3(f_{75})_8 \\
\psi_i(x, y, \zeta_2, \zeta_3) &= q_i(x, y) - q_i(x_7, y_7)\psi_7(x, y, \zeta_2, \zeta_3) \\
&\quad - q_i(x_8, y_8)\psi_8(x, y, \zeta_2, \zeta_3), \qquad i = 1, 2, \ldots, 6
\end{aligned}\right\} \qquad (2.7.14)$$

The coordinate transformation (2.7.12) is bilinear, so the final basis (2.7.14) is a cubic polynomial in ζ_2, ζ_3. However, the basis in the (x, y)-plane is now of degree 2 (complete quadratics), whereas the previous parametric basis is only of degree 1 (complete linears).

2.7.3 Unification

The rational and high-order transformation bases are closely related. Let us briefly summarize the constructions in a general form that permits direct comparison of the schemes.

Rational Basis: Let the boundary curve of element Ω_e be denoted Γ_m and consist of m continuous curves (sides). Define the "opposite factor" associated with node i to be the product

$$O_i(x, y) = \prod_{j \neq s(i)} f_j(x, y) \tag{2.7.15}$$

where $f_j(x, y) = 0$ is the equation of the curved side j of Ω_e, and $j \neq s(i)$ implies that those sides containing node i [here denoted $s(i)$] are excluded from the product. Hence, $O_i(x, y)$ is zero on all sides of Ω_e except those containing node i.

To satisfy the desired interpolation properties, the basis functions should also be zero at the remaining nodes on the side or sides containing node i. We construct an "adjacent function" $A_i(x, y)$ which is zero at all nodes on this side or sides except node i.

Combining the functions above, we obtain

$$N_i(x, y) = O_i(x, y)A_i(x, y) \tag{2.7.16}$$

This establishes the numerator expression for the rational basis. The denominator $D(x, y)$ is the unique algebraic curve defined by the exterior intersection points of the extended sides and in algebraic geometry is called the "adjoint" polynomial. Thus, the rational basis functions have the form

$$r_i(x, y) = \frac{C_i O_i(x, y) A_i(x, y)}{D(x, y)} \tag{2.7.17}$$

where C_i is a normalizing constant chosen so that $r_i(x_i, y_i) = 1$.

The denominator $D(x, y)$ is necessary because of the unwanted external intersections of the element sides. $D(x, y)$ is of degree $M - 3$, where M is the sum of the degrees of the element sides. Thus, for the triangle with one quadratic side in Fig. 2.26, $M = 4$, so that $D(x, y)$ is a straight line. If the extended conic intersects the straight sides at infinity, $D(x, y) = 1$ and the rational basis degenerates to a polynomial basis. Observe that the simpler the shape of the element, the fewer the number of external points and thus the simpler the form of $D(x, y)$; the complexity of $O_i(x, y)$ increases with the distortion and number of sides.

Higher-Order Transformation: Introduce a parametric map from a master element $\hat{\Omega}$ to Ω_e. On $\hat{\Omega}$ we can repeat the steps described above to construct a rational basis. Let (ξ, η) be the coordinates in the master-element frame. The rational basis on $\hat{\Omega}$ becomes, according to (2.7.17),

$$\hat{r}_i(\xi, \eta) = \frac{C_i O_i(\xi, \eta) A_i(\xi, \eta)}{D(\xi, \eta)} \qquad (2.7.18)$$

In the standard case of the master triangle or square, $D(\xi, \eta) = 1$ and the familiar polynomial bases on $\hat{\Omega}$ are recovered.

The general form of the higher-order transformation basis is qualitatively the same as in (2.7.17), with the notable exception that $A_i(\xi, \eta)$ is modified to a new function $A_i(\xi, \eta, x, y)$, so that

$$\hat{\psi}_i(\xi, \eta) = \frac{C_i O_i(\xi, \eta) A_i(\xi, \eta, x, y)}{D(\xi, \eta)} \qquad (2.7.19)$$

The modifications to A_i are given below and result from the need to impose $(k + 1)(k + 2)/2$ linear conditions which guarantee an approximability of degree k in x and y. To do this we introduce additional side nodes (x_i, y_i) in Ω_e and $\hat{\Omega}$ and as in (2.7.16) construct $A_i(\xi, \eta, x, y)$ to be the adjacent factor for the corresponding rational basis. This produces a basis function which is a polynomial of degree k on the corresponding side since $O_i(x, y)/D(x, y)$ is linear modulo that side and $A_i(x, y)$ is of degree $k - 1$. As the curve approaches a straight line, singularities may occur due to the fact that the adjoint curve is receding toward infinity. Other choices of higher-order transformation bases that avoid the singular limit behavior can be devised (McLeod and Murphy [1979]). If i is a vertex node, $A_i(x, y)$ can be chosen to have simple intersections in ξ and η at the corresponding adjacent nodes and no other intersections with these sides (see the example of Fig. 2.26). These intersections, together with the intersections from the opposite factor $O_i(x, y)$ produce the correct behavior on the adjacent sides.

EXERCISES

2.7.1 Consider a triangle, denoted 1–2–3, and perturb side 2–3 to form a quadrilateral. Construct rational functions on the quadrilateral and show that in the limit, as the quadrilateral degenerates to the triangle, the appropriate polynomial shape functions are recovered.

2.7.2 Construct a similar argument to that of Exercise 2.7.1 for a quadrilateral deforming to a rectangle.

2.7.3 Show that the basis of rational functions for the quadrilateral defined by (2.7.2) contains the linears 1, x, and y.

2.7.4 Construct a rational basis for the pentagon shown here with linear (degree $k = 1$) approximation. Include midside nodes for the degree 2 approximation. Verify that representative rational basis functions are

$$r_1(x, y) = C_1 \frac{f_2(x, y) f_3(x, y) f_4(x, y) f_{6-10}(x, y)}{q(x, y)}$$

$$r_6(x, y) = C_6 \frac{f_2(x, y) f_3(x, y) f_4(x, y) f_5(x, y)}{q(x, y)}$$

where $q(x, y)$ is the unique conic containing points TM, $f_i(x, y) = 0$ is the equation to side i and $f_{6-10}(x, y) = 0$ is the equation to the straight line joining points 6 and 10.

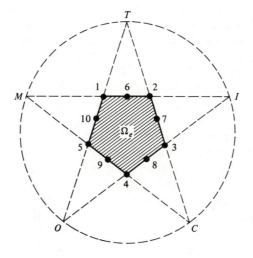

2.8 HIERARCHIC ELEMENT FAMILIES

2.8.1 The Hierarchic Concept

In the development of the C^0-Lagrange bases we have seen that the Pascal triangle and tetrahedron provide a framework for defining one form of natural hierarchy: the number and position of the nodes of successively higher-degree elements are nested in successive levels of the Pascal pattern (recall Figs. 2.5, 2.6, 2.8, and 2.9). For example, the linear, quadratic, cubic, quartic, and quintic triangles are deduced directly from the pattern in Fig. 2.5.

In this instance the hierarchic nature is confined to the form of the element only, and only implicitly to the shape functions themselves. By this we mean that the quadratic C^0-Lagrange basis contains, of course, the linears but not in the explicit form in which they are constructed for the linear triangle. Recall from equations (2.3.13) and (2.3.16) in this case that we have for the linear element

$$\hat{\psi}_1(\zeta) = \zeta_1, \quad \hat{\psi}_2(\zeta) = \zeta_2, \quad \hat{\psi}_3(\zeta) = \zeta_3 \qquad (2.8.1)$$

and for the quadratic element

$$\hat{\psi}_1(\zeta) = \zeta_1(2\zeta_1 - 1), \quad \hat{\psi}_2(\zeta) = \zeta_2(2\zeta_2 - 1), \quad \hat{\psi}_3(\zeta) = \zeta_3(2\zeta_3 - 1),$$
$$\hat{\psi}_4(\zeta) = 4\zeta_1\zeta_2, \qquad \hat{\psi}_5(\zeta) = 4\zeta_2\zeta_3, \qquad \hat{\psi}_6(\zeta) = 4\zeta_3\zeta_1 \qquad (2.8.2)$$

where ζ_i, $i = 1, 2, 3$, are the area coordinates for the element. Evidently, even though the basis in (2.8.2) is complete for quadratics, the precise form of the linears in (2.8.1) does not appear explicitly as a subset in (2.8.2).

The idea of nesting the shape functions so that lower-degree bases are explicitly contained in successively higher-degree bases has led to a different form of hierarchic structure, and such families have been termed *hierarchic element families* (see Szabo [1979]). The main motivation for considering such an alternative classification is the desire to embed the element calculations in a similar manner: a nesting of element matrix and vector calculations is thereby achieved and this can be exploited in computation. For instance, rather than refining the mesh to improve accuracy, the mesh can be held fixed and the degree of the local polynomial can be increased. As the degree p of the element is increased, previously computed arrays are used and only new terms involving the added polynomials of degree p need be computed. Such strategies have been shown in practice to be more efficient than standard (fixed degree) finite element methods for some applications. The study of convergence of these adaptive algorithms has been termed "p-convergence" to distinguish it from more typical "h-convergence" using mesh refinement (Babuška et al. [1979]). In Volume III we consider the techniques for formulating element contributions and implementing this scheme in a program and we state error estimates. Here we restrict our treatment to the development of the nested hierarchic shape functions for standard elements in one, two, and three dimensions.

2.8.2　One-Dimensional Hierarchies

The main idea can be easily explained by examining linear and quadratic interpolation in one dimension. The usual Lagrange shape functions $\hat{\psi}_j$ are shown in Fig. 2.1. Consider the quadratic curve on the master element $\hat{\Omega}$ sketched in Fig. 2.27. The nodes have been numbered as shown for conven-

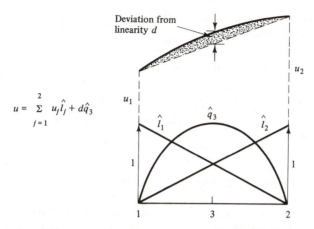

FIGURE 2.27 *Interpolation using hierarchic basis of linears and quadratic deviation from linearity.*

ience in "nesting" the hierarchic bases. If we linearly interpolate the endpoint values, the dashed line is obtained. The actual quadratic curve can be recovered by adding to the linear curve the *deviation* from linearity represented by the difference between the linear and quadratic curves.

This observation suggests the alternative form of the expansion

$$\hat{U}(\xi) = u_1 \hat{l}_1(\xi) + u_2 \hat{l}_2(\xi) + d\hat{q}_3(\xi) \tag{2.8.3}$$

where $d = u_3 - (u_1 + u_2)/2$ is the departure from linearity at the midpoint, $\hat{l}_1$ and $\hat{l}_2$ are linear Lagrange polynomials, and $\hat{q}_3$ is the quadratic Lagrange polynomial associated with the midpoint node 3.

An alternative form is obtained as follows: Differentiating $\hat{U}$ twice with respect to ξ, we obtain

$$\hat{U}_{\xi\xi} = -2d \tag{2.8.4}$$

so that the degree of freedom at node 3 can equivalently be taken as the curvature $(\hat{U}_{\xi\xi})_3$, whence $d = -(u_{\xi\xi})_3/2$ in (2.8.3).

The expansion (2.8.3) becomes

$$\hat{U}(\xi) = \tfrac{1}{2}(1 - \xi)u_1 + \tfrac{1}{2}(1 + \xi)u_2 - \tfrac{1}{2}(1 - \xi^2)(u_{\xi\xi})_3 \tag{2.8.5}$$

The nodal degrees of freedom are then u_1, u_2 for linears and u_1, u_2, $(u_{\xi\xi})_3$ for quadratics. The corresponding element shape functions are

Linears: $\tfrac{1}{2}(1 - \xi)$, $\tfrac{1}{2}(1 + \xi)$

Quadratics: $\tfrac{1}{2}(1 - \xi)$, $\tfrac{1}{2}(1 + \xi)$, $-\tfrac{1}{2}(1 - \xi^2)$ (2.8.6)

and the hierarchic form is established.

The generalization to higher degree is immediate. The polynomial expansion of degree k becomes

$$\hat{U}(\xi) = u_1 \hat{\psi}_1(\xi) + u_2 \hat{\psi}_2(\xi) + \sum_{j=3}^{k+1} u_3^{(j-1)} \hat{\psi}_j(\xi) \qquad (2.8.7)$$

where $\hat{\psi}_1(\xi) = \frac{1}{2}(1 - \xi)$, $\hat{\psi}_2(\xi) = \frac{1}{2}(1 + \xi)$, and the remaining basis functions are constructed such that the derivatives $u_3^{(j-1)}$ with respect to ξ at midpoint node 3 are degrees of freedom.

2.8.3 Hierarchic Simplex Elements

The same procedure can be followed in developing C^0-simplex elements with hierarchic bases. On any side the problem reduces to a one-dimensional problem, so that the tangential derivatives of second order and higher are now the corresponding degrees of freedom.

Consider the quadratic triangle and let ζ_i, $i = 1, 2, 3$, again be the area coordinates. The linear shape functions for C^0-interpolation are $\hat{\psi}_i(\zeta) = \zeta_i$, $i = 1, 2, 3$, as before. Applying the argument for one dimension on the sides, the tangential derivatives are the quadratic degrees of freedom and the hierarchic quadratic shape functions are

$$\begin{array}{lll}
\hat{\psi}_1(\zeta) = \zeta_1, & \hat{\psi}_2(\zeta) = \zeta_2, & \hat{\psi}_3(\zeta) = \zeta_3, \\
\hat{\psi}_4(\zeta) = -\frac{1}{2}\zeta_1\zeta_2, & \hat{\psi}_5(\zeta) = -\frac{1}{2}\zeta_2\zeta_3, & \hat{\psi}_6(\zeta) = -\frac{1}{2}\zeta_3\zeta_1
\end{array} \qquad (2.8.8)$$

The C^0-cubic shape functions can be obtained by including the third tangential derivatives at the midsides and an interior nodal degree of freedom. The additional shape functions associated with the sides are

$$\hat{\psi}_7(\zeta) = \tfrac{1}{12}\zeta_1\zeta_2(\zeta_1 - \zeta_2), \quad \hat{\psi}_8(\zeta) = \tfrac{1}{12}\zeta_2\zeta_3(\zeta_2 - \zeta_3),$$
$$\hat{\psi}_9(\zeta) = \tfrac{1}{12}\zeta_3\zeta_1(\zeta_3 - \zeta_1) \qquad (2.8.9)$$

and in the interior

$$\hat{\psi}_{10}(\zeta) = \zeta_1\zeta_2\zeta_3 \qquad (2.8.10)$$

A specified degree of freedom need not be associated with the interior node. This internal degree of freedom can be removed by pre-elimination (condensation). Alternatively, all the shape functions can be modified by adding $C\hat{\psi}_{10}(\zeta)$, where C is an arbitrary constant [since $\hat{\psi}_{10}(\zeta)$ is zero on the sides] and this additional degree of freedom can be eliminated in advance.

Hierarchic basis functions for the tetrahedron are constructed in a like manner. Let ζ_i, $i = 1, 2, 3, 4$, be volume coordinates for the tetrahedral element. The linear C^0 basis is

$$\hat{\psi}_i(\zeta) = \zeta_i, \qquad i = 1, 2, 3, 4 \qquad (2.8.11)$$

Adding the tangential edge derivatives with the shape functions

$$\hat{\psi}_5(\zeta) = -\tfrac{1}{2}\zeta_1\zeta_2, \quad \hat{\psi}_6 = -\tfrac{1}{2}\zeta_2\zeta_3, \quad \ldots, \quad \hat{\psi}_{10} = -\tfrac{1}{2}\zeta_1\zeta_4 \qquad (2.8.12)$$

determines the quadratic.

These results for the C^0-triangle and tetrahedron are summarized in Figs. 2.28 and 2.29 for polynomial degree up to cubic.

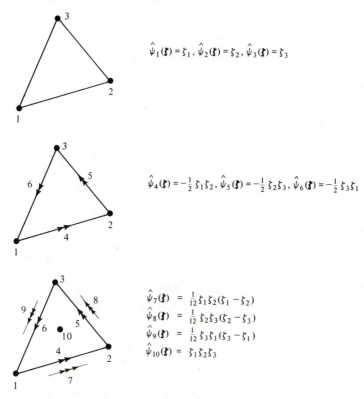

FIGURE 2.28 *Hierarchic C^0-family up to cubic degree for the triangle. Arrows indicate tangential derivatives as degrees-of-freedom.*

2.8.4 Tensor-Product Hierarchic Bases

The construction of tensor-product hierarchic bases follows immediately from the one-dimensional case. For the bilinear-biquadratic pair on a quadrilateral the degrees of freedom are the corner node values, the second tangen-

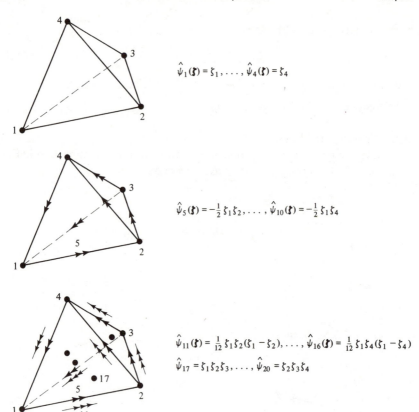

$$\hat{\psi}_1(\mathbf{\zeta}) = \zeta_1, \ldots, \hat{\psi}_4(\mathbf{\zeta}) = \zeta_4$$

$$\hat{\psi}_5(\mathbf{\zeta}) = -\tfrac{1}{2}\zeta_1\zeta_2, \ldots, \hat{\psi}_{10}(\mathbf{\zeta}) = -\tfrac{1}{2}\zeta_1\zeta_4$$

$$\hat{\psi}_{11}(\mathbf{\zeta}) = \tfrac{1}{12}\zeta_1\zeta_2(\zeta_1 - \zeta_2), \ldots, \hat{\psi}_{16}(\mathbf{\zeta}) = \tfrac{1}{12}\zeta_1\zeta_4(\zeta_1 - \zeta_4)$$

$$\hat{\psi}_{17} = \zeta_1\zeta_2\zeta_3, \ldots, \hat{\psi}_{20} = \zeta_2\zeta_3\zeta_4$$

FIGURE 2.29 *Hierarchic C^0-family up to cubic degree for the tetrahedron.*

tial derivatives at the midside nodes, and an interior nodal value. Recalling the one-dimensional basis in (2.8.6), the tensor-product basis on the master square is

$$\hat{\psi}_{ij}(\xi, \eta) = \hat{\psi}_i(\xi)\hat{\psi}_j(\eta) \tag{2.8.13}$$

whence

$$\begin{aligned}
\hat{\psi}_{11}(\xi, \eta) &= \tfrac{1}{4}(1 - \xi)(1 - \eta), \quad \ldots, \quad \hat{\psi}_{22}(\xi, \eta) = \tfrac{1}{4}(1 + \xi)(1 + \eta) \\
\hat{\psi}_{13}(\xi, \eta) &= -\tfrac{1}{4}(1 - \xi)(1 - \eta^2), \ldots, \quad \hat{\psi}_{32}(\xi,\eta) = -\tfrac{1}{4}(1 - \xi^2)(1 + \eta) \\
\hat{\psi}_{33}(\xi, \eta) &= \tfrac{1}{4}(1 - \xi^2)(1 - \eta^2)
\end{aligned} \tag{2.8.14}$$

Once again, a nodal degree of freedom need not be associated with the interior node. The development of higher-degree tensor-product bases is left as an exercise.

We remark that hierarchic C^1-simplex elements can also be developed but the construction is quite complex. More significantly, C^1-elements of minimal complexity are already of high degree, so that the hierarchic concept is less practical in this instance.

EXERCISES

2.8.1 Follow the approach described for developing the one-dimensional hierarchic bases to construct the hierarchic quadratic and cubic shape functions on a triangle.

2.8.2 Construct the following tensor-product hierarchic bases:

(a) Bicubics for a quadrilateral

(b) Biquadratics for a "quadrilateral brick"

2.8.3 Show that the hierarchic quartics for the tetrahedron are obtained by including the following additional shape functions:

(a) At the midedge nodes,

$$\psi_{21}(\zeta) = -\tfrac{1}{48}\zeta_1\zeta_2(\zeta_1^2 + \zeta_2^2), \quad \ldots, \quad \psi_{26} = -\tfrac{1}{48}\zeta_1\zeta_4(\zeta_1^2 + \zeta_4^2)$$

(b) On the faces,

$$\psi_{27}(\zeta) = \zeta_1^2\zeta_2\zeta_3, \quad \ldots, \quad \psi_{34}(\zeta) = \zeta_2\zeta_3^2\zeta_4$$

(c) In the interior,

$$\psi_{35}(\zeta) = \zeta_1\zeta_2\zeta_3\zeta_4$$

3

MIXED METHODS, HYBRID METHODS,

AND PENALTY METHODS

3.1 THE STANDARD FINITE ELEMENT METHOD AND MINIMIZATION PROBLEMS

3.1.1 Minimization of Functionals

In many variants of the finite element method, it is informative to view the boundary-value problem to be approximated as one of finding the minimizer of a certain energy functional. For instance, consider the second-order boundary-value problem

$$\sum_{i,j=1}^{3} -\frac{\partial}{\partial x_i}\left[a_{ij}(\mathbf{x})\frac{\partial u(\mathbf{x})}{\partial x_j}\right] + b(\mathbf{x})u(\mathbf{x}) = f(\mathbf{x}) \quad \text{in} \quad \Omega$$

$$u(\mathbf{x}) = 0 \quad \text{on} \quad \partial\Omega$$

(3.1.1)

where $\mathbf{x} = (x_1, x_2, x_3) = (x, y, z)$ is a point in Ω and the coefficients a_{ij} and b are given functions of position $\mathbf{x}$ in Ω, $\Omega \subset \mathbb{R}^3$. We shall assume that the coefficients a_{ij} and b are smooth (e.g., twice differentiable) functions of $\mathbf{x}$ with $b(\mathbf{x}) \geq 0$ and such that for any three real numbers ξ_i,

$$\sum_{i,j=1}^{3} a_{ij}(\mathbf{x})\xi_i\xi_j \geq a_0 \sum_{i=1}^{3} \xi_i^2 \quad \forall \mathbf{x} \in \Omega$$

(3.1.2)

where a_0 is a positive constant. When the coefficients satisfy a condition such as (3.1.2), the governing operator in problem (3.1.1) is referred to as *strongly elliptic* and (3.1.1) is a (strongly) elliptic boundary-value problem. We assume that $f \in L^2(\Omega)$ for the present.

A direct variational statement of (3.1.1) is: find $u \in H_0^1(\Omega)$ such that

$$\int_\Omega \left[\sum_{i,j=1}^3 a_{ij} \frac{\partial u}{\partial x_i} \frac{\partial v}{\partial x_j} + buv \right] dx = \int_\Omega fv \, dx \tag{3.1.3}$$

$$\forall v \in H_0^1(\Omega)$$

Recall that this variational problem is the basis from which our usual Galerkin finite element methods are developed.

In the special case in which the coefficients appearing in (3.1.1) are symmetric, that is, when

$$a_{ij}(\mathbf{x}) = a_{ji}(\mathbf{x}), \quad 1 \le i, j \le 3, \quad \mathbf{x} \in \Omega$$

problem (3.1.3) is equivalent to finding the minimum of the functional

$$J(v) = \frac{1}{2} \int_\Omega \left(\sum_{i,j=1}^3 a_{ij} \frac{\partial v}{\partial x_i} \frac{\partial v}{\partial x_j} + bv^2 - 2fv \right) dx \tag{3.1.4}$$

over all $v \in H_0^1(\Omega)$. We refer to J in (3.1.4) as a functional to emphasize that its domain is the space $H_0^1(\Omega)$ of admissible functions, and its range (i.e., its set of values) is a set of real numbers (i.e., a subset of $\mathbb{R}$). We express these facts symbolically by writing

$$J : H_0^1(\Omega) \longrightarrow \mathbb{R}$$

which is read "J maps the space of admissible functions $H_0^1(\Omega)$ into the real numbers $\mathbb{R}$."

We now get a hint as to why we refer to (3.1.3) as a "variational" problem. Following standard arguments of variational calculus, we may determine how much J varies at a "point" $u \in H_0^1(\Omega)$ as we move a distance ϵv from u, ϵ being an arbitrary positive number and v an arbitrary member of $H_0^1(\Omega)$ (i.e., ϵv is an arbitrary *admissible variation* in u, so that $u + \epsilon v$ is a nearby admissible function for ϵ small). The value of J at $u + \epsilon v$ is

$$J(u + \epsilon v) = \frac{1}{2} \int_\Omega \left[\sum_{i,j=1}^3 a_{ij} \left(\frac{\partial u}{\partial x_i} + \epsilon \frac{\partial v}{\partial x_i} \right) \left(\frac{\partial u}{\partial x_j} + \epsilon \frac{\partial v}{\partial x_j} \right) \right.$$
$$\left. + b(u + \epsilon v)^2 - 2f(u + \epsilon v) \right] dx$$

Expanding and collecting terms, we have

$$J(u + \epsilon v) = J(u) + \epsilon \langle \delta J(u), v \rangle + \tfrac{1}{2} \epsilon^2 \langle \delta^2 J(u)v, v \rangle \tag{3.1.5}$$

where

$$\langle \delta J(u), v \rangle \equiv \int_{\Omega} \left(\sum_{i,j=1}^{3} a_{ij} \frac{\partial u}{\partial x_i} \frac{\partial v}{\partial x_j} + buv - fv \right) dx \tag{3.1.6}$$

$$\langle \delta^2 J(u)v, v \rangle \equiv \int_{\Omega} \left(\sum_{i,j=1}^{3} a_{ij} \frac{\partial v}{\partial x_i} \frac{\partial v}{\partial x_j} + bv^2 \right) dx \tag{3.1.7}$$

The quantity $\langle \delta J(u), v \rangle$ is called the *first variation* in J; it is linear in v [and in $\delta J(u)$]. Under our assumptions on the coefficients a_{ij} and b, expression (3.1.6) defines a continuous linear functional on $H_0^1(\Omega)$. Hence, $\delta J(u)$ can be regarded as a member of the dual space $H^{-1}(\Omega)$ of $H_0^1(\Omega)$ and the symbols $\langle \cdot, \cdot \rangle$ may then denote a "duality pairing" on $H_0^1(\Omega)$ and its dual. If v is very smooth, an integration of (3.1.6) by parts reveals that $\delta J(u)$ is *formally*

$$\delta J(u) = - \sum_{i,j=1}^{3} \frac{\partial}{\partial x_i} \left(a_{ij} \frac{\partial u}{\partial x_j} \right) + bu - f, \qquad u = 0 \quad \text{on} \quad \partial\Omega \tag{3.1.8}$$

We remark that the first two terms in the first variation (3.1.6) define the bilinear form $B(\cdot, \cdot)$ for this problem as discussed in Chapter 1 [recall (1.3.2)]. Hence, for problem (3.1.1),

$$\begin{aligned} B(u, v) &= \int_{\Omega} \left(\sum_{i,j=1}^{3} a_{ij} \frac{\partial u}{\partial x_i} \frac{\partial v}{\partial x_j} + buv \right) dx \\ &= \langle \delta J(u), v \rangle + \int_{\Omega} fv \, dx \end{aligned} \tag{3.1.9}$$

A function $u \in H_0^1(\Omega)$ is called a *critical point* of J if the first variation of J vanishes at u, that is, if

$$\langle \delta J(u), v \rangle = 0 \qquad \forall v \in H_0^1(\Omega) \tag{3.1.10}$$

The values of J at critical points are called *stationary values* of the functional. Equation (3.1.10) is equivalent to

$$\delta J(u) = 0 \quad \text{in} \quad (H_0^1(\Omega))' = H^{-1}(\Omega) \tag{3.1.11}$$

Comparing (3.1.10) with (3.1.3), we see that the *variational boundary-value problem (3.1.3) is equivalent to the problem of finding critical points of the functional J. Moreover, in view of* (3.1.11), *the variational problems (3.1.3) and (3.1.10) are equivalent to (3.1.1) if we interpret (3.1.1) as a problem set in the dual space $H^{-1}(\Omega)$ of the space of admissible functions.* Any classical solution to (3.1.1) is also a solution to (3.1.10) or (3.1.11); any solution of (3.1.10) is also a solution of (3.1.1) provided that (3.1.1) is interpreted in the sense of linear functionals as in (3.1.11).

It is clear from (3.1.5) that the first variation in J can be computed from the formula

$$\langle \delta J(u), v \rangle = \lim_{\epsilon \to 0} \frac{\partial}{\partial \epsilon} J(u + \epsilon v) \qquad (3.1.12)$$

Whenever this limit exists and $\delta J(u)$ is in the dual space of the space of admissible functions on which it is defined, the functional is said to be *differentiable* (or differentiable "in the sense of Gâteaux") at u.

The quantity $\langle \delta^2 J(u)v, v \rangle$ defined in (3.1.7) is known as the *second variation* of J. If the functions a_{ij} satisfy (3.1.2), then (3.1.7) looks very much like a norm on $H_0^1(\Omega)$. Indeed, in such cases, there exists a positive constant C depending only on the constant a_0 [in (3.1.2)] and b such that, for all v in $H_0^1(\Omega)$,

$$\langle \delta^2 J(u)v, v \rangle \geq C \| v \|_1^2 \qquad (3.1.13)$$

We may use this property to characterize the critical point. Suppose that (3.1.13) holds and that u is a critical point of J. Take $\epsilon = \sqrt{2}$ and let $w = u + \epsilon v$ be an arbitrary element of $H_0^1(\Omega)$. Then, in view of (3.1.11) and (3.1.13), (3.1.5) yields

$$J(w) = J(u) + \langle \delta^2 J(u)v, v \rangle \geq J(u) + C \| v \|_1^2$$

whence

$$J(u) \leq J(w) \qquad \forall w \in H_0^1(\Omega) \qquad (3.1.14)$$

Thus, in this case the solution u of our problem is a *minimizer* of J; the functional J assumes its smallest possible value when evaluated at the solution u of the variational boundary-value problem (3.1.3).

3.1.2   Finite Element Approximations

The finite element approximation of this minimization problem is nothing more than a simple variant of the classical Rayleigh–Ritz method. The value of J over a finite element mesh is clearly

$$J(v_h) = \sum_{e=1}^{E} J_e(v_h^e) \qquad (3.1.15)$$

where

$$J_e(v_h^e) = \frac{1}{2} \int_{\Omega_e} \left(\sum_{i,j=1}^{3} a_{ij} \frac{\partial v_h^e}{\partial x_i} \frac{\partial v_h^e}{\partial x_j} + b v_h^{e2} - 2 f v_h^e \right) dx \qquad (3.1.16)$$

Here the mesh contains E elements Ω_e and v_h^e is the restriction of a finite element test function v_h [belonging to a finite-dimensional subspace H^h

of $H_0^1(\Omega)$] to element Ω_e. Suppose that v_h^e is expressed in the form

$$v_h^e(\mathbf{x}) = \sum_{l=1}^{N_e} v_l^e \psi_l^e(\mathbf{x}), \qquad v_l^e = v_h^e(\mathbf{x}_l), \quad \mathbf{x}_l \in \bar{\Omega}_e \qquad (3.1.17)$$

where $\{\mathbf{x}_l\}_{l=1}^{N_e}$ are nodes in $\bar{\Omega}_e$ and ψ_l^e are the usual element shape functions. Then (3.1.16) reduces to

$$J_e(v_h^e) = \tfrac{1}{2}\left(\sum_{m,l=1}^{N_e} v_m^e k_{ml}^e v_l^e - 2 \sum_{m=1}^{N_e} f_m^e v_m^e \right) \qquad (3.1.18)$$

where k_{ml}^e and f_m^e are entries in the element stiffness matrix and load vector, respectively:

$$k_{ml}^e = \int_{\Omega_e} \left(\sum_{i,j=1}^{3} a_{ij} \frac{\partial \psi_m^e}{\partial x_i} \frac{\partial \psi_l^e}{\partial x_j} + b\psi_m^e \psi_l^e \right) dx$$

$$f_m^e = \int_{\Omega_e} f\psi_m^e \, dx \qquad (3.1.19)$$

Introducing (3.1.18) into (3.1.15) gives

$$J(v_h) = \tfrac{1}{2}\mathbf{v}^T \mathbf{K} \mathbf{v} - \mathbf{v}^T \mathbf{F} \qquad (3.1.20)$$

wherein $\mathbf{v}$ is the vector of nodal values v_j of v_h, $\mathbf{K}$ is the final (global) stiffness matrix, and $\mathbf{F}$ is the final load vector.

Thus, the finite element approximation of the minimization problem is one of seeking a minimizer to the functional $J: H^h \longrightarrow \mathbb{R}$ given by (3.1.20). The first variation of J, restricted to the finite-dimensional space H^h, is clearly

$$\langle \delta J(u_h), v_h \rangle = \sum_{j=1}^{M} \frac{\partial J(u_h)}{\partial v_j} v_j \qquad (3.1.21)$$

where the sum is taken over all M degrees of freedom of the model. The approximation u_h is a critical point if $\partial J(u_h)/\partial v_j = 0$ for all j, $1 \le j \le M$, and, in view of (3.1.20), this happens if and only if

$$\mathbf{Ku} = \mathbf{F} \qquad (3.1.22)$$

This, of course, is the final system of linear equations for our finite element model. The analysis is completed by solving (3.1.22) for the nodal values u_j and thereby determining the local approximations $u_h^e(\mathbf{x}) = \sum_{j=1}^{N_e} u_j^e \psi_j^e(\mathbf{x})$.

3.1.3 Conditions for a Minimum

Let us now take a more detailed look at some concrete conditions which are sufficient to guarantee the existence of a unique minimizer of a given functional J. We will discover that such conditions also provide useful information on the acceptability of various approximations of J.

We first consider a rather abstract situation in which J is a functional defined on a general real Hilbert space H [of which $H^m(\Omega)$, $H_0^m(\Omega)$, etc., are typical examples]. Thus, H is endowed with an inner product $(\cdot, \cdot)_H$ and norm $\| \cdot \|_H = \sqrt{(\cdot, \cdot)_H}$ and we wish to find $u \in H$ such that

$$J(u) \leq J(v) \quad \forall v \in H \quad (J: H \longrightarrow \mathbb{R}) \quad (3.1.23)$$

The minimization arguments made previously can be shown to apply whenever J has the following properties:

1. *J is differentiable.* In other words, J is such that it is always possible to compute its first variation,

$$\langle \delta J(u), v \rangle = \lim_{\epsilon \to 0} \frac{\partial}{\partial \epsilon} J(u + \epsilon v)$$

   Here $\langle \cdot, \cdot \rangle$ denotes duality pairing on $H' \times H$; that is, if H' is the dual of H, then $\delta J(u) \in H'$ and $\delta J(u)(v) \equiv \langle \delta J(u), v \rangle$.

2. *J is strictly convex.* That is, if u and v are arbitrary elements of the space H on which J is defined and θ is a real number, $0 < \theta < 1$, then

$$J(\theta u + (1 - \theta)v) < \theta J(u) + (1 - \theta)J(v)$$

   whenever $u, v \neq 0$, $u \neq v$.

3. *J is coercive.* This means that if $\| \cdot \|_H$ is an appropriate norm on H, the magnitude of the value of J at u grows with increasing values of $\| u \|_H$; that is,

$$\lim_{\| u \|_H \to \infty} J(u) = +\infty$$

Under these conditions, we can easily prove the following minimization theorem:*

Theorem 3.1.1.   Let H be a Hilbert space with inner product $(\cdot, \cdot)_H$ and norm $\| \cdot \|_H = \sqrt{(\cdot, \cdot)_H}$. Let $J: H \to \mathbb{R}$ be a real functional satisfying conditions 1 through 3 above. Then there exists a unique element u in H which minimizes J over all of H:

$$J(u) \leq J(v) \quad \text{for all } v \in H$$

Moreover, the minimizer is characterized by the variational (Euler–Lagrange) equation

$$\langle \delta J(u), v \rangle = 0 \quad \forall v \in H \qquad \square$$

---

* Significantly weaker conditions can be given to guarantee the existence of a minimizer, but these are standard and suit our present purposes. For more general theorems, see, for example, Vainberg [1973].

As a sample application of this theorem, consider the energy functional for Poisson's problem in the plane:

$$J: H_0^1(\Omega) \longrightarrow \mathbb{R}, \qquad J(v) = \tfrac{1}{2} \int_\Omega (v_x^2 + v_y^2 - 2fv)\, dx\, dy \qquad (3.1.24)$$

where $v_x = \partial v/\partial x$, $v_y = \partial v/\partial y$, and f is given in $L^2(\Omega) = H^0(\Omega)$. We easily verify that

$$\langle \delta J(u), v \rangle = \lim_{\epsilon \to 0} \frac{\partial}{\partial \epsilon} J(u + \epsilon v)$$

$$= \int_\Omega (u_x v_x + u_y v_y - fv)\, dx\, dy \qquad (3.1.25)$$

$$= B(u, v) - \int_\Omega fv\, dx\, dy$$

where, in this case [as in (1.3.20)],

$$B(u, v) = \int_\Omega \left(\frac{\partial u}{\partial x}\frac{\partial v}{\partial x} + \frac{\partial u}{\partial y}\frac{\partial v}{\partial y} \right) dx\, dy \qquad (3.1.26)$$

A necessary condition that u be a critical point of J is that

$$\langle \delta J(u), v \rangle = B(u, v) - \int_\Omega fv\, dx\, dy = 0$$
$$\forall v \in H_0^1(\Omega) \qquad (3.1.27)$$

so that if the critical point u is sufficiently smooth, the solution to (3.1.27) is also a solution of the Dirichlet problem

$$-\Delta u = f \quad \text{in} \quad \Omega, \qquad u = 0 \quad \text{on} \quad \partial\Omega$$

where $\Delta = \partial^2/\partial x^2 + \partial^2/\partial y^2$ is the Laplacian operator.

We easily verify that J is strictly convex: For $0 < \theta < 1$,

$$J(\theta u + (1 - \theta)v) = \tfrac{1}{2} \int_\Omega \{[\theta u_x + (1 - \theta)v_x]^2$$

$$+ [\theta u_y + (1 - \theta)v_y]^2 - 2f[\theta u + (1 - \theta)v]\}\, dx\, dy$$

$$= \tfrac{1}{2} \int_\Omega \{\theta^2(u_x^2 + u_y^2) + (1 - \theta)^2(v_x^2 + v_y^2)$$

$$+ 2\theta(1 - \theta)(u_x v_x + u_y v_y)\}\, dx\, dy$$

$$- \int_\Omega [\theta fu + (1 - \theta)fv]\, dx\, dy$$

But since

$$\int_\Omega 2(u_x v_x + u_y v_y)\, dx\, dy < \int_\Omega (u_x^2 + v_x^2 + u_y^2 + v_y^2)\, dx\, dy$$

we must have strict convexity of J,

$$J[\theta u + (1 - \theta)v] < \theta J(u) + (1 - \theta)J(v)$$

Finally, J is coercive with respect to the norm

$$\| u \|_H = \left[\int_\Omega (u_x^2 + u_y^2)\, dx\, dy \right]^{1/2}$$

Indeed, since*

$$J(u) = \tfrac{1}{2} \| u \|_H^2 - \int_\Omega fu\, dx\, dy$$

$$\geq \tfrac{1}{2} \| u \|_H^2 - c \| f \|_0 \| u \|_H$$

where c is a constant, we have

$$\lim_{\| u \|_H \to \infty} J(u) = +\infty$$

Hence, from Theorem 3.1.1, J has a unique minimizer in $H_0^1(\Omega)$ and that minimizer is characterized as a solution of the variational boundary-value problem (3.1.27).

We can now ask if the conditions in Theorem 3.1.1 are transferred to the discrete minimization problem,

$$J(u_h) \leq J(v_h) \qquad \forall\, v_h \in H^h; \qquad J(v_h) = \tfrac{1}{2} \mathbf{v}^T \mathbf{K} \mathbf{v} - \mathbf{v}^T \mathbf{F} \qquad (3.1.28)$$

where J is now understood to be restricted to the finite-dimensional space H^h spanned by the finite element basis functions. J is clearly differentiable on H^h [indeed, $\langle \delta J(u_h), v_h \rangle = (\mathbf{Ku} - \mathbf{F})^T \mathbf{v}$]. Since $\mathbf{K}$ is symmetric and the stiffnesses K_{ij} are bounded (being finite real numbers), we also verify that J is strictly convex by essentially following the same steps used in proving convexity of J on $H_0^1(\Omega)$. Finally, if $\mathbf{K}$ is strongly positive definite in the sense that a constant $c_1 > 0$ exists such that

$$\mathbf{v}^T \mathbf{K} \mathbf{v} \geq c_1 \mathbf{v}^T \mathbf{v} \qquad (3.1.29)$$

---

* Here we use Schwarz's inequality: $\int_\Omega fu\, dx\, dy \leq \left(\int_\Omega f^2\, dx\, dy \right)^{1/2} \left(\int_\Omega u^2\, dx\, dy \right)^{1/2}$ $= \| f \|_{H^0(\Omega)} \| u \|_{H^0(\Omega)}$ and Poincaré's inequality, $\| u \|_0 \leq c \| u \|_H$. Recall equation (1.2.19).

then J is coercive on H^h. To interpret (3.1.29), let $\{\lambda_i\}_{i=1}^M$ and $\{w_i\}_{i=1}^M$ denote eigenvalues and eigenvectors of $\mathbf{K}$ (i.e., $\mathbf{K}w_i = \lambda_i w_i$, $1 \leq i \leq M$). Then

$$w_i^T \mathbf{K} w_i = \lambda_i w_i^T w_i = \lambda_i \| w_i \|_{\mathbb{R}^M}^2$$

where $\| w_i \|_{\mathbb{R}^M} = \sqrt{w_i^T w_i}$ is the $\mathbb{R}^M$-Euclidean norm of the vector w_i. Then

$$c_1 = \min_{v \in \mathbb{R}^M} \frac{v^T \mathbf{K} v}{v^T v} = \min_{1 \leq i \leq M} \lambda_i$$

Thus, if $\mathbf{K}$ is nonsingular with a minimum eigenvalue $\lambda_{\min} > 0$, J is coercive and the discrete problem has a unique mimimizer in H^h.

EXERCISES

3.1.1 Describe the variational boundary-value problems and the formal differential equations and boundary conditions corresponding to stationary values of the following functionals:

(a) $J_1(v) = \frac{1}{2} \int_0^1 [k(v'')^2 + p(v')^2 - 2fv] \, dx$

    for k, p, and f given functions of x, $0 \leq x \leq 1$;

$$J_1 : H_0^2(0, 1) \longrightarrow \mathbb{R}$$

(b) $J_2(v) = \frac{1}{2} \int_\Omega (\nabla v \cdot \nabla v + v^2 - 2fv) \, dx \, dy - \oint_{\partial\Omega} gv \, ds$ for $\Omega \subset \mathbb{R}^2$, g

    given data on the boundary $\partial\Omega$ and f a given function of position in Ω;

$$J_2 : H^1(\Omega) \longrightarrow \mathbb{R}$$

(c) $J_3(v) = \frac{1}{2} \int_\Omega (\nabla v \cdot \nabla v + v^2 - 2fv) \, dx \, dy$ for $\Omega \subset \mathbb{R}^2$, and f a given

    function of position in Ω;

$$J_3 : H_0^1(\Omega) \longrightarrow \mathbb{R}$$

(d) $J_4(v)$ is of the same form as $J_3(v)$ except that

$$J_4 : H^1(\Omega) \longrightarrow \mathbb{R}$$

(e) $J_5(v) = \frac{1}{2} \int_0^1 [k(v'')^2 - 2fv] \, dx - v'(0)M$ with k, M = constants, and

    f a given function of x;

$$J_5 : H \longrightarrow \mathbb{R}$$
$$H = \{v \in H^2(0, 1): v(0) = v'(1) = v(1) = 0\}$$

3.1.2 Consider the four functionals

$$J_1(v) = \tfrac{1}{2}\int_0^1 [(v'')^2 + (v')^2 + v^2 - 2xv \, dx]; \qquad J_1: H_0^2(0,1) \longrightarrow \mathbb{R}$$

$$J_2(v) = \frac{1}{2}\int_0^1 \left[(2v')^2 + \left(1 - \frac{1}{x}\right)v^2 - 2v \sin x\right] dx; \qquad J_2: H_0^1(0,1) \longrightarrow \mathbb{R}$$

$$J_3(v) = \frac{1}{2}\int_\Omega \left[\left(\frac{\partial v}{\partial x}\right)^2 + \left(\frac{\partial v}{\partial y}\right)^2 + v^2 - 2v \cos x \cos y\right] dx \, dy$$

$$+ \oint_{\partial\Omega} gv \, ds; \qquad J_3: H^1(\Omega) \longrightarrow \mathbb{R} \ (\Omega \subset \mathbb{R}^2)$$

$$J_4(v) = \frac{1}{2}\int_\Omega \left\{\left(\frac{\partial^2 v}{\partial x^2}\right)^2 + \left(\frac{\partial^2 v}{\partial y^2}\right)^2 - 2\left[\frac{\partial^2 v}{\partial x^2}\frac{\partial^2 v}{\partial y^2} - \left(\frac{\partial^2 v}{\partial x \, \partial y}\right)^2\right]\right\} dx \, dy$$

$$- \int_\Omega fv \, dx \, dy; \qquad J_4: H_0^2(\Omega) \longrightarrow \mathbb{R} \ (\Omega \subset \mathbb{R}^2)$$

where $f \in L^2(\Omega)$, $g \in L^2(\partial\Omega)$.

(a) Show that each of the functionals has a unique minimizer in the space of admissible functions on which it is defined. (In the case of J_3 some additional conditions on the data g are required; what are they?)

(b) Derive the boundary-value problems that formally characterize the minimizers in each case.

3.1.3 Using two finite elements, construct a piecewise-cubic C^1-approximation of the functional

$$J: H_0^2(0,1) \longrightarrow \mathbb{R}; \qquad J(v) = \tfrac{1}{2}\int_0^1 [(v'')^2 + v^2 - 2v] \, dx$$

Carry out all the computations, including the solution of the stiffness equations, and compare the approximate minimizer with the exact minimizer of J.

3.1.4 Suppose that, instead of (3.1.13), we have $\langle \delta^2 J(u)v, v \rangle \geq 0$, $\forall v \in H_0^1(\Omega)$, at a critical point u. Show that u is still a minimizer of J but that it may not be unique.

3.2 LAGRANGE MULTIPLIERS AND MIXED FINITE ELEMENT METHODS

A wide variety of finite element methods can be developed using the classical notion of Lagrange multipliers. We will survey some of the basic ideas behind these methods and discuss some specific applications.

3.2.1 Constrained Minimization Problems

The minimization problems discussed in the preceding section are un-constrained; that is, we seek the minimum of J over an entire space H of admissible functions without any side conditions imposed on the minimizer. We consider next a more difficult class of minimization problems in which we demand that the minimum of J be obtained on those subsets of functions in H that satisfy side conditions,* such as

$$Bu = g \qquad (3.2.1)$$

where B is a given linear operator from H into another Hilbert space Q and g represents given data in Q. For instance,

$$Bu = \frac{\partial u}{\partial x}, \qquad g = 1$$

is an example of a constraint one might impose on minimizers of functionals on spaces such as $H^1(\Omega)$, and B maps $H^1(\Omega)$ into $H^0(\Omega)$.

The subset K of H consisting of all elements that satisfy the constraint (3.2.1) is referred to as the *constraint set* or the *set of admissible functions:*

$$K = \{v \in H \mid Bv = g\} \qquad (3.2.2)$$

The minimization problem now consists of seeking specific u in K such that

$$J(u) \leq J(v) \qquad \forall v \in K \qquad (3.2.3)$$

The problem of finding a minimum of J in a constraint set K is consider-ably more difficult than the unconstrained minimization problem discussed earlier. Of particular concern here is the fact that the elements of K may be difficult or impossible to approximate.

The method of Lagrange multipliers provides a powerful alternative which enables us to seek a minimizer in the whole space H, not just in the constraint set K. In the method of Lagrange multipliers, we introduce a second space of functions Q', assumed here to be a Hilbert space and called the *space of Lagrange multipliers*, which is the dual of the space Q which contains the elements Bu; that is, B maps H into Q as asserted earlier, and we introduce a functional L on $H \times Q'$ into $\mathbb{R}$, defined by

$$L(v, q) = J(v) + [q, Bv - g] \qquad (3.2.4)$$

---

* The constraint need not be an equality constraint. We consider problems involving inequality constraints in Volume V.

where $[\cdot, \cdot]$ denotes duality pairing on $Q' \times Q$. In other words, Q is a space of functions containing the given data g in the constraint $Bu = g$, and Q' is its dual (i.e., Q' is the space of continuous linear functionals on Q). Thus, if q is an element of Q', we have

$$\left.\begin{array}{c} q(g) \equiv [q, g] \in \mathbb{R}; \qquad g \in Q \\ q(\alpha g_1 + \beta g_2) = \alpha[q, g_1] + \beta[q, g_2], \qquad \forall \alpha, \beta \in \mathbb{R} \\ \forall g_1, g_2 \in Q \end{array}\right\} \tag{3.2.5}$$

The members $q \in Q'$ are *multipliers* (of the constraint) and the functional L in (3.2.4) is called the *Lagrangian* corresponding to J and the constraint set K. Its first variation at $(u, p) \in H \times Q'$ is

$$\begin{aligned} \langle \delta L(u, p), (v, q) \rangle_{H \times Q'} &= \lim_{\epsilon \to 0} \frac{\partial}{\partial \epsilon} L(u + \epsilon v, p + \epsilon q) \\ &= \langle \delta J(u), v \rangle + [p, Bv] + [q, Bu - g] \end{aligned} \tag{3.2.6}$$

where $\langle \cdot, \cdot \rangle_{H \times Q'}$ denotes duality pairing on $(H \times Q')' \times (H \times Q')$. Here $\langle \delta J(u), v \rangle$ is the first variation of J evaluated at u and, since B is assumed to be linear, we have used the fact that

$$\lim_{\epsilon \to 0} \frac{\partial [p + \epsilon q, B(u + \epsilon v) - g]}{\partial \epsilon} = [p, Bv] + [q, Bu - g]$$

Note that the transpose B^* of B has the property that B^* maps Q' into H' and for arbitrary $q \in Q', v \in H$,

$$[q, Bv] = \langle B^*q, v \rangle \tag{3.2.7}$$

where $\langle \cdot, \cdot \rangle$ denotes duality pairing on $H' \times H$. Thus, (3.2.6) can also be written

$$\langle \delta L(u, p), (v, q) \rangle_{H \times Q'} = \langle \delta J(u), v \rangle + \langle B^*p, v \rangle + [q, Bu - g] \tag{3.2.8a}$$

or

$$\begin{aligned} \langle \delta L(u, p), (v, q) \rangle_{H \times Q'} &= \langle \delta J(u), v \rangle + \langle B^*p, v \rangle \\ &\quad + \langle B^*q, u \rangle - [q, g] \end{aligned} \tag{3.2.8b}$$

By demanding that (u, p) be such that $\langle \delta L(u, p), (v, q) \rangle_{H \times Q'}$ vanish for arbitrary v in H and q in Q', we arrive at the variational boundary-value problem

$$\begin{aligned} \langle \delta J(u), v \rangle + [p, Bv] &= 0 \qquad \forall v \in H \\ [q, Bu - g] &= 0 \qquad \forall q \in Q' \end{aligned} \tag{3.2.9}$$

Under certain conditions on J and B, the first component u of the solution of (3.2.9) is precisely the solution of the constrained minimization problem (3.2.3).

Example 3.2.1   Boundary Conditions as Constraints

As a first example, we consider the problem of minimizing the *Dirichlet integral*

$$J(v) = \tfrac{1}{2} \int_\Omega |\nabla v|^2 \, dx - \int_\Omega fv \, dx, \qquad \Omega \subset \mathbb{R}^2$$

$$J : H^1(\Omega) \longrightarrow \mathbb{R}, \qquad f \in L^2(\Omega)$$

(3.2.10)

subject to the constraint that the minimizer satisfy the Dirichlet boundary condition

$$u(s) = \hat{u}(s), \qquad s \in \partial\Omega \tag{3.2.11}$$

where $\hat{u}$ is a smooth function given on the boundary.

In this case we can take

$$Bu = u|_{\partial\Omega}, \qquad B : H^1(\Omega) \longrightarrow Q$$

$$Q = L^2(\partial\Omega) = Q'$$

(3.2.12)

$$L(v, q) = J(v) + \int_{\partial\Omega} q(u - \hat{u}) \, ds \tag{3.2.13}$$

The functional L attains a stationary value at any pair of functions (u, p) satisfying

$$\int_\Omega \nabla u \cdot \nabla v \, dx + \int_{\partial\Omega} pv \, ds = \int_\Omega fv \, dx \qquad \forall v \in H^1(\Omega)$$

$$\int_{\partial\Omega} q(u - \hat{u}) \, ds = 0 \qquad \forall q \in L^2(\partial\Omega)$$

(3.2.14)

Example 3.2.2   Stokes' Problem

One of the most important and popular examples of a constrained minimization problem that fits naturally in the framework established above is the classical problem of Stokes describing the slow flow of an incompressible viscous fluid. Let us consider an example in which the domain Ω is a smooth, bounded, two-dimensional region, and the total energy is given by the functional

$$J(\mathbf{v}) = \frac{1}{2} \int_\Omega \left[\nu \sum_{i,j=1}^2 \frac{\partial v_i}{\partial x_j} \frac{\partial v_i}{\partial x_j} - 2\mathbf{f} \cdot \mathbf{v} \right] dx \tag{3.2.15}$$

where v is the viscosity of the fluid, a positive constant, and $v_i = v_i(\mathbf{x})$, $i = 1, 2$ [$\mathbf{x} = (x_1, x_2)$, $dx = dx_1 \, dx_2$], are the components of the velocity vector,

$$\mathbf{v}(\mathbf{x}) = (v_1(\mathbf{x}), v_2(\mathbf{x})), \qquad \mathbf{x} \in \Omega$$

The vector $\mathbf{f} = \mathbf{f}(\mathbf{x})$ represents the body force per unit volume (area) of fluid and is given as part of the data.

For simplicity, let us assume that

$$v_i = 0 \quad \text{on} \quad \partial\Omega \tag{3.2.16}$$

and that $f_i \in H^0(\Omega) \, [= L^2(\Omega)]$. Then, in this case, the space H of admissible velocities is

$$H = H_0^1(\Omega) \times H_0^1(\Omega) = \{\mathbf{v} = (v_1, v_2) | v_i \in H_0^1(\Omega), \quad i = 1, 2\} \tag{3.2.17}$$

and

$$\| \mathbf{v} \|_H = \left\{ \int_\Omega \left[\left(\frac{\partial v_1}{\partial x_1}\right)^2 + \left(\frac{\partial v_1}{\partial x_2}\right)^2 + \left(\frac{\partial v_2}{\partial x_1}\right)^2 + \left(\frac{\partial v_2}{\partial x_2}\right)^2 \right] dx \right\}^{1/2} \tag{3.2.18}$$

Since the fluid is assumed to be incompressible, all flows are subjected to the constraint

$$-\operatorname{div} \mathbf{v} = 0 \quad \text{in} \quad \Omega \tag{3.2.19}$$

where, of course, $\operatorname{div} \mathbf{v} \equiv \partial v_1/\partial x_1 + \partial v_2/\partial x_2$. This means that, in this case, the constraint operator B is given by

$$B: H \longrightarrow H^0(\Omega), \qquad B\mathbf{v} = -\operatorname{div} \mathbf{v} \tag{3.2.20}$$

Thus, the constraint set K is given by

$$K = \{\mathbf{v} \in H | \operatorname{div} \mathbf{v} = 0 \quad \text{in} \quad \Omega\} \tag{3.2.21}$$

Then we can take

$$Q = Q' = H^0(\Omega) = L^2(\Omega)$$
$$[q, B\mathbf{v}] = \int_\Omega q \operatorname{div} \mathbf{v} \, dx = (q, B\mathbf{v}) \tag{3.2.22}$$

where $(\cdot, \cdot)$ is the inner product on $L^2(\Omega)$. Formally, if q is sufficiently smooth,

$$[q, B\mathbf{v}] = \langle B^*q, \mathbf{v} \rangle = \int_\Omega \nabla q \cdot \mathbf{v} \, dx \tag{3.2.23}$$

so that (formally) $B^*q = \operatorname{grad} q$ (plus boundary conditions).

A variational statement of Stokes' problem amounts to seeking a velocity **u** which minimizes J over all velocities that satisfy the incompressibility condition (3.2.19):

$$J(\mathbf{u}) \leq J(\mathbf{v}) \qquad \forall \mathbf{v} \in K \tag{3.2.24}$$

To obtain a Lagrange-multiplier formulation of this problem, we introduce the Lagrangian

$$L : (H_0^1(\Omega) \times H_0^1(\Omega)) \times L^2(\Omega) \longrightarrow \mathbb{R}$$
$$L(\mathbf{v}, q) = J(\mathbf{v}) + [q, B\mathbf{v}] = \tfrac{1}{2}B(\mathbf{v}, \mathbf{v}) - (\mathbf{f}, \mathbf{v}) + [q, B\mathbf{v}] \tag{3.2.25}$$

where

$$B(\mathbf{u}, \mathbf{v}) = \int_\Omega \nu \, \nabla\mathbf{u} : \nabla\mathbf{v} \, dx \equiv \int_\Omega \nu \sum_{i,j=1}^2 \frac{\partial u_i}{\partial x_j} \frac{\partial v_i}{\partial x_j} \, dx$$
$$(\mathbf{f}, \mathbf{v}) = \int_\Omega \mathbf{f} \cdot \mathbf{v} \, dx \tag{3.2.26}$$

Then, in accordance with (3.2.9), a stationary value of L is achieved at the point $(\mathbf{u}, p) \in H \times Q$ satisfying the system

$$B(\mathbf{u}, \mathbf{v}) - \int_\Omega p \, \mathrm{div}\, \mathbf{v} \, dx = (\mathbf{f}, \mathbf{v}) \qquad \forall \mathbf{v} \in H$$
$$\int_\Omega q \, \mathrm{div}\, \mathbf{u} \, dx = 0 \qquad \forall q \in Q \tag{3.2.27}$$

The variational boundary-value problem (3.2.27) thus characterizes a solution $(\mathbf{u}, p)$ of a two-dimensional Stokes problem; this solution also provides a stationary value of the Lagrangian L of (3.2.25).

We remark that if **u** and p are sufficiently smooth, the following Green's formula holds:

$$\int_\Omega \sum_{i,j=1}^2 \frac{\partial u_i}{\partial x_j} \frac{\partial v_i}{\partial x_j} \, dx = \int_\Omega - \sum_{i,j=1}^2 \frac{\partial}{\partial x_j}\left(\frac{\partial u_i}{\partial x_j}\right) v_i \, dx$$
$$+ \int_{\partial\Omega} \sum_{i,j=1}^2 \frac{\partial u_i}{\partial x_j} n_j v_i \, ds \tag{3.2.28}$$

and this can be used to verify that the classical boundary-value problem corresponding to (3.2.27) is

$$\left.\begin{array}{r} -\nu \, \Delta\mathbf{u} + \nabla p = \mathbf{f} \\ \mathrm{div}\, \mathbf{u} = 0 \end{array}\right\} \ \text{in } \Omega \\ \mathbf{u} = 0 \qquad \text{on } \partial\Omega \left.\begin{array}{r} \\ \\ \\ \end{array}\right\} \tag{3.2.29}$$

where $\Delta u = \partial^2 u/\partial x_1^2 + \partial^2 u/\partial x_2^2$. Clearly, the Lagrange multiplier p has an important physical interpretation: it is the *pressure* in the fluid. Note that p is determinable only to within an arbitrary constant.

3.2.2 Conditions for Saddle Points

Let us return to the general functional L of (3.2.4) and list some basic properties of Lagrange multiplier formulations.

Saddle Point Problem: The introduction of the Lagrange functional L of the form (3.2.4) transforms the minimization problem (3.2.3) into a *saddle point problem*. A *saddle point* of the functional $L : H \times Q' \to \mathbb{R}$ is a pair (u, p) with $u \in H$ and $p \in Q'$ such that

$$L(u, q) \leq L(u, p) \leq L(v, p) \qquad \forall v \in H, \quad \forall q \in Q' \qquad (3.2.30)$$

Roughly speaking, for fixed q, $L(v, q)$ is a minimum at $u \in H$; and for fixed v, $L(v, q)$ is a maximum at $p \in Q'$. Thus, $L(v, q)$ can be imagined as a surface over the (v, q)-plane which resembles a saddle with double curvature, convex in the direction of u and concave in the direction of p.

Conditions for the Existence of Saddle Points: These conditions are similar to those listed in Theorem 3.1.1:

(a) For fixed $q_0 \in Q'$, $L(v, q_0)$ is strictly convex and differentiable with respect to v; that is, the limit $\langle \delta_v L(u, q_0), (v, q_0) \rangle_{H \times Q'} = \lim_{\epsilon \to 0} \partial L(u + \epsilon v, q_0)/\partial \epsilon$ exists at all u in H and, for $u, v \neq 0$, $L(\theta u + (1 - \theta)v, q_0) < \theta L(u, q_0) + (1 - \theta)L(v, q_0)$, where $0 < \theta < 1$.

(b) For fixed $v_0 \in H$, $L(v_0, q)$ is concave and differentiable with respect to q; that is, the limit $\langle \delta_q L(v_0, p), (v_0, q) \rangle_{H \times Q'} = \lim_{\epsilon \to 0} \partial L(v_0, p + \epsilon q)/\partial \epsilon$ exists at all p in Q' and, for $p, q \neq 0$, $L(v_0, \theta p + (1 - \theta)q) \geq \theta L(v_0, p) + (1 - \theta)L(v_0, q)$, where $0 \leq \theta \leq 1$.

(c) One can find a $q_0 \in Q'$ such that $L(v, q_0)$ is coercive with respect to v in the sense that $\lim_{\|v\|_H \to \infty} L(v, q_0) = +\infty$.

(d) One can find a $v_0 \in H$ such that $L(v_0, q)$ is (negatively) coercive with respect to q in the sense that $\lim_{\|q\|_{Q'} \to \infty} L(v_0, q) = -\infty$.

$$(3.2.31)$$

It can be shown (see, e.g., Ekeland and Temam [1976]) that whenever conditions (3.2.31) hold, there exists at least one saddle point $(u, p) \in H \times Q'$ of $L(\cdot, \cdot)$ and that (u, p) is a solution of the variational boundary-value problem (3.2.9).

Additional Assumptions: Let us apply conditions (3.2.31) to the functional $L(\cdot, \cdot)$ defined in (3.2.4), assuming that the following conditions hold:

(a)   The functional $J : H \to \mathbb{R}$ is continuous and satisfies the conditions of Theorem 3.1.1; in particular, suppose that

$$J(v) \geq C_0 \|v\|_H^2 - C_1(f) \|v\|_H$$

for all v in H, where C_0 and C_1 are positive constants and C_1 depends on the data f.

(b)   The constraint operator $B : H \to Q$ is a bounded linear operator; that is, a constant $C_B > 0$ exists such that for every $v \in H$,

$$\|Bv\|_Q \leq C_B \|v\|_H$$

(c)   The variational operator $\delta J : H \to H'$ is bounded; that is, if $\|v\|_H \leq C_1 = $ constant, then $\|\delta J(v)\|_{H'} \leq C_2$ = constant.

(3.2.32)

Under conditions (a) and (b) above, we easily verify that conditions (a) and (b) of (3.2.31) hold. Also, since

$$|[q, Bv - g]| \leq \|q\|_{Q'}(\|Bv\|_Q + \|g\|_Q)$$
$$\leq \|q\|_{Q'}(C_B \|v\|_H + \|g\|_Q)$$

we have

$$L(v, q) = J(v) + [q, Bv - g]$$
$$\geq C_0 \|v\|_H^2 - [C_1(f) + C_B \|q\|_{Q'}] \|v\|_H - \|q\|_{Q'} \|g\|_Q$$

As the positive term grows quadratically as $\|v\|_H \to \infty$, whereas the negative terms depend only linearly on $\|v\|_H$, we see that $L(v, q) \to \infty$ as $\|v\|_H \to \infty$ for any fixed q. Hence, condition (c) of (3.2.31) also holds.

Unfortunately, *we cannot always verify that the final condition of (3.2.31) holds without imposing an additional condition.* Thus, conditions (a) through (c) in (3.2.32) are insufficient to guarantee the existence of a saddle point of L.

Perturbed Lagrangian: In order to satisfy all of the conditions (3.2.31), we introduce a *perturbed Lagrangian* L_ϵ defined by

$$L_\epsilon(v, q) = L(v, q) - \frac{\epsilon}{2} \| q \|_{Q'}^2 \qquad (3.2.33)$$

where L is given by (3.2.4) and ϵ is an arbitrary positive number. Since $L_\epsilon(v_0, q) \to -\infty$, as $\| q \|_{Q'} \to \infty$, all of the conditions in (3.2.31) hold. Thus, there exists a saddle point (u_ϵ, p_ϵ) of L_ϵ in $H \times Q'$ for every $\epsilon > 0$. Moreover, (u_ϵ, p_ϵ) is a solution of the variational boundary-value problem*

$$\begin{aligned} \langle \delta J(u_\epsilon), v \rangle + [p_\epsilon, Bv] &= 0 & \forall v \in H \\ [q, Bu_\epsilon] - \epsilon[q, p_\epsilon] &= [q, g] & \forall q \in Q' \end{aligned} \qquad (3.2.34)$$

Weak Convergence of Perturbed Saddle Points: The next question is whether or not the perturbed Lagrangian method (3.2.33) will "work." We know that under the conditions listed earlier, we can always solve (3.2.34) for (u_ϵ, p_ϵ) for each choice of $\epsilon > 0$, but will those perturbed solutions converge to a solution of the original problem as ϵ tends to zero?

To resolve this issue, we must first recognize that there are two types of convergence of sequences $\{u_\epsilon\}$, $\{p_\epsilon\}$ of solutions that should be considered for problems of this type: *weak convergence* in H and *strong convergence* in H.

By weak convergence, of a sequence $\{u_\epsilon\}$ to u in H, we mean that for any linear functional σ in the dual space H',

$$\lim_{\epsilon \to 0} \langle \sigma, u_\epsilon \rangle = \langle \sigma, u \rangle$$

For example, u_ϵ is said to converge weakly to u in $L^2(\Omega)$ if, for all f in $L^2(\Omega)$,

$$\lim_{\epsilon \to 0} \int_\Omega fu_\epsilon \, dx = \int_\Omega fu \, dx$$

(see Exercise 3.2.6). It is a fundamental and extremely useful property of Hilbert spaces H that *whenever a sequence* $\{u_\epsilon\}$ *is "bounded in norm,"* that is, whenever a constant C independent of ϵ exists such that

$$\| u_\epsilon \|_H \leq C$$

---

* We assume here, for simplicity, that $\lim_{\epsilon \to 0} \partial \| p + \epsilon q \|_{Q'}^2 / \partial \epsilon = 2[p, q]$, which applies, for example, when $Q = L^2(\Omega)$. Much more general situations could be handled, but these involve technicalities unwarranted in this introductory discussion. For additional details, see Oden [1978] or Kikuchi [1979].

there always exists a subsequence $\{u_{\epsilon'}\}$ *of* $\{u_{\epsilon}\}$ *that converges weakly to an element u in H.*

By strong convergence of a sequence $\{u_{\epsilon}\}$ to u in H, we mean convergence with respect to the norm defined on H; that is,

$$\lim_{\epsilon \to 0} \| u_{\epsilon} - u \|_H = 0$$

If a sequence converges weakly, it does not necessarily converge strongly. To ensure strong convergence of $\{(u_{\epsilon}, p_{\epsilon})\}$ to a solution of the original constrained problem, additional (stronger) conditions on the functional J must be imposed. We will comment further on this point later.

Without loss in generality, we shall hereafter take $g \equiv 0$ for simplicity so that our constraint simply becomes

$$Bu = 0 \quad (\text{in } Q)$$

The development of parallel results for the case $Bu = g \neq 0$ is left as an exercise.

Returning to (3.2.34), suppose that we calculate a sequence $\{(u_{\epsilon}, p_{\epsilon})\}$ of solutions of this problem for various values of ϵ as ϵ tends to zero. Each pair $(u_{\epsilon}, p_{\epsilon})$ is, by construction, the saddle point of the perturbed functional L_{ϵ}. Hence, for any q in Q' and v in H,

$$J(u_{\epsilon}) + [q, Bu_{\epsilon}] - \frac{\epsilon}{2} \| q \|_{Q'}^2 \leq J(v) + [p_{\epsilon}, Bv] - \frac{\epsilon}{2} \| p_{\epsilon} \|_{Q'}^2$$

Picking $q = 0$, $v = 0$, we have

$$J(u_{\epsilon}) \leq J(0) - \frac{\epsilon}{2} \| p_{\epsilon} \|_{Q'}^2 \leq J(0) = \text{constant}$$

and, from property (a) in (3.2.32),

$$C_0 \| u_{\epsilon} \|_H^2 - C_1(f) \| u_{\epsilon} \|_H \leq J(0) = \text{constant}$$

From this* it follows that

$$\| u_{\epsilon} \|_H \leq C \qquad (3.2.35)$$

where C is a constant, depending on J and f, but independent of ϵ. Hence, *there must exist a sequence* $\epsilon' \to 0$ *and an element u such that* $\{u_{\epsilon'}\}$ *converges*

---

* In particular, since $J(v)$ has the property (coerciveness) that it approaches $+\infty$ as $\| v \|_H \to +\infty$ and since $J(u_{\epsilon}) < \text{constant}$, we must have $\| u_{\epsilon} \|_H \leq \text{constant}$.

weakly to u in H. We have some hope, therefore, that this weak limit u may be a solution to our constrained minimization problem.

If only it were possible to show that the sequence $\{p_\epsilon\}$ of multipliers of the perturbed Lagrangian were also bounded in Q', we could be assured that there might be some hope of extracting a solution (u, p) of problem (3.2.9) as the limit of solutions (u_ϵ, p_ϵ) of (3.2.34) obtained as $\epsilon \to 0$. In particular, if a constant C_1 independent of ϵ exists such that

$$\|p_\epsilon\|_{Q'} \leq C_1 \qquad (3.2.36)$$

then we are, in fact, guaranteed the existence of an element $p \in Q'$ and a subsequence $\{p_{\epsilon'}\}$ such that $p_{\epsilon'}$ converges weakly to p in Q' as ϵ' tends to zero. Moreover, *if conditions (a) through (c) of (3.2.32) and both (3.2.35) and (3.2.36) hold, one can show that a sequence $\{(u_\epsilon, p_\epsilon)\}$ of solutions to (3.2.34) exists which converges to a point (u, p) where u is a solution of the original boundary-value problem,*

$$\begin{aligned}
\langle \delta J(u), v \rangle + [p, Bv] = 0 \qquad \forall v \in H \\
[q, Bu] = 0 \qquad \forall q \in Q'
\end{aligned} \qquad (3.2.37)$$

We leave the proof of this result as an exercise (see Exercise 3.2.3).

Unfortunately, the key condition (3.2.36) cannot be shown to hold under the assumptions laid down thus far; an additional and fundamental condition is needed to complete our theory.

The Babuška–Brezzi Condition: Recall assumption (c) of (3.2.32). Clearly, if $\|u_\epsilon\|_H \leq C_1$, we must have $\langle \delta J(u_\epsilon), v \rangle \leq \|\delta J(u_\epsilon)\|_{H'} \|v\|_H \leq C_2 \|v\|_H$ for all $v \in H$. Thus, the first equation in (3.2.34) implies

$$|[p_\epsilon, Bv]| = |\langle \delta J(u_\epsilon), v \rangle| \leq C_2 \|v\|_H$$

for arbitrary test functions $v \in H$. Thus, if a constant $\beta > 0$ exists such that

$$\beta \|p_\epsilon\|_{Q'} \|v\|_H \leq |[p_\epsilon, Bv]| \leq C_2 \|v\|_H \qquad (3.2.38)$$

for arbitrary v in H, we would have

$$\|p_\epsilon\|_{Q'} \leq \frac{C_2}{\beta}$$

That is, (3.2.36) would then hold and the existence of a solution to our saddle-point problem (3.2.9) [or (3.2.34)] would be guaranteed.

In view of (3.2.38), we introduce the following additional condition into our theory:*

There exists a constant $\beta > 0$ such that

$$\beta \, \| q \|_{Q'} \leq \sup_{\substack{v \in H \\ (v \neq 0)}} \frac{|[q, Bv]|}{\| v \|_H} \qquad \forall q \in Q'. \qquad (3.2.39)$$

Clearly, whenever (3.2.39) holds, we are guaranteed the existence of a $p \in Q'$ to which a sequence of multipliers p_ϵ, which are solutions of the perturbed problem (3.2.34), converges weakly in Q'.

Conditions such as (3.2.39) play a key role in the study of elliptic boundary-value problems with constraints. Such conditions were first proposed and studied for linear elliptic problems and their approximation by Babuška (e.g., Babuška and Aziz [1972], Babuška [1973a,b], 1977]) and by Brezzi [1972] in his important work on saddle point problems. We shall refer to such conditions as *Babuška–Brezzi conditions*. We summarize the main results in the following theorem.

Theorem 3.2.1.   Let conditions (a) through (c) in (3.2.32) and condition (3.2.39) hold. Then there exists a unique saddle point $(u, p) \in H \times Q'$ of the Lagrangian L of (3.2.9) and this saddle point is also the solution of problem (3.2.37). $\square$

There are still some generalizations of (3.2.39) that are necessary for more general linear elliptic problems with constraints. For example, the fact that the solution (u, p) of (3.2.37) is unique under the conditions of Theorem 3.2.1 suggests that in many cases (3.2.39) may be too strong. For instance, if there exist nonzero functions $q_0 \in Q'$ such that

$$B^* q_0 = 0 \qquad (3.2.40)$$

then

$$[q_0, Bv] = \langle B^* q_0, v \rangle = 0 \qquad \forall v \in H$$

and we would necessarily have $\beta = 0$ in (3.2.39). We overcome this difficulty by simply eliminating the elements in Q' which satisfy (3.2.40). To accomplish this, we first denote

$$\ker B^* = \{ q \in Q' \,|\, \langle B^* q, v \rangle = 0 \qquad \forall v \in H \} \qquad (3.2.41)$$

---

* In (3.2.39), "sup" denotes the supremum or least upper bound. Condition (3.2.39) is then a special case of the "inf-sup" condition in the existence theorems for linear elliptic problems discussed in Volume IV, a condition that plays a fundamental role in the analysis of convergence and stability of finite element methods. A detailed discussion of this and similar conditions is given in Chapters 4 and 5 of Volume IV.

We can then proceed to reformulate the saddle point problem so that the multipliers are defined in the space Q' *modulo elements in* ker B^* (written Q'/ker B^*). The elements in the space Q'/ker B^* are then the "cosets"

$$[q] = \{\bar{q} \in Q' \,|\, q - \bar{q} \in \text{ker } B^*\}$$

and the norm on Q'/ker B^* is defined by

$$\| [q] \|_{Q'/\text{ker } B^*} = \min_{\bar{q} \in \text{ker} B^*} \| q + \bar{q} \|_{Q'} \qquad (3.2.42)$$

In other words, we treat as unknowns in the problem the minimizer $u \in H$ of J and the coset $[q] \in Q'$/ker B^*, so that the saddle point (u, p) is determined only to within an arbitrary element in ker B^*. In this case, we impose, instead of (3.2.39), the Babuška–Brezzi condition: There exists a constant $\beta > 0$ such that

$$\beta \min_{\bar{q} \in \text{ker} B^*} \| q + \bar{q} \|_{Q'} \; (\equiv \| [q] \|_{Q'/\text{ker} B^*}) \leq \sup_{\substack{v \in H \\ (v \neq 0)}} \frac{|[q, Bv]|}{\| v \|_H} \qquad \forall q \in Q' \quad (3.2.43)$$

Then, as a corollary to Theorem 3.2.1, we may establish that when conditions (a) through (c) of (3.2.32) and (3.2.43) hold, a solution (u, p) to (3.2.37) exists *with p determined up to an arbitrary element in* ker B^*. Still other alternative forms of the Babuška–Brezzi condition may be imposed, depending on the smoothness of the functions v and q.

Strong Convergence: We now show that under mild additional conditions, the perturbed Lagrangian solutions $\{(u_\epsilon, p_\epsilon)\}$, converge *strongly* to a solution (u, p) of (3.2.9) as ϵ tends to zero. Let us set the stage for such a demonstration by assuming that the functional J is of the form

$$J(v) = \tfrac{1}{2} B(v, v) - \langle f, v \rangle$$

where B is a symmetric bilinear form on $H \times H$ of the type discussed earlier which is continuous and H-elliptic, that is, constants M and α exist such that

$$B(u, v) \leq M \| u \|_H \| v \|_H; \qquad B(v, v) \geq \alpha \| v \|_H^2 \qquad (3.2.44)$$

for all $u, v \in H$. We shall also assume for simplicity that $Q = Q'$, ker $B^* = \{0\}$, and that the relations

$$[p, Bv] = (p, Bv); \qquad (p, p) = \| p \|_Q^2$$
$$(p, q) \leq \| p \|_{Q'} \| q \|_{Q'} \qquad (3.2.45)$$

hold for any $p, q \in Q'$, where $(\cdot, \cdot) = (\cdot, \cdot)_{Q'}$.

In this case, the saddle points (u_ϵ, p_ϵ) of the perturbed Lagrangian are characterized by

$$B(u_\epsilon, v) + (p_\epsilon, Bv) = (f, v) \qquad \forall v \in H$$
$$p_\epsilon = +\epsilon^{-1} Bu_\epsilon \quad \text{in} \quad Q' \tag{3.2.46}$$

whereas the saddle point (u, p) of L satisfies

$$B(u, v) + (p, Bv) = (f, v) \qquad \forall v \in H$$
$$Bu = 0 \quad \text{in} \quad Q' \tag{3.2.47}$$

We also must have the Babuška–Brezzi condition,

$$\beta \| q \|_{Q'} \le \sup_{\substack{v \in H \\ (v \ne 0)}} \frac{|(q, Bv)|}{\| v \|_H} \qquad \text{for all } q \in Q' \tag{3.2.48}$$

With these conventions now in force, we begin the analysis by observing that when the first expression in (3.2.46) is subtracted from the first expression in (3.2.47), we obtain the orthogonality condition,

$$B(u - u_\epsilon, v) + (p - p_\epsilon, Bv) = 0 \qquad \forall v \in H \tag{3.2.49}$$

Thus, according to the second expression in (3.2.44),

$$\alpha \| u - u_\epsilon \|_H^2 \le B(u - u_\epsilon, u - u_\epsilon) = (p_\epsilon - p, Bu - Bu_\epsilon) \tag{3.2.50}$$

In addition, the Babuška–Brezzi condition (3.2.48), the orthogonality condition (3.2.49), and the first expression in condition (3.2.44) yield

$$\beta \| p - p_\epsilon \|_{Q'} \le \sup_{\substack{v \in H \\ v \ne 0}} \frac{|(p - p_\epsilon, Bv)|}{\| v \|_H}$$

$$= \sup_{\substack{v \in H \\ v \ne 0}} \frac{|B(u - u_\epsilon, v)|}{\| v \|_H}$$

$$\le M \| u - u_\epsilon \|_H$$

Thus,

$$\| p - p_\epsilon \|_{Q'} \le \frac{M}{\beta} \| u - u_\epsilon \|_H \tag{3.2.51}$$

Moreover,

$$[p_\epsilon - p, B(u - u_\epsilon)] = (p - p_\epsilon, Bu_\epsilon)$$

$$= \epsilon(p - p_\epsilon, p_\epsilon)$$

$$= \epsilon(p - p_\epsilon, p_\epsilon - p) + \epsilon(p - p_\epsilon, p)$$

$$= -\epsilon \| p - p_\epsilon \|_{Q'}^2 + \epsilon(p - p_\epsilon, p)$$

$$\le \epsilon \| p - p_\epsilon \|_{Q'} \| p \|_{Q'}$$

Therefore, combining this last result with (3.2.50) and (3.2.51), we have

$$\| u - u_\epsilon \|_H \leq \left(\frac{M}{\alpha\beta} \| p \|_{Q'} \right) \epsilon \tag{3.2.52}$$

and, according to (3.2.51),

$$\| p - p_\epsilon \|_{Q'} \leq \left(\frac{M^2}{\alpha\beta^2} \| p \|_{Q'} \right) \epsilon \tag{3.2.53}$$

Thus, *under conditions* (3.2.44), (3.2.45), u_ϵ *converges strongly to u in H and* p_ϵ *converges strongly to p in Q′ as ϵ approaches zero.* Note that, once again, the Babuška–Brezzi condition plays a fundamental role in establishing this convergence property.

Example 3.2.3   Pressure Conditions in Stokes' Problems

In the case of Stokes' problem (3.2.27) or (3.2.29), we have

$$\ker B^* = \ker (\text{grad})$$
$$= \{\text{space of constant functions defined on } \Omega\} \tag{3.2.54}$$

Then (3.2.43) becomes*

$$\beta \min_{c \in R} \left(\int_\Omega | q + c |^2 \, dx \right)^{1/2} \leq \sup_{\substack{v \in H \\ v \neq 0}} \frac{\left| \int_\Omega q \, \text{div} \, v \, dx \right|}{\| v \|_H} \tag{3.2.55}$$

for all $q \in L^2(\Omega)$ ($v \neq 0$) where

$$\| v \|_H^2 = \int_\Omega \sum_{i,j=1}^2 \frac{\partial v_i}{\partial x_j} \frac{\partial v_i}{\partial x_j} \, dx$$

Note that if **f** and $\partial\Omega$ are smooth, the variational form (3.2.27) of Stokes' problem can be replaced by

$$\int_\Omega \nu \, \mathbf{V}\mathbf{u} : \mathbf{V}\mathbf{v} \, dx + \int_\Omega \mathbf{V}p \cdot \mathbf{v} \, dx = \int_\Omega \mathbf{f} \cdot \mathbf{v} \, dx \qquad \forall \mathbf{v} \in H$$
$$\int_\Omega \mathbf{V}q \cdot \mathbf{u} \, dx = 0 \qquad\qquad \forall q \in Q' \tag{3.2.56}$$

where H is given by (3.2.17), and we can now take

$$Q' = H^1(\Omega), \qquad \| q \|_{Q'/R} = \| \mathbf{V}q \|_0 \tag{3.2.57}$$

---

* This condition was apparently first derived in an alternative but equivalent form by Ladyzhenskaya [1969].

In this case, we might expect a stronger condition to hold, such as

$$\beta \, \| \nabla q \|_0 \leq \sup_{\mathbf{v} \in H} \frac{\left| \int_\Omega \nabla q \cdot \mathbf{v} \, dx \right|}{\| \mathbf{v} \|_0} \tag{3.2.58}$$

for all $q \in Q'$ (see Chapter 4, Vol. IV).

3.2.3 Mixed Finite Element Approximations

Finite element methods based on Lagrange multiplier formulations of constrained problems are called *mixed* finite element methods. To develop such methods, we follow the usual procedure and construct finite-dimensional subspaces H^h of H and Q^h of Q' spanned by polynomial basis functions $\{\phi_i(\mathbf{x})\}_{i=1}^N$ and $\{\chi_j(\mathbf{x})\}_{j=1}^M$, respectively. The domain Ω is, as usual, partitioned into a collection of finite elements and the global basis functions ϕ_i and χ_j are constructed by patching together local shape functions defined over each finite element. The only peculiarity that we face in the present formulation is that now we must approximate two dependent variables over each element: the solution u and the Lagrange multiplier p.

Algebraically, the procedure is straightforward. To provide some detail, let us consider the case in which

$$L(v, q) = J(v) + \int_\Omega q(Bv - g) \, dx$$
$$= \tfrac{1}{2} B(v, v) - \int_\Omega fv \, dx + \int_\Omega q(Bv - g) \, dx \tag{3.2.59}$$

where $B(\cdot, \cdot)$ is a bilinear form of the type described in Chapter 1 [recall (1.3.2)] and $Bv = g$ represents the constraint. [Here $Q = Q' = L^2(\Omega)$.]

A conforming finite element approximation of (3.2.59) (for the case $\Omega_h = \Omega$) consists of seeking $u_h \in H^h$ and $p_h \in Q^h$ such that

$$B(u_h, v_h) + \int_\Omega p_h Bv_h \, dx = \int_\Omega fv_h \, dx \qquad \forall v_h \in H^h$$
$$\int_\Omega q_h Bu_h \, dx = \int_\Omega gq_h \, dx \qquad \forall q_h \in Q^h \tag{3.2.60}$$

For a typical finite element Ω_e in the mesh, the weighted residual statement $\langle \delta L(u, p), (v, q) \rangle_{H \times Q'} = 0$ leads to the pair of variational equations

$$B_e(u, v) + \int_{\Omega_e} pBv \, dx = \int_{\Omega_e} fv \, dx - \sigma_e \qquad \forall v \in H$$
$$\int_{\Omega_e} q(Bu - g) \, dx = 0 \qquad \forall q \in Q' \tag{3.2.61}$$

where σ_e is a boundary flux term arising from "integration by parts" and containing contour integrals which sum (in the absence of line or point sources, etc.) to zero on element interfaces upon assembling the elements. To obtain a local finite element approximation of (3.2.61), we introduce the expansions

$$u_h^e(\mathbf{x}) = \sum_{m=1}^{M_e} u_m^e \psi_m^e(\mathbf{x}), \qquad p_h^e(\mathbf{x}) = \sum_{i=1}^{N_e} p_i \omega_i^e(\mathbf{x}) \qquad (3.2.62)$$

where shape functions ψ_m^e contain polynomials of degree $k \geq 1$, the shape functions ω_i^e contain polynomials of degree $r \geq 0$, and

$$\psi_m^e(\mathbf{x}_n^e) = \delta_{mn}, \qquad 1 \leq m, n \leq M_e$$
$$\omega_i^e(\mathbf{y}_j^e) = \delta_{ij}, \qquad 1 \leq i, j \leq N_e$$

wherein $\{\mathbf{x}_n^e\}_{1 \leq n \leq M}$ and $\{\mathbf{y}_j^e\}_{1 \leq j \leq N}$ are collections of nodal points corresponding to the degrees of freedom assigned to u_h^e and p_h^e, respectively. We emphasize that the choice of polynomials and the corresponding location of nodal points for representing u_h^e and p_h^e are, at this stage, independent. Thus, we might choose quadratic elements for u_h^e and linear or constant elements for p_h^e. We will provide some criteria for these choices later.

The use of (3.2.62) in (3.2.61) yields the following element stiffness relations:

$$\mathbf{k}^e \mathbf{u}^e + \mathbf{c}^e \mathbf{p}^e = \mathbf{f}^e - \boldsymbol{\sigma}^e$$
$$\mathbf{c}^{eT} \mathbf{u}^e = \mathbf{g}^e \qquad (3.2.63)$$

where $\boldsymbol{\sigma}^e$ is the vector of element boundary fluxes and

$$\left.\begin{aligned}
\mathbf{k}^e = [k_{mn}^e]; \qquad & k_{mn}^e = B_e(\psi_m^e, \psi_n^e), \quad 1 \leq m, n \leq M_e \\
\mathbf{c}^e = [c_{ni}^e]; \qquad & c_{ni}^e = \int_{\Omega_e} B \psi_n^e \omega_i^e \, dx, \quad 1 \leq n \leq M_e, \quad 1 \leq i \leq N_e \\
\mathbf{f}^e = \{f_n^e\}; \qquad & f_n^e = \int_{\Omega_e} f \psi_n^e \, dx, \quad 1 \leq n \leq M_e \\
\mathbf{g}^e = \{g_i^e\}; \qquad & g_i^e = \int_{\Omega_e} g \omega_i^e \, dx, \quad 1 \leq i \leq N_e \\
\mathbf{u}^e = \{u_1, u_2, \ldots, u_{M_e}\}^T; \qquad & \mathbf{p}^e = \{p_1, p_2, \ldots, p_{N_e}\}^T
\end{aligned}\right\} \qquad (3.2.64)$$

If conforming finite element approximations are used, it is not, in general, possible to eliminate the vector $\mathbf{p}^e$ of Lagrange multipliers from (3.2.63) at the element level. Thus, upon assembling the elements, we obtain a global system of equations of the form

$$\begin{bmatrix} \mathbf{K} & \mathbf{C} \\ \mathbf{C}^T & \mathbf{O} \end{bmatrix} \begin{bmatrix} \mathbf{U} \\ \mathbf{P} \end{bmatrix} = \begin{bmatrix} \mathbf{F} \\ \mathbf{G} \end{bmatrix} \qquad (3.2.65)$$

where $\mathbf{K}$ is the global stiffness matrix, $\mathbf{C}$ is the constraint matrix, $\mathbf{U}$ and $\mathbf{P}$ are the global vectors of nodal values of u_h and p_h, and so on. The system (3.2.65) is symmetric, and upon applying boundary conditions, invertible, so this system can be solved for the vector $\{\mathbf{U}, \mathbf{P}\}$. We discuss methods for solving such linear systems of equations in Volume III.

3.2.4 Selection of Shape Functions

There are some pitfalls in the procedure just described, and care must be taken in selecting the local shape function ψ_m^e and ω_i^e if a stable and convergent method is to be obtained. To illustrate some of the considerations that must be taken into account, let us examine finite element approximations of the Stokes problem (3.2.56). Since we expect to use *continuous* pressure approximations p_h, this particular formulation seems to be a natural one.

Our finite element approximation of (3.2.56) consists of seeking $\mathbf{u}_h \in H^h \subset H_0^1(\Omega) \times H_0^1(\Omega)$ and $p_h \in Q^h \subset H^1(\Omega)$ (with $\Omega = \Omega_h$) such that

$$\int_\Omega \nu \nabla \mathbf{u}_h : \nabla \mathbf{v}_h \, dx + \int_\Omega \nabla p_h \cdot \mathbf{v}_h \, dx = \int_\Omega \mathbf{f} \cdot \mathbf{v}_h \, dx \qquad \forall \, \mathbf{v}_h \in H^h$$

$$\int_\Omega \nabla q_h \cdot \mathbf{u}_h \, dx = 0 \qquad \forall \, q_h \in Q^h$$

(3.2.66)

The first and most important question that must be asked is: *Are our choices of elements for constructing H^h and Q^h such that the discrete problem (3.2.66) is well posed and well behaved as $h \rightarrow 0$?* To answer this question, we turn to the general list of conditions in Theorem 3.2.1 for the existence of solutions to such problems and make two key observations: (1) all the conditions in (3.2.31) [which were shown to hold for the actual "continuous" problem (3.2.56)] are preserved in passing to the discrete problem except condition (3.2.39); (2) for the discrete problem, we must impose, instead of (3.2.39), *a discrete Babuška–Brezzi condition* [analogous to (3.2.39) or, if appropriate, (3.2.43)]: There exists a constant $\beta_h > 0$ such that

$$\beta_h \| q_h \|_{Q'} \leq \sup_{\substack{\mathbf{v}_h \in H^h \\ (\mathbf{v}_h \neq 0)}} \frac{\int_\Omega q_h \, \mathrm{div} \, \mathbf{v}_h \, dx}{\| \mathbf{v}_h \|_H} \qquad \text{for all } q_h \in Q^h \qquad (3.2.67)$$

Thus, this discrete Babuška–Brezzi condition emerges as a critical requirement in the success (specifically, the *stability*) of finite element methods such as (3.2.66).

We remark that a discrete analogue of condition (3.2.58) can be shown to hold under certain special conditions:

$$\beta_h \| \nabla q_h \|_0 \leq \sup_{\substack{v_h \in H^h \\ v_h \neq 0}} \frac{\left| \int_\Omega \nabla q_h \cdot v_h \, dx \right|}{\| v_h \|_0} \qquad \forall \, q_h \in Q^h \qquad (3.2.68)$$

Here

$$\| v_h \|_0 = \left(\int_\Omega v_h \cdot v_h \, dx \right)^{1/2}; \qquad v_h \in H^h$$

If a β_h satisfying (3.2.67) exists at all, then, from what has been established thus far, we know that we can at least solve the discrete problem (3.2.66). However, this is not enough! If β_h depends on the mesh parameter h, it could happen that $\beta_h \to 0$ as $h \to 0$ (or, practically speaking, β_h becomes very small for fine meshes). This is, in general, unacceptable. In the very best circumstances, a β_h dependent on h results in a low (suboptimal) rate of convergence of the method; in the worst circumstances, it characterizes a scheme that is *numerically unstable* since unbounded sequences of pressure approximations p_h can result.

REMARK: We noted earlier that the pressures are determined within an arbitrary constant since ker $B^* = \mathbb{R}$, the constant functions. A similar observation applies to the discrete gradient operator B_h^* defined by the approximate formulation. However, for some elements, the approximation may permit additional functions (called "modes" because of the connection with the discrete eigenvalue problem) in ker B_h^*. For example, an alternating piecewise-constant "checkerboard" mode has been observed in computations with C^0 bilinear velocity and discontinuous piecewise-constant pressure approximations. Similar difficulties arise in connection with penalty methods (see Section 3.5).

In view of these difficulties, one must ask which finite element methods work for these problems. As examples, we note that Bercovier and Pironneau [1977] have shown* that condition (3.2.68) is satisfied for β_h independent of h whenever: (1) $\Omega = \Omega_h$ is a polygon in $\mathbb{R}^2$; (2) each finite element in the mesh has no more than one node on the boundary $\partial\Omega$; and (3) the approximations of velocities v_h and hydrostatic pressures p_h are constructed according to the following rules:

1. Triangular or rectangular elements $\Omega_e, e = 1, 2, \ldots, E$, are used in the construction of the mesh.

2. Conforming (C^0) approximations are used for both the velocities and the pressures.

3. (a) $(v_h^e)_i \, (i = 1, 2; e = 1, 2, \ldots, E)$ is a complete quadratic in x_1 or x_2 if Ω_e is a triangle, and p_h^e is then linear; or $(v_h^e)_i$ is a tensor product of quadratics if Ω_e is rectangular and p_h^e is then bilinear.

---

* We outline portions of their proofs of these results in Volume IV.

(b) $(v_h^e)_i$ is piecewise linear over each of four subtriangles if Ω_e is
      a triangle, and p_h^e is then linear over the composite triangle;
      or $(v_h^e)_i$ is piecewise bilinear over each of four subrectangles if
      Ω_e is rectangular and p_h^e is then bilinear over the composite
      rectangle.

Illustrations of these cases are given in Fig. 3.1. In these examples, one
can also show that rates of convergence are of optimal order. Indeed, if the
interpolation properties (2.2.1) hold for H^h and Q^h, then constants C_1 and C_2
independent of h exist such that

$$\| \nabla(\mathbf{u} - \mathbf{u}_h) \|_0 \leq C_1 h^k (\| \mathbf{u} \|_{k+1} + \| p \|_{k/\mathbb{R}})$$
$$\| \nabla(p - p_h) \|_0 \leq C_2 h^{k-1} (\| \mathbf{u} \|_{k+1} + \| p \|_{k/\mathbb{R}}) \tag{3.2.69}$$

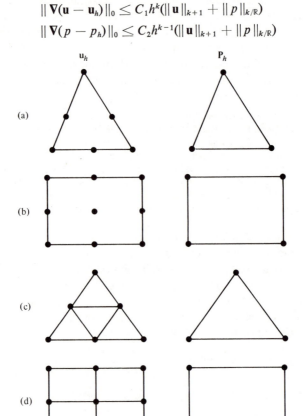

FIGURE 3.1 *Examples of finite element approximations of
the velocity* **u** *and the pressure* p *which satisfy the
Babuška–Brezzi stability condition* (3.2.68) *with* β_h
*independent of h, provided that no more than one side per
triangle or two per rectangle lie on the boundary* $\partial\Omega$.

where $k = 2$ in case 3(a) (quadratic velocities) and $k = 1$ in case 3(b). Here $\| \nabla u \|_0 = \{\int_\Omega \sum_{i,j=1}^2 (\partial u_i / \partial x_j \, \partial u_i / \partial x_j) \, dx\}^{1/2}$. The quotient norms $\| \cdot \|_{k/R}$ are defined as in (3.2.42). If $k = 1$ [case 3(b)], $\| \nabla (p - p_h) \|_0 = O(1)$ and the pressures may not converge in the norm $\| \nabla q \|_0$. However, in this case, if Ω and the data are sufficiently smooth, one can expect that $\| p - p_h \|_0 = O(h)$.

EXERCISES

3.2.1   Show that problem (3.2.27) characterizes a saddle point $(\mathbf{u}, p)$ of the Lagrangian L defined in (3.2.25).

3.2.2   (a)   Using the Green–Gauss divergence theorem, derive the Green's formula (3.2.28).

(b)   Show that any sufficiently smooth solution of (3.2.27) is also a solution of the classical problem (3.2.29).

3.2.3   Let $\{u_\epsilon\}$ and $\{p_\epsilon\}$ be sequences of solutions of problem (3.2.34), and suppose that $\| u - u_\epsilon \|_H \longrightarrow 0$ as $\epsilon \longrightarrow 0$ and $\| p_\epsilon \|_{Q'} < C = \text{constant}$. Under these conditions, show that the limit u of the sequence $\{u_\epsilon\}$ is, in fact, a minimizer of J in the constraint set K.

[*HINT:*   Use the fact that (u_ϵ, p_ϵ) is a saddle point of L_ϵ and that $Bv - g = 0$ if $v \in K$.]

3.2.4   Let J_1, J_2, J_3, and J_4 be the functionals described in Exercise 3.1.2. Derive Lagrangians for the following constraint conditions:

(a) J_1: $v(\tfrac{1}{2}) = 5$

(b) J_2: $H^1(0, 1) \longrightarrow \mathbb{R}$; $v(0) = v(1) = 3$

(c) J_3: $v = xy$

(d) J_4: $H^2(\Omega) \times H_0^1(\Omega) \longrightarrow \mathbb{R}$
$$\frac{\partial v}{\partial n} = 3 \quad \text{on} \quad \partial \Omega$$

3.2.5   Develop variational boundary-value problems characterizing saddle points of the Lagrangians constructed in Exercise 3.2.4.

3.2.6   Prove that conditions (a) and (b) of (3.2.32) imply that conditions (a) and (b) of (3.2.31) are satisfied. Then use these results and Theorem 3.1.1 to establish that problem (3.2.34) always has a solution (u_ϵ, p_ϵ) for each $\epsilon > 0$.

3.2.7   Consider the classical Fourier series representation of a function $f \in L^2(0, 1)$:

$$f(x) = \sum_{n=1}^\infty f_n \phi_n(x), \qquad \phi_n(x) = \sqrt{2} \, \sin n\pi x$$

$$f_n = (f, \phi_n)_{L^2(0, 1)} = \int_0^1 f \phi_n \, dx$$

(a)   Discuss what is meant by the symbolism "$f = \sum_{n=1}^{\infty} f_n \phi_n$ in $L^2(0, 1)$" in terms of strong convergence in $L^2(0, 1)$.

[**HINT:** Construct the Nth partial sum, $f^N = \sum_{n=1}^{N} f_n \phi_n$ and observe that the Fourier series represents f if this sequence of partial sums converges strongly to f in $L^2(0, 1)$ as $N \longrightarrow \infty$.]

(b)   Since f_n converges to zero as n tends to infinity, establish that the sequence of functions $\{\phi_n\}$ converges weakly to zero but *not* strongly to zero in $L^2(0, 1)$.

3.2.8   The purpose of this exercise is to establish that if $\{(u_\epsilon, p_\epsilon)\}$ is a sequence of solutions of the perturbed saddle point problem (3.2.34) such that $\{u_\epsilon\}$ converges weakly to u in H and p_ϵ converges weakly to p in Q', then (u, p) is a saddle point of the functional L in (3.2.4) and, therefore, a solution of problem (3.2.37). First note that since (u_ϵ, p_ϵ) is a saddle point of L_ϵ,

$$J(u_\epsilon) + [q, Bu_\epsilon] - \frac{\epsilon}{2}\|q\|_{Q'}^2 \leq J(u_\epsilon) + [p_\epsilon, Bu_\epsilon] - \frac{\epsilon}{2}\|p_\epsilon\|_{Q'}^2, \qquad (1)$$

for all $q \in Q'$ and

$$J(u_\epsilon) + [p_\epsilon, Bu_\epsilon] - \frac{\epsilon}{2}\|p_\epsilon\|_{Q'}^2 \leq J(v) + [p_\epsilon, Bv] - \frac{\epsilon}{2}\|p_\epsilon\|_{Q'}^2, \qquad (2)$$

for all $v \in H$. Also, use the fact that under the assumptions given in the text,

$$\underline{\lim_{\epsilon \to 0}}\, L_\epsilon(u_\epsilon, q) \geq L(u, q) \quad \forall\, q \in Q' \qquad (3)$$

$$\overline{\lim_{\epsilon \to 0}}\, L_\epsilon(v, p_\epsilon) \leq L(v, p) \quad \forall\, v \in H \qquad (4)$$

(a)   Use conditions (1)–(4) to prove that the weak limit (u, p) of the perturbed solutions (u_ϵ, p_ϵ) is a solution of (3.2.37).

(b)   Prove that if $\|p_\epsilon\|_{Q'} \leq C =$ constant, the weak limit u of $\{u_\epsilon\}$ satisfies the constraint $Bu = 0$.

[**HINT:** Let $\bar{q}$ be an arbitrary element of Q' and set $q = \bar{q} + p_\epsilon$ in (1) to obtain $\lim_{\epsilon \to 0} [\bar{q}, Bu_\epsilon] \leq 0$, $\forall\, \bar{q} \in Q'$; then deduce the assertion.]

3.2.9   In each of the examples of Exercises 3.2.4 and 3.2.5, write down the finite element approximation to the variational problem. Introduce global basis functions as in the text examples to determine the finite element system in integral form.

3.2.10   In each of the cases in Exercise 3.2.9, derive representative element matrix and vector contributions. Leave your answers in integral form.

3.2.11   Select any example from those of Exercise 3.2.9 and indicate how the element matrices are to be computed in practice and how boundary conditions are applied. Discuss the properties of the final system and compare them with the systems arising for problems without constraints.

3.3 OTHER MIXED METHODS

Up to this point, we have described mixed finite element methods for elliptic problems with constraints in which the constraint entered naturally into the formulation of the problem (e.g., incompressibility in the Stokes problem), and in which the Lagrange multiplier was an intrinsic and physically important variable of the problem (e.g., the hydrostatic pressure). There are, however, popular mixed methods in which a constraint is introduced somewhat artificially for the purpose of producing a formulation which leads to computational methods which may have advantages over conventional methods. Most frequently, these constraints manifest themselves in the decomposition of a given operator into products of lower-order operators so that independent approximations of the solution and its various derivatives can be made.

For example, the second-order mixed boundary-value problem

$$\left.\begin{aligned} -\nabla \cdot (a\, \nabla u) = f \quad &\text{in} \quad \Omega \\ u = 0 \quad &\text{on} \quad \partial\Omega_1 \\ a(\mathbf{x})\, \nabla u \cdot \mathbf{n} = g \quad &\text{on} \quad \partial\Omega_2 \end{aligned}\right\} \qquad (3.3.1a)$$

can be decomposed into the equivalent system

$$\left.\begin{aligned} \left.\begin{aligned} a\, \nabla u = \boldsymbol{\sigma} \\ -\nabla \cdot \boldsymbol{\sigma} = f \end{aligned}\right\} \quad &\text{in} \quad \Omega \\ u = 0 \quad &\text{on} \quad \partial\Omega_1 \\ \boldsymbol{\sigma} \cdot \mathbf{n} = g \quad &\text{on} \quad \partial\Omega_2 \end{aligned}\right\} \qquad (3.3.1b)$$

Clearly, problems (3.3.1a) and (3.3.1b) are equivalent, but in the formulation (3.3.1b), we seek a pair of functions $(u, \boldsymbol{\sigma})$, whereas in (3.3.1a) a single dependent variable u is sought. Thus, in developing finite element approximations of (3.3.1b), we can employ independent approximations of u and $\boldsymbol{\sigma} = a\, \nabla u$. Such "mixed formulations" are still basically Lagrange multiplier methods since (3.3.1b) can be regarded as the problem of solving $-\nabla \cdot \boldsymbol{\sigma} = f$ subject to the constraints $\boldsymbol{\sigma} - a\, \nabla u = 0$ in Ω and $u = 0$ on $\partial\Omega_1$.

Similarly, for higher-order problems such as

$$\frac{d^m u}{dx^m} = f \qquad (3.3.2a)$$

a decomposition such as

$$\frac{du}{dx} = v_1, \quad \frac{dv_1}{dx} = v_2, \quad \frac{dv_2}{dx} = v_3, \ldots, \quad \frac{dv_{m-1}}{dx} = f \qquad (3.3.2b)$$

127

makes it possible to use independent polynomial approximations of each component of the vector $(u, v_1, v_2, \ldots, v_m)$. Similarly, in the case of the biharmonic equation

$$\Delta^2 u \equiv \frac{\partial^4 u}{\partial x^4} + 2\frac{\partial^4 u}{\partial x^2 \partial y^2} + \frac{\partial^4 u}{\partial y^4} = f \qquad (3.3.3a)$$

we can consider the equivalent pair of second-order equations

$$\begin{aligned} -\Delta u &= w \\ -\Delta w &= f \end{aligned} \qquad (3.3.3b)$$

Clearly, mixed methods based on decompositions such as these may have some significant computational advantages over standard finite element methods since the usual variational formulation of, say, (3.3.3a), suggests that we employ C^1-finite element approximations of u, whereas a pair of C^0-finite elements can be employed to approximate (3.3.3b).

The idea of using mixed finite elements of this type was first proposed by Hermann [1966], and extensive applications of mixed methods have been reported in the engineering literature. For additional references, see the bibliography of Norrie and de Vries [1976]. Mathematical properties of mixed methods, including convergence criteria and error estimates, were studied by Oden [1973, 1974], Johnson [1973], Oden and Reddy [1976a,b], Sheu [1978], Babuška et al. [1977, 1978], Mansfield [1976], Falk and Osborn [1979], Girault and Raviart [1979], and many others. Our objective in this section is to outline some typical mixed methods and to list some of their properties.

3.3.1 A Mixed Method for a Fourth-Order Problem

As an example of a popular application of mixed methods, consider the fourth-order boundary-value problem

$$\Delta^2 u = f \quad \text{in} \quad \Omega$$

$$u = \frac{\partial u}{\partial n} = 0 \quad \text{on} \quad \partial\Omega \qquad (3.3.4)$$

where Δ^2 is the biharmonic operator in (3.3.3a) and Ω is a convex polygonal domain in $\mathbb{R}^2$. As indicated in (3.3.3a), we can recast this problem as the system of second-order equations (3.3.3b) together with the boundary conditions indicated in (3.3.4). A variational statement of this system of reduced order is as follows: find a pair of functions $(u, w) \in H_0^1(\Omega) \times H^1(\Omega)$ such that

$$\int_\Omega wv \, dx - \int_\Omega \nabla u \cdot \nabla v \, dx = 0, \qquad \forall v \in H^1(\Omega)$$

$$\int_\Omega \nabla w \cdot \nabla q \, dx = \int_\Omega fq \, dx, \qquad \forall q \in H_0^1(\Omega)$$

(3.3.5)

Here we assume that f is square-integrable over Ω.

It is interesting to note that this system of equations describes a stationary value of the functional

$$L : H^1(\Omega) \times H_0^1(\Omega) \longrightarrow \mathbb{R}$$

$$L(v, z) = \int_\Omega \left(\tfrac{1}{2} v^2 - \nabla z \cdot \nabla v + zf \right) dx$$

(3.3.6)

The finite element approximation of this variational boundary-value problem is constructed in the usual way; however, since the mixed variational principle (3.3.5) involves only H^1-functions, there is now no longer the need to use C^1-elements. Indeed, we can choose standard C^0-elements of the types commonly employed in second-order problems.

Let us assume that

$$\left.\begin{aligned}
Q^h = \{v_h \in C^0(\Omega) \,|\, &v_h^e \text{ contains complete polynomials} \\
&\text{of degree} \leq k, \, k \geq 2; \, 1 \leq e \leq E\} \\
H^h = \{v_h \in Q^h \,|\, &v_h = 0 \quad \text{on} \quad \partial\Omega\}
\end{aligned}\right\}$$

(3.3.7)

Then a mixed approximation of (3.3.5) consists of seeking $(u_h, w_h) \in H^h \times Q^h$ such that

$$\left.\begin{aligned}
\int_\Omega w_h v_h \, dx - \int_\Omega \nabla u_h \cdot \nabla v_h \, dx = 0 \qquad &\forall v_h \in H^h \\
\int_\Omega \nabla w_h \cdot \nabla q_h \, dx = \int_\Omega fq_h \, dx \qquad &\forall q_h \in Q^h
\end{aligned}\right\}$$

(3.3.8)

The corresponding discrete system assumes the form

$$\left.\begin{aligned}
\mathbf{Kw} - \mathbf{Cu} = \mathbf{0} \\
-\mathbf{C}^T\mathbf{w} = \mathbf{f}
\end{aligned}\right\}$$

(3.3.9)

Here $\mathbf{w}$ and $\mathbf{u}$ are the vectors of nodal values of w_h and u_h, respectively. Notice that in this particular formulation, the matrix $\mathbf{K}$ is merely the Gram or mass matrix corresponding to the basis functions ϕ_i of H^h,

$$K_{ij} = \int_\Omega \phi_i \phi_j \, dx, \qquad i, j = 1, 2, \ldots, N$$

(3.3.10)

whereas the constraint matrix $\mathbf{C}$ is simply a matrix for standard conforming approximations of Laplace's equation,

$$C_{ij} = \int_{\Omega} \nabla \phi_i \cdot \nabla \phi_j \, dx, \qquad i, j = 1, 2, \ldots, N \qquad (3.3.11)$$

This remarkably simple formulation suggests that codes developed for the solution of Laplace's equation might be adapted to an iterative solution of the biharmonic problem with minor modifications. There are other advantages of such formulations. In Volume IV (Section 4.5) we prove that the scheme (3.3.8) is stable. Moreover, when the solution u of (3.3.4) is in $H^r(\Omega)$, $r \geq 3$, and when the usual interpolation properties of the spaces H^h and Q^h are in force as $h \to 0$, the errors in the finite element approximation (3.3.8) satisfy bounds of the type

$$\left.\begin{array}{c} \| w - w_h \|_s \leq C_s h^{\mu-s} \| u \|_\mu, \qquad s = 0, 1 \\[2mm] \| u - u_h \|_1 \leq C_2 h^{\mu-1} \| u \|_\mu \\[2mm] \mu = \min (r, k + 1) \end{array}\right\} \qquad (3.3.12)$$

where C_0, C_1, C_2 are constants independent of h. This means that if, for example, we use six-node quadratic triangles to approximate both u and w, and if u is smooth, then the error in $\| u - u_h \|_1 = O(h^2)$ *and* $\| w - w_h \|_1 = O(h^2)$. This is the best possible (optimal) rate of convergence for this method and these results show that such mixed methods produce approximations of the Laplacian $w = \Delta u$ of the solution to an equal order of accuracy as the approximation of u itself in $H^1(\Omega)$.

3.3.2 A Mixed Method for a Model Second-Order Problem

As a second example, consider the second-order problem (3.3.1) for the case in which Ω is an open bounded two-dimensional domain, $\partial\Omega = \overline{\partial\Omega_1} \cup \overline{\partial\Omega_2}, f$ and g are given data, and, as usual, $\mathbf{n}$ is a unit vector normal to $\partial\Omega$. To develop a finite element approximation of (3.3.1b), a variational principle of the *Hellinger–Reissner* type* is needed. In such principles, we consider a pair H, S of spaces of admissible functions and an energy functional L defined on $H \times S$. For example, suppose that the coefficient $a = a(x_1, x_2) \geq$

---

* Mixed variational principles of this type were first proposed by Hellinger [1914] and Reissner [1948, 1953] and are well known in the literature on solid mechanics. For an extensive list of such mixed variational principles for problems in solid and fluid mechanics, see Oden and Reddy [1976b].

$a_0 > 0$ for all points (x_1, x_2) in Ω and consider the functional

$$L(v, \tau) = \int_\Omega \left(\tau \cdot \nabla v - \frac{1}{2a} \tau \cdot \tau - fv \right) dx - \int_{\partial\Omega_2} gv \, ds \quad (3.3.13)$$

with

$$\left. \begin{array}{l} L : H \times S \longrightarrow \mathbb{R} \\ H = \{v \in H^1(\Omega) | v = 0 \quad \text{on} \quad \partial\Omega_1\} \\ S = \{\tau = (\tau_1, \tau_2) \in (H^0(\Omega))^2 \,|\, \nabla \cdot \tau \in H^0(\Omega)\} \end{array} \right\} \quad (3.3.14)$$

We easily verify that the functional L achieves a stationary value at a point $(u, \sigma) \in H \times S$ whenever

$$\int_\Omega \left[\left(\nabla u - \frac{1}{a} \sigma \right) \cdot \tau + \sigma \cdot \nabla v \right] dx = \int_\Omega fv \, dx + \int_{\partial\Omega_2} gv \, ds \quad (3.3.15)$$

$$\forall (v, \tau) \in H \times S$$

which, of course, is equivalent to the pair of variational equations

$$\left. \begin{array}{ll} \int_\Omega \sigma \cdot \nabla v \, dx = \int_\Omega fv \, dx + \int_{\partial\Omega_2} gv \, ds & \forall v \in H \\ \int_\Omega \left(\nabla u - \frac{1}{a} \sigma \right) \cdot \tau \, dx = 0 & \forall \tau \in S \end{array} \right\} \quad (3.3.16)$$

The finite element analysis of a mixed problem such as (3.3.16) is non-standard and deserves a more detailed examination. To construct finite element approximations of (3.3.16), we proceed as usual and partition Ω into a collection $\{\Omega_e\}_{1 \le e \le E}$ of finite elements and introduce over each element polynomial approximations u_h^e of u and σ_h^e of σ. For representing u_h^e, we use local shape functions ψ_m^e, $1 \le m \le M_e$, which contain complete polynomials of degree k, and for representing $\sigma_h^e = [(\sigma_h^e)^1, (\sigma_h^e)^2]$, we use local shape functions ω_n^e, $1 \le n \le N_e$, which contain complete polynomials of degree t. Thus, over a typical finite element Ω_e we have

$$\left. \begin{array}{ll} u_h^e(\mathbf{x}) = \sum_{m=1}^{M_e} u_m^e \psi_m^e(\mathbf{x}); & (\sigma_h^e(\mathbf{x}))^i = \sum_{n=1}^{N_e} \sigma_n^{ei} \omega_n^e(\mathbf{x}), \quad i = 1, 2 \\ & \sigma_h^e(\mathbf{x}) = ((\sigma_h^e)^1, (\sigma_h^e)^2) \end{array} \right\} \quad (3.3.17)$$

Upon assembling a collection of mixed finite elements of the type described above, we effectively patch together the interpolation functions $\{\psi_m^e(\mathbf{x})\}$ and $\{\omega_n^e(\mathbf{x})\}$ to produce collections of global basis functions $\{\phi_i(\mathbf{x})\}$ and $\{\chi_j(\mathbf{x})\}$ which provide bases for finite-dimensional spaces H^h and S^h, respectively. If the boundary $\partial\Omega$ is matched exactly by that of the finite element mesh and if $u_h = 0$ on $\partial\Omega_1$, it is possible to construct H^h and S^h so that they are linear subspaces of H and S, respectively.

Over each element Ω_e, the variational problem (3.3.16) assumes the form

$$
\left.
\begin{aligned}
&\int_{\Omega_e} (\nabla u_h^e - a^{-1}\sigma_h^e) \cdot \tau_h^e \, dx = 0 \\
&\int_{\Omega_e} \sigma_h^e \cdot \nabla v_h^e \, dx = \int_{\Omega_e} f v_h^e \, dx + \int_{\partial\Omega_e \cap \partial\Omega_2} g v_h^e \, ds \\
&\qquad\qquad + \int_{\partial\Omega_e - (\partial\Omega_2 \cap \partial\Omega_e)} \sigma \cdot n_e v_h^e \, ds
\end{aligned}
\right\}
\tag{3.3.18}
$$

for all admissible τ_h^e and v_h^e. Thus, we obtain the following local approximation by substituting (3.3.17) into (3.3.18):

$$
\left.
\begin{aligned}
\mathbf{Au} - \mathbf{Hs} &= \mathbf{0} \\
\mathbf{A}^T \mathbf{s} &= \mathbf{f} + \bar{\boldsymbol{\sigma}}
\end{aligned}
\right\}
\tag{3.3.19}
$$

Here we have, for the moment, eliminated the element label e for clarity, and have denoted

$$
\left.
\begin{aligned}
&\mathbf{u} = \{u_1^e, u_2^e, \ldots, u_{M_e}^e\}^T \\
&\mathbf{s} = \{\sigma_1^1, \sigma_2^1, \ldots, \sigma_{N_e}^1, \sigma_1^2, \sigma_2^2, \ldots, \sigma_{N_e}^2\}^T \\
&\mathbf{A} = [A_{nm}^e]; \qquad 1 \le n \le 2N_e, \quad 1 \le m \le M_e \\
&A_{nm}^e = \begin{cases} \displaystyle\int_{\Omega_e} \omega_n^e \frac{\partial \psi_m^e}{\partial x_1} \, dx; & 1 \le n \le N_e \quad 1 \le m \le M_e \\[2ex] \displaystyle\int_{\Omega_e} \omega_{n-N_e}^e \frac{\partial \psi_m^e}{\partial x_2} \, dx; & \begin{array}{l} N_e + 1 \le n \le 2N_e, \\ 1 \le m \le M_e \end{array} \end{cases} \\
&\mathbf{H} = [H_{ij}^e]; \qquad 1 \le i, j \le 2N_e \\
&H_{ij}^e = \begin{cases} \displaystyle\int_{\Omega_e} a^{-1}\omega_i^e \omega_j^e \, dx; & 1 \le i, j \le N_e \\[1.5ex] 0; & N_e + 1 \le i \le 2N_e, \quad 1 \le j \le N_e \\[1.5ex] 0; & 1 \le i \le N_e, \quad N_e + 1 \le j \le 2N_e \\[1.5ex] \displaystyle\int_{\Omega_e} a^{-1}\omega_{i-N_e}^e \omega_{j-N_e}^e \, dx; & N_e + 1 \le i, j \le 2N_e \end{cases} \\
&\mathbf{f} = \{f_1^e, f_2^e, \ldots, f_{M_e}^e\}^T \\
&f_m^e = \int_{\Omega_e} f \psi_m^e \, dx + \int_{\partial\Omega_e \cap \partial\Omega_2} g \psi_m^e \, ds \\
&\bar{\boldsymbol{\sigma}} = \{\bar{\sigma}_1^e, \bar{\sigma}_2^e, \ldots, \bar{\sigma}_{M_e}^e\}^T \\
&\bar{\sigma}_m^e = \int_{\partial\Omega_e - (\partial\Omega_2 \cap \partial\Omega_e)} \sigma \cdot n_e \psi_m^e \, ds
\end{aligned}
\right\}
\tag{3.3.20}
$$

Here **u** and **s** are vectors of nodal values of u_h^e and σ_h^e, respectively, and $\bar{\sigma}$ is the usual generalized force vector which (in the absence of point or line forces) sums to zero when the elements are assembled to form the global model.

The matrix **H** in (3.3.19) is symmetric, positive definite, and thus invertible. Thus, we can eliminate **s** to obtain

$$\mathbf{s} = \mathbf{H}^{-1}\mathbf{A}\mathbf{u} \tag{3.3.21}$$

and then

$$\mathbf{k}^e \mathbf{u}^e = \mathbf{f}^e + \bar{\sigma}^e \tag{3.3.22}$$

where $\mathbf{k}^e$ is the local stiffness matrix for the mixed finite element method:

$$\mathbf{k}^e = \mathbf{A}^T \mathbf{H}^{-1} \mathbf{A} \tag{3.3.23}$$

Having reduced the local construction of a mixed finite element method to the standard stiffness matrix formulation (3.3.22), the remainder of the analysis is identical to that for conventional finite element methods discussed in Volume I. Once the nodal values **u** are determined, $u_h^e = \sum_{m=1}^{M_e} u_m^e \psi_m^e$ is known, but in mixed methods we do not use ∇u_h as an approximation of the gradient of u. Instead, we hope that $a^{-1}\sigma_h$ is a better approximation of ∇u than ∇u_h.

3.3.3 Remarks on Stability and Convergence

One must also be concerned about the numerical stability of mixed methods and whether or not a form of the Babuška–Brezzi condition holds for the approximating spaces H^h and S^h. One form of the condition on the compatibility of the spaces S^h and $\nabla H^h = \{\mathbf{w}_h = (w_{1h}, w_{2h}); \mathbf{w}_h = \nabla v_h, v_h \in H^h\}$ is: If P_h denotes the projection of ∇H^h into S^h, the mixed method is stable whenever the constant α_h, given by*

$$\alpha_h = \min_{v_h \in H^h} \frac{\| P_h \nabla v_h \|_0}{\| \nabla v_h \|_0} \tag{3.3.24}$$

is positive and independent of h, where $\| \nabla v_h \|_0^2 = \int_\Omega \nabla v_h \cdot \nabla v_h \, dx$. If α_h depends on h and is positive for a given h, the approximate problem has a unique solution (u_h, σ_h) for that mesh; however, the solutions may become numerically unstable as $h \to 0$. Condition (3.3.24) is a convenient alternative form related to the Babuška–Brezzi condition, which can be more easily verified for certain classes of problems.

---

* See, for example, Oden and Lee [1978] and Babuška et al. [1977].

If condition (3.3.24) is to hold, so also must a condition on the rank of the assembled form $\bar{\mathbf{A}}$ of matrix $\mathbf{A}$ appearing in (3.3.20). Then an α_h exists such that (3.3.24) holds whenever

$$\bar{\mathbf{A}}\mathbf{u} = \mathbf{0} \qquad \text{implies that} \qquad \mathbf{u} = \mathbf{0} \qquad\qquad (3.3.25)$$

This means that we should have dim $\mathbf{S}^h \geq$ dim $\boldsymbol{\nabla} H^h$.

The mixed methods remain a delicate and intriguing class of finite element methods for linear elliptic boundary problems. Very robust mixed methods can be developed which are insensitive to irregularities in the solution but which may converge at a suboptimal rate of convergence. Still other mixed methods may be unstable, in the sense that parameter α_h appearing in the Babuška–Brezzi condition [or equivalently, (3.3.24)] is $O(h^\sigma)$ for some $\sigma \geq 1$, yet *appear* to produce very acceptable results when the actual solution is very smooth.

One property of mixed methods that illustrates this delicate behavior arises in the case in which polynomials of equal degree are used to approximate u_h and the approximation $\boldsymbol{\sigma}_h$ of $\boldsymbol{\nabla} u$. Consider again problem (3.3.1a) with $a = 1$ and $\partial\Omega = \partial\Omega_1$ for simplicity (i.e., Laplace's equation with homogeneous Dirichlet boundary conditions). For the approximation spaces, take

$$
\begin{aligned}
H^h &\sim \text{spanned by conforming piecewise polynomials} \\
&\quad \text{of degree} \leq k \\
\mathbf{S}^h &\sim \text{spanned by conforming piecewise polynomials} \\
&\quad \text{of degree} \leq r
\end{aligned}
\qquad (3.3.26)
$$

Then, if $k = r$ Oden and Lee [1978] have shown that these mixed methods are, in general, unstable in that the stability parameter α_h tends to zero as $h \longrightarrow 0$. In particular, for $\Omega \subset \mathbb{R}^2$ and $k = r = 1$ they have shown that

$$\alpha_h = O(h)$$

This is unacceptable, since the convergence of these methods is typically governed by error estimates of the type

$$
\begin{aligned}
\| u &- u_h \|_1 + \| \boldsymbol{\sigma} - \boldsymbol{\sigma}_h \|_0 \\
&\leq \frac{c}{\alpha_h} [\inf_{v_h \in H^h} \| u - v_h \|_1 + \inf_{\tau_h \in \mathbf{S}^h} \| \boldsymbol{\sigma} - \boldsymbol{\tau}_h \|_0]
\end{aligned}
\qquad (3.3.27)
$$

Thus, if $k = r = 1$, it is known that $\inf_{v_h \in H^h} \| u - v_h \|_1 \leq c(u)h$ for regular refinements of the finite element mesh. Thus, the error in (3.3.27) is $O(1)$; that is, the method may not be convergent!

Nevertheless, these unstable methods appear to "work" in special cases in which the solution is very smooth. For example, in the special case of a one-dimensional linear boundary-value problem, Mansfield [1980] has shown that special superconvergence properties of mixed methods may come into play that lead to estimates of the type

$$\| e \|_W \le ch^\gamma, \qquad \gamma = \min{(k, r)} \tag{3.3.28}$$

where the spaces (3.3.26) are used and

$$\| e \|_W = \left[\| u - u_h \|_1^2 + \| \sigma - \sigma_h \|_0^2 + \left\| \frac{d}{dx}(\sigma - \sigma_h) \right\|_0^2 \right]^{1/2} \tag{3.3.29}$$

This result has been proved only for second-order two-point problems with smooth solutions and may not hold for more general problems in higher dimensions. Numerical experiments suggest that similar results may hold for fourth-order two-point problems. In any case, these results are very special in that they assume that the solution is very regular. In the presence of irregular solutions, the instabilities in these methods suggested by stability conditions such as (3.3.24) should be quite conspicuous as h is decreased.

3.3.4 Some Numerical Experiments

To illustrate the peculiar convergence and stability properties of certain mixed finite element methods described in the preceding subsection, we consider two simple one-dimensional boundary-value problems:

1. $-u''(x) = x^3, \quad 0 < x < 1$
   $u(0) = u(1) = 0$

2. $u^{(\mathrm{iv})}(x) = x^3, \quad 0 < x < 1$
   $u(0) = u(1) = u''(0) = u''(1) = 0$

The experiments we describe here were performed by Sheu [1978]. We use a uniform mesh containing elements of length h. For problem 1, we use the decomposition

$$u' = \sigma, \qquad -\sigma' = x^3; \qquad u(0) = u(1) = 0$$

and for the fourth-order problem 2 we use

$$u'' = \sigma, \qquad u(0) = u(1) = 0$$
$$\sigma'' = x^3, \qquad \sigma(0) = \sigma(1) = 0$$

We are interested in the behavior of the errors $e_u = u - u_h$ and $e_\sigma = \sigma -$

σ_h in the norms $\|v\|_1 = \left[\int_0^1 (v'^2 + v^2)\, dx\right]^{1/2}$ and $\|\tau\|_0 = \left[\int_0^1 \tau^2\, dx\right]^{1/2}$ for various choices of polynomials of degree k for u_h and r for σ_h. For $k = r = 3$, Hermite polynomials are used. Results of numerical experiments are shown in Fig. 3.2 for various choices of h, k, and r, and are summarized in the table

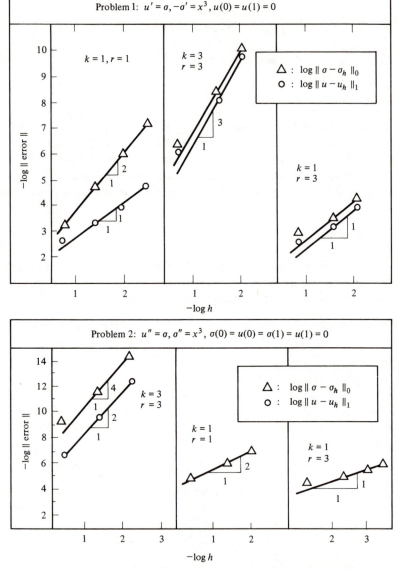

FIGURE 3.2 *Computed convergence rates for two model problems.*

below. Note that we obtain the rate of convergence of the method by calculating the norm of the error for each h, plotting $\log \| \text{error} \|$ versus $\log h$, and calculating the slope of this line. The methods tested here are basically unstable in the sense that (3.3.24) is not satisfied by an α_h independent of h. The solution, however, is a C^∞-function and the computed rates of convergence agree with the estimate (3.3.28). Note that in the fourth-order example the method fails because the rank condition (3.3.25) is violated and this is a necessary condition for the success of the method.

Problem 1: $\quad u' = \sigma, \; -\sigma' = x^3, \; u(0) = u(1) = 0$

Degree k	*Degree r*	$\| u - u_h \|_1$	$\| \sigma - \sigma_h \|_0$
1	1	$O(h)$	$O(h^2)$
3	3	$O(h^3)$	$O(h^3)$
1	3	$O(h)$	$O(h)$
3	1	Method fails	

Problem 2: $\quad u'' = \sigma, \; \sigma'' = x^3, \; \sigma(0) = \sigma(1) = u(0) = u(1) = 0$

Degree k	*Degree r*	$\| u - u_h \|_1$	$\| \sigma - \sigma_h \|_0$
1	1	$O(h^0)$	$O(h^2)$
3	3	$O(h^3)$	$O(h^4)$
1	3	$O(h)$	$O(h)$
3	1	Method fails	

EXERCISES

3.3.1 Show that any sufficiently smooth solution of (3.3.5) is a classical solution of the fourth-order problem (3.3.4).

3.3.2 Show that saddle points of the functional L of (3.3.6) can be characterized as solutions of (3.3.5).

3.3.3   Verify that if (u, σ) is a smooth solution of (3.3.15), u is a solution of (3.3.1).

3.3.4   Develop mixed variational principles for the following boundary-value problems:

(a) $u^{(iv)} + u'' + u = f, \quad 0 < x < 1$

$$u''(0) = M_0, \quad u''(1) = M_1 \quad (M_0, M_1 = \text{constants})$$
$$u(0) = 0, \qquad u(1) = 0$$

(b) $-u'' + u' + u = x, \quad 0 < x < 1$

$$u(0) = 1, \quad u(1) = 0$$

3.3.5   For polynomials of degree k for u and r for σ as approximations of the boundary-value problems in Exercise (3.3.4), develop all of the local stiffness equations for an element of length h for the cases:

(a) $k = 3, r = 3$

(b) $k = 1, r = 1$

(c) $k = 2, r = 1$

3.4   HYBRID METHODS

We recall that in conforming finite element methods, the global finite element approximations are required to satisfy certain continuity requirements at interelement boundaries. For example, in second-order problems, C^0-continuity of the global basis functions is required; for fourth-order problems, C^1-elements must be used, and for $2m$th-order problems, C^{m-1}-elements are necessary. Hybrid finite element methods are special mixed methods obtained by regarding these continuity requirements as constraints and using the method of Lagrange multipliers to enforce the constraints in an "average" (variational) sense. The result of such a formulation is that independent approximations of the solution on the interior of an element and the values of the solution and its derivatives on the boundary of an element can be made.

Hybrid finite elements were developed by Pian and Tong and their associates, and have been applied to a wide range of physical problems. See, for instance, Pian [1964, 1966], Pian and Tong [1969], Tong [1970], and Atluri [1971] and the references therein. Mathematical properties of hybrid finite element methods for second-order problems were studied by Babuška et al. [1977, 1978], Oden and Lee [1978], Lee [1976], Raviart and Thomas [1976], and Thomas [1975], and theories of hybrid methods for fourth-order problems were developed by Brezzi [1974, 1975] and Brezzi and Marini [1974]. We will give a brief description of a typical hybrid finite element method for a model second-order elliptic boundary-value problem.

3.4.1 Hybrid Variational Principles

To illustrate the major ideas, let us consider a representative second-order problem of finding u such that

$$\left.\begin{array}{rl} -\Delta u + u = f & \text{in} \quad \Omega \\ u = 0 & \text{on} \quad \partial\Omega \end{array}\right\} \tag{3.4.1}$$

where Ω is a domain in $\mathbb{R}^2$ with a smooth boundary $\partial\Omega$, and f represents the given data. We now construct a special hybrid variational principle for this problem.

Our first step is to divide the domain Ω into a collection of subdomains $\{\Omega_e\}$, $1 \le e \le E$, as shown in Fig. 3.3. At this point in our analysis, these

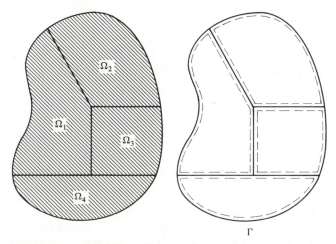

FIGURE 3.3 *Subdivision of a domain Ω into four subdomains, and the connected interdomain boundary Γ.*

domains are arbitrary and have nothing to do with the finite element approximations that are to come later. Over each domain Ω_e we define a functional

$$J_e(v, \eta) = \tfrac{1}{2}\int_{\Omega_e} (\nabla v \cdot \nabla v + v^2 - 2fv)\, dx + \oint_{\partial\Omega_e} v\eta\, ds \tag{3.4.2}$$

where η is a function defined only on the boundary $\partial\Omega_e$. By summing such functionals defined on all of the subdomains, we obtain a functional J defined on a class $\mathscr{X}$ of functions whose domain is the entire region Ω:

$$J: \mathscr{X} \longrightarrow \mathbb{R}, \; J(v, \eta) = \sum_{e=1}^{E} J_e(v, \eta) \tag{3.4.3}$$

The functions in the class $\mathfrak{X}$ are made up of pairs (v, η), where v is a function defined on the interiors of the subdomains Ω_e and η is a function defined on the connected interdomain boundary $\Gamma = \bigcup_{e=1}^{E} \partial\Omega_e$, as shown in Fig. 3.3.

The first variation in J at (u, η) is

$$\langle \delta J(u, \eta), (\bar{u}, \bar{\eta}) \rangle_{\mathfrak{X}} = \sum_{e=1}^{E} \left[\int_{\Omega_e} (\nabla u \cdot \nabla \bar{u} + u\bar{u} - f\bar{u})\, dx + \oint_{\partial\Omega_e} (u\bar{\eta} + \eta\bar{u})\, ds \right]$$

$$(3.4.4)$$

for arbitrary $(\bar{u}, \bar{\eta})$ in $\mathfrak{X}$. The hybrid variational problem is to find (u, η) such that

$$\langle \delta J(u, \eta), (\bar{u}, \bar{\eta}) \rangle_{\mathfrak{X}} = 0 \qquad \forall\, (\bar{u}, \bar{\eta}) \in \mathfrak{X} \qquad (3.4.5)$$

By integrating (3.4.4) by parts and using (3.4.5), we easily verify that (3.4.5) is formally equivalent to the system of Euler–Lagrange equations

$$\left.\begin{array}{ll}
\text{(a)} & -\Delta u + u = f \quad \text{in} \quad \Omega_e \\[6pt]
\text{(b)} & \dfrac{\partial u}{\partial n_e} = -\eta \quad \text{on} \quad \partial\Omega_e
\end{array}\right\} e = 1, 2, \ldots, E \\[12pt]
\text{(c)} \quad \sum_{e=1}^{E} \oint_{\partial\Omega_e} u_e \bar{\eta}_e\, ds = 0 \left.\right\}$$

$$(3.4.6)$$

Here $\mathbf{n}_e$ is the unit outward normal to $\partial\Omega_e$. Clearly, η is a Lagrange multiplier associated with the continuity condition "u_e is continuous across interdomain boundaries," which is implied by condition (c) of (3.4.6). Mathematically, η corresponds to $-\partial u/\partial n_e$. Since any solution of (3.4.1) also satisfies (3.4.6), the variational principle (3.4.5) can be used as a basis for approximations of (3.4.1). It is fundamental to note that this particular principle allows us to construct independent approximations of u on the interior of each subdomain and $\eta = -\partial u/\partial n_e$ on the boundary of each subdomain.

The variational principle (3.4.5) is by no means the only hybrid formulation possible for problem (3.4.1). For example, the Euler–Lagrange equations for the functional

$$G(u, \boldsymbol{\sigma}, w) = \sum_{e=1}^{E} \left[\tfrac{1}{2} \int_{\Omega_e} (u^2 - \boldsymbol{\sigma} \cdot \boldsymbol{\sigma} - u\nabla \cdot \boldsymbol{\sigma} - 2fu)\, dx + \oint_{\partial\Omega_e} \mathbf{n} \cdot \boldsymbol{\sigma} w\, ds \right]$$

$$(3.4.7)$$

are (formally)

$$
\begin{aligned}
&\text{(a)} \quad \left.\begin{array}{r} \sigma - \nabla u = 0 \\ -\nabla \cdot \sigma + u = f \end{array}\right\} \text{ in } \Omega_e \\
&\text{(b)} \qquad\qquad w = u \quad \text{ on } \partial\Omega_e \\
&\text{(c)} \quad \sum_{e=1}^{E} \oint_{\partial\Omega_e} \sigma \cdot \mathbf{n}_e \, ds = 0
\end{aligned} \left.\right\} e = 1, 2, \ldots, E \tag{3.4.8}
$$

which are also satisfied by solutions of (3.4.1). In this case, w is a Lagrange multiplier associated with the constraint of continuity of $\partial u/\partial n$ across interdomain boundaries. Clearly, (3.4.7) can serve as a basis for "mixed" hybrid approximations, since independent approximations of u and σ can be made on the interior of each element and of the values w on the boundaries of each element.

For higher-order problems (e.g., $\Delta^2 u = f$), many different hybrid variational principles can be formulated depending on which type of interelement continuity requirement is viewed as a constraint. For additional examples of such principles, see Pian and Tong [1969], Atluri [1971], Oden and Reddy [1976b], and the references therein.

3.4.2 Hybrid Finite Elements

As representative examples of hybrid finite elements, we will consider the two types of approximations suggested in Fig. 3.4. The first is based on the hybrid principle (3.4.5) and involves the use of polynomials of degree k for u and of degree t for $-\eta = \partial u/\partial n$. The second is based on the principle associated with the functional G in (3.4.7) and involves the use of polynomials of degree k for u, r for σ, and t for the values of u on the boundary of each element. These elements were studied by Lee [1976], Oden and Lee [1978], and Babuška et al. [1977]. Now, of course, the subdomains Ω_e are regarded as finite elements. We refer to the element in Fig. 3.4a as a *primal hybrid element* and that in Fig. 3.4b as a *dual-mixed-hybrid element*.

Primal Hybrid Element: We use as local approximations,

$$
u_h^e = \sum_{i=1}^{I} u_i^e \psi_i^e(\mathbf{x}), \qquad \eta_h^e = \sum_{j=1}^{T} \eta_j^e \chi_j^e(s) \tag{3.4.9}
$$

where ψ_i^e are shape functions containing polynomials of degree k, and χ_j^e are *piecewise* polynomials of degree t on $\partial\Omega_e$. Since η corresponds to $\partial u/\partial n$, *we do not place nodal points at the corners of elements* (since $\partial u/\partial n$ is not

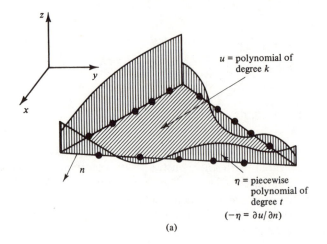

(a)

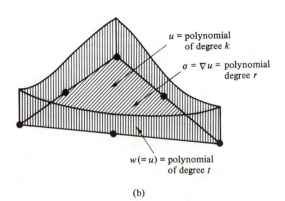

(b)

FIGURE 3.4 *Two types of hybrid finite elements.*

uniquely defined at a corner). Substitution of (3.4.9) into (3.4.5) yields the linear system of equations,

$$\left.\begin{array}{c} \mathbf{k}^e\mathbf{u}^e + \mathbf{c}^e\boldsymbol{\eta}^e = \mathbf{f}^e + \bar{\boldsymbol{\sigma}}^e \qquad \text{for element } \Omega_e \\[2mm] \sum_{e=1}^{E} (\mathbf{c}^T\mathbf{u})^e = 0 \end{array}\right\} \qquad (3.4.10)$$

where

$$\left.\begin{array}{ll} \mathbf{k}^e = [k^e_{ij}]; & k^e_{ij} = \displaystyle\int_{\Omega_e} (\nabla\psi^e_i \cdot \nabla\psi^e_j + \psi^e_i\psi^e_j)\, dx, \quad 1 \le i,j \le I \\[3mm] \mathbf{c}^e = [c^e_{is}]; & c^e_{is} = \displaystyle\sum_{l=1}^{N_l} \int_{\partial\Omega^l_e} \psi^e_i \chi^e_s\, ds, \quad 1 \le i \le I, 1 \le s \le T \\[3mm] \mathbf{u}^e = \{u_1, u_2, \ldots, u_I\}^T; & \mathbf{f}^e = \left\{\displaystyle\int_{\Omega_e} f\psi^e_i\, dx\right\}, \quad 1 \le i \le I \\[3mm] \boldsymbol{\eta}^e = \{\eta_1, \eta_2, \ldots, \eta_T\}^T; & T = N_l(t+1) \end{array}\right\} \qquad (3.4.11)$$

Here N_t is the number of sides of the element ($N_t = 3$ in Fig. 3.4a), and the vectors $\bar{\boldsymbol{\sigma}}^e$ sum to zero upon assembly of the elements. The final system of global equations is then (symbolically)

$$\sum_{e=1}^{E} [\mathbf{c}^T \mathbf{k}^{-1}(\mathbf{f} - \mathbf{c}\boldsymbol{\eta})]^e = 0 \qquad (3.4.12)$$

In other words, the nodal values $\boldsymbol{\eta}$ are taken as the unknowns. Observe that the global equations (3.4.12) involve the *inverse of a stiffness matrix;* hence, this particular formulation is of the "direct flexibility" type rather than a stiffness formulation.

Dual-Mixed-Hybrid Element: In this case, we have locally

$$\left.\begin{array}{ll} u_h^e(\mathbf{x}) = \sum_{i=1}^{I} u_i^e \psi_i^e(\mathbf{x}), & \sigma_h^e(\mathbf{x}) = \sum_{s=1}^{S} [\sigma_1^s \omega_s(\mathbf{x})\mathbf{e}_1 + \sigma_2^s \omega_s(\mathbf{x})\mathbf{e}_2] \\[2mm] w_h^e(s) = \sum_{l=1}^{L} w_l^e \zeta_l(s), & L = N_t t \end{array}\right\} \quad (3.4.13)$$

where $\mathbf{e}_1$ and $\mathbf{e}_2$ are orthonormal base vectors for $\mathbb{R}^2$. In this case, corner nodal points are used and the approximation of $w = u$ on $\partial\Omega_e$ is continuous over $\partial\Omega_e$.

Substitution of (3.4.13) into the variational equations associated with the functional G in (3.4.7) yields the system

$$\left.\begin{array}{l} (\mathbf{Mu} - \mathbf{A}_1^T\boldsymbol{\sigma}_1 - \mathbf{A}_2^T\boldsymbol{\sigma}_2)^e = \mathbf{f}^e + \bar{\boldsymbol{\sigma}}^e \\[1mm] (-\mathbf{A}_1\mathbf{u} - \mathbf{H}\boldsymbol{\sigma}_1 + \mathbf{C}_1\mathbf{w})^e = 0 \\[1mm] (-\mathbf{A}_2\mathbf{u} - \mathbf{H}\boldsymbol{\sigma}_2 + \mathbf{C}_2\mathbf{w})^e = 0 \\[1mm] \sum_{e=1}^{E} (\mathbf{C}_1^T\boldsymbol{\sigma}_1 + \mathbf{C}_2^T\boldsymbol{\sigma}_2)^e = 0 \end{array}\right\} \begin{array}{l} \text{for element } \Omega_e \end{array} \quad (3.4.14)$$

where (omitting the element label e to simplify the notation),

$$\left.\begin{array}{ll} M_{ij} = \int_{\Omega_e} \psi_i \psi_j \, dx; & f_i = \int_{\Omega_e} f\psi_i \, dx, \quad 1 \leq i, j \leq I \\[3mm] A_{\alpha,si} = \int_{\Omega_e} \frac{\partial \omega_s}{\partial x_\alpha} \psi_i \, dx; & \alpha = 1, 2, \quad 1 \leq s \leq S \quad 1 \leq i \leq I \\[3mm] H_{st} = \int_{\Omega_e} \omega_s \omega_t \, dx, & 1 \leq s, t \leq S \\[3mm] C_{\alpha,sl} = \oint_{\partial\Omega_e} \omega_s n_\alpha \zeta_l \, ds; & \alpha = 1, 2, \quad 1 \leq s \leq S, \quad 1 \leq l \leq L \\[3mm] \boldsymbol{\sigma}_\alpha = \{\sigma_\alpha^1, \sigma_\alpha^2, \dots, \sigma_\alpha^S\}^T, & \alpha = 1, 2; \quad \mathbf{u} = \{u^1, u^2, \dots, u^I\}^T \\[2mm] \mathbf{w} = \{w_1, w_2, \dots, w_L\}^T; & \text{etc.} \end{array}\right\} \quad (3.4.15)$$

As in the case of mixed methods, we can eliminate $\boldsymbol{\sigma}_\alpha$ to obtain

$$\boldsymbol{\sigma}_\alpha = \mathbf{H}^{-1}(-\mathbf{A}_\alpha\mathbf{u} + \mathbf{C}_\alpha\mathbf{w}), \qquad \alpha = 1, 2 \qquad (3.4.16)$$

so that for element e,

$$\mathbf{ku} = \mathbf{Dw} + \mathbf{f} + \bar{\boldsymbol{\sigma}} \tag{3.4.17}$$

where

$$\begin{aligned}
\mathbf{k} &= \mathbf{A}_1^T\mathbf{H}^{-1}\mathbf{A}_1 + \mathbf{A}_2^T\mathbf{H}^{-1}\mathbf{A}_2 + \mathbf{M} \\
\mathbf{D} &= \mathbf{A}_1^T\mathbf{H}^{-1}\mathbf{C}_1 + \mathbf{A}_2^T\mathbf{H}^{-1}\mathbf{C}_2
\end{aligned} \tag{3.4.18}$$

Thus, a stiffness relation involving only $\mathbf{w}$ is obtained if $\mathbf{k}$ is invertible; for example,

$$\sum_{e=1}^{E} (\mathbf{Kw} - \mathbf{F})^e = 0 \tag{3.4.19}$$

where

$$\begin{aligned}
\mathbf{K} &= \mathbf{C}_1^T\mathbf{H}^{-1}\mathbf{C}_1 + \mathbf{C}_2^T\mathbf{H}^{-1}\mathbf{C}_2 - \mathbf{D}^T\mathbf{k}^{-1}\mathbf{D} \\
\mathbf{F} &= \mathbf{D}^T\mathbf{k}^{-1}\mathbf{f}
\end{aligned} \tag{3.4.20}$$

Once $\mathbf{w}$ is obtained, $\mathbf{u}$ and $\boldsymbol{\sigma}$ are calculated using (3.4.17) and (3.4.16) for each finite element.

3.4.3 Some Properties of Hybrid Elements

Since hybrid elements are based on a special Lagrange-multiplier formulation, it should be no surprise that certain conditions must be satisfied by the approximations if stable and consistent schemes are to be obtained. In the case of hybrid methods, necessary conditions for these stability requirements to be satisfied manifest themselves in the form of so-called *rank conditions** similar to (3.3.25).

As an example, in the case of the primal model (3.4.9), the rank condition takes the following form:

$$\left. \begin{array}{l} \text{A necessary condition† for the primal hybrid method} \\ \text{(3.4.9) to lead to a solvable system of equations is that} \\[2mm] \qquad \displaystyle\oint_{\partial\Omega_e} \eta_h v_h \, ds = 0 \qquad \forall v_h \in H^h \\[2mm] \text{implies that } \eta_h = 0. \end{array} \right\} \tag{3.4.21}$$

---

* As usual, a Babuška–Brezzi condition of the type (3.2.39) must hold for all hybrid methods. These conditions involve a number of technical details and we refer the reader to Volume IV for a more detailed discussion.

† The rank conditions are necessary conditions for the existence of a positive constant

This condition is, in effect, a condition on the rank of the matrix c^e in (3.4.10): c^e must be of full rank. It can be shown that condition (3.4.21) holds for triangular elements for which u_h is a polynomial of degree k and η_h is a piecewise polynomial of degree t if

and only if
$$\left. \begin{array}{ll} t \leq k - 1 & \text{if } k \text{ is odd} \\[2ex] t \leq k - 2 & \text{if } k \text{ is even} \end{array} \right\} \qquad (3.4.22)$$

In the case of the dual-mixed-hybrid method, the rank condition takes the form:

$$\left. \begin{array}{l} \text{If the dual-mixed-hybrid equations (3.4.14) are to} \\ \text{have a unique solution, it is necessary that} \\[2ex] \displaystyle\sum_{e=1}^{E} \int_{\partial\Omega_e} \bar{\boldsymbol{\sigma}}_h \cdot \mathbf{n}_e w^e \, ds = 0 \qquad \forall \bar{\boldsymbol{\sigma}}_h \in \mathbf{S}^h \\[2ex] \text{implies that } w_h = 0. \end{array} \right\} \qquad (3.4.23)$$

If we solve $-\Delta u = f$ instead of $-\Delta u + u = f$, we also need the stability condition for mixed methods:

$$\left. \begin{array}{l} \text{It is necessary that condition (3.4.23) holds} \\ \textit{and} \text{ that} \\[2ex] \displaystyle\int_{\Omega_e} u_h \nabla \cdot \bar{\boldsymbol{\sigma}}_h \, dx = 0 \qquad \forall \bar{\boldsymbol{\sigma}}_h \in \mathbf{S}^h \\[2ex] \text{implies that } u_h = 0, \text{ if (3.4.14) is to have a} \\ \text{unique solution.} \end{array} \right\} \qquad (3.4.24)$$

Again, (3.4.23) is equivalent to the condition that the matrices $\mathbf{C}_\alpha$ in (3.4.14) have full rank, while (3.4.24) asserts that the matrices $\mathbf{A}_\alpha$ of (3.4.14) must have full rank (i.e., $\mathbf{C}_\alpha \mathbf{w} = \mathbf{0}$ implies that $\mathbf{w} = \mathbf{0}$ and $\mathbf{A}_\alpha \mathbf{u} = \mathbf{0}$ implies that $\mathbf{u} = \mathbf{0}$, $\alpha = 1, 2$).

Rank conditions such as these must be checked for all hybrid finite element methods.

β_h in Babuška–Brezzi conditions for hybrid elements, but as we have seen, the existence of $\beta_h > 0$ does not guarantee stability since β_h may depend on h. Thus, the rank conditions may not be sufficient conditions to guarantee convergence of hybrid methods.

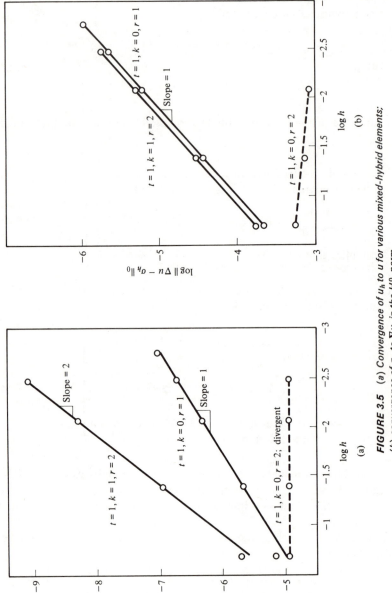

FIGURE 3.5 (a) Convergence of u_h to u for various mixed-hybrid elements; (b) convergence of σ_h to ∇u, in the H^0-norm.

3.4.4 Numerical Experiments

As a representative numerical example, we will outline some results of Lee [1976] for dual-mixed-hybrid approximations of the problem

$$-\Delta u = f \quad \text{in} \quad \Omega$$
$$u = 0 \quad \text{on} \quad \partial\Omega$$

where $\Omega = (0, 1) \times (0, 1)$ and

$$f(x, y) = 2x^2(1 - 3y)(x - 1) + 2y^2(1 - 3x)(y - 1)$$

The exact solution is $u = x^2y^2(1 - x)(1 - y)$.

Numerical results were obtained by Lee for the following choices of local approximations on a uniform mesh of square elements:

u_h	σ_h	w_h
1. Constant ($k = 0$)	Linear ($r = 1$)	Linear ($t = 1$)
2. Constant ($k = 0$)	Quadratic ($r = 2$)	Linear ($t = 1$)
3. Linear ($k = 1$)	Quadratic ($r = 2$)	Linear ($t = 1$)

Case (2) violates the stability condition (3.4.23) and should not converge.

Some typical numerical results on the rates of convergence are given in Fig. 3.5, where we have plotted $\log \| u - u_h \|_0$ and $\log \| \nabla u - \sigma_h \|_0$ versus $\log h$. As expected, case (2) is divergent, while the most rapid convergence* for u_h is obtained in case (3). Additional results of this type can be found in the work of Lee [1976] and Oden and Lee [1978].

EXERCISES

3.4.1 Give a finite element analysis of the boundary-value problem

$$-u'' + u = x^3, \qquad 0 < x < 1$$
$$u(0) = u(1) = 0$$

---

* We derive error estimates for several types of mixed and hybrid methods in Volume IV.

147

using a mixed hybrid finite element in which u is approximated on the interior of each element using polynomials of degree $k = 2$, $u = w = $ constant at the ends (boundaries) of the element, and u' is approximated on the interior using polynomials of degree $r = 1$. Carry out all computations, displaying all matrices, for a model consisting of at least three finite elements. Discuss the rank conditions (3.4.23) and (3.4.24) for this model.

3.5 PENALTY METHODS, PERTURBED LAGRANGIANS, AND REDUCED INTEGRATION

Penalty methods arose from the theory of constrained minimization outlined in Section 3.2 (but they are not limited to such problems). Suppose that we wish to minimize a functional J defined on a Hilbert space H subject to the constraint $Bu = g$, where B is an operator taking H into another Hilbert space Q, as before. We recall from previous discussions that this problem can be viewed as one of finding the minimum of J in some constraint set $K = \{v \in H \mid Bv = g\}$. The method of Lagrange multipliers enabled us to seek solutions to this problem in the whole space H, provided that we were willing to pay the price of including the Lagrange multiplier p as an additional unknown in the formulation. Penalty methods, on the other hand, also provide for minimizing functionals on all of H, but they possess the significant advantage of not requiring the inclusion of an additional unknown. In many instances, this property can lead to finite element formulations which have significantly fewer unknowns than those obtained using Lagrange multipliers and that are easier to implement in existing finite element programs.

3.5.1 Penalty Functionals

The idea behind penalty methods for constrained minimization problems is to append to the functional being minimized a *penalty term*, which gets larger in magnitude the more severely the constraint $Bu = g$ is violated. In other words, if a function v is tested as a candidate for a minimizer of J, the further v is away from satisfying the constraint, the greater the penalty that we must pay.

Let K be the constraint set,

$$K = \{v \in H \mid Bv = 0\} \tag{3.5.1}$$

A functional P is called a *penalty functional* if it satisfies the following conditions:

(a) $P: H \rightarrow \mathbb{R}$ is differentiable [i.e., its first
     variation, $\lim_{\epsilon \to 0} \partial P(u + \epsilon v)/\partial \epsilon = \langle \delta P(u), v \rangle_H$
     exists].*

(b) $P(v) \geq 0$ for all v in H.

(c) $P(v) = 0$ if $v \in K$ and $P(v) > 0$
     if $v \notin K$.

(d) P is convex;† that is, $P[\theta u + (1 - \theta)v]$
     $\leq \theta P(u) + (1 - \theta)P(v), 0 \leq \theta \leq 1.$

(3.5.2)

Once a functional P with properties (3.5.2) is available, the problem of finding the minimum of J in K can be solved by the penalty method in the following steps:

1. Let J be a functional to be minimized subject to a constraint such as (3.5.1). Let P be any functional satisfying conditions (3.5.2). Our first step is to introduce a new functional J_ϵ defined by

$$J_\epsilon(v) = J(v) + \frac{1}{\epsilon} P(v), \qquad v \in H \qquad (3.5.3)$$

    where ϵ is an arbitrary positive number.

2. If J satisfies the conditions of Theorem 3.1.1 and P satisfies conditions (3.5.2), we easily verify by Theorem 3.1.1 that J_ϵ has a minimizer $u_\epsilon \in H$ for each fixed $\epsilon > 0$. Of course, u_ϵ will not, in general, satisfy the constraint condition (i.e., $u_\epsilon \notin K$). However, u_ϵ will be a solution of the variational problem

$$\langle \delta J(u_\epsilon), v \rangle_H + \frac{1}{\epsilon} \langle \delta P(u_\epsilon), v \rangle_H = 0 \qquad \forall v \in H \qquad (3.5.4)$$

3. Let $\{u_\epsilon\}$ be a sequence of solutions of (3.5.4) obtained as ϵ approaches zero. The assumed coerciveness of J (recall condition 3 of Theorem 3.1.1) implies the existence of a subsequence of $\{u_\epsilon\}$ which converges weakly to an element u in H.

4. Remarkably, the limit u of the subsequence $\{u_\epsilon\}$ of solutions to (3.5.4)

---

* This requirement can be weakened; penalty methods are not limited to equations involving operators derived from a potential energy functional.

† This condition can also be relaxed. What is generally needed is that P be "weakly lower semicontinuous": that is, if $\{u_n\}$ is a sequence in H that converges weakly to u in H, then $\lim \inf_{n \to \infty} P(u_n) \geq P(u)$. In the examples with which we will be concerned, it is sufficient to replace this condition with the requirement that P be a convex differentiable functional. See, for example, Oden [1978, 1981].

described in step 3 is a solution of our problem; that is, u is a mini-mizer of the functional J in the constraint set K.

3.5.2 Lagrange Multipliers by Penalty Methods

In many physical problems involving constraints, Lagrange multipliers appear as important dependent variables in the formulation of the governing equations. For example, in the study of incompressible fluid flow or the deformation of incompressible elastic bodies, the Lagrange multiplier associated with the constraint of incompressiblity represents physically the hydro-static pressure. The question then arises as to how to compute Lagrange multipliers using penalty methods.

To resolve this question, recall that if we use the *perturbed Lagrangian* method (3.2.33), we are led to a system of variational equations (3.2.34):*

$$\langle \delta J(u_\epsilon), v \rangle_H + [p_\epsilon, Bv] = 0 \qquad \forall v \in H \left. \right\} \tag{3.5.5}$$
$$[q, Bu_\epsilon] - \epsilon[q, p_\epsilon] = [q, g] \qquad \forall q \in Q \left. \right\}$$

Here ϵ is an arbitrary positive number and (u_ϵ, p_ϵ) is a saddle point of the perturbed Lagrangian L_ϵ in (3.2.33). We can solve the last equation in (3.5.5) for the perturbed Lagrange multiplier p_ϵ:

$$p_\epsilon = \frac{1}{\epsilon}(Bu_\epsilon - g) \tag{3.5.6}$$

Thus, u_ϵ must be such that

$$\langle \delta J(u_\epsilon), v \rangle_H + \frac{1}{\epsilon}[Bu_\epsilon - g, Bv] = 0 \qquad \forall v \in H \tag{3.5.7}$$

The form of this variational boundary-value problem is very close to that of the penalty method (3.5.4). Indeed, if we have

$$P(v) \equiv \tfrac{1}{2}\| Bv - g \|_Q^2 \left. \right\} \tag{3.5.8}$$
$$\langle \delta P(u), v \rangle_H = [Bu - g, Bv] \left. \right\}$$

then the penalty method is equivalent to the perturbed Lagrangian method.†

---

* Here we have effectively identified Q with its dual Q' and treated $[\cdot, \cdot]$ as an inner product on Q; a more general situation would be to replace $[q, p_\epsilon]$ by, for example, $(j^{-1}p_\epsilon, q)_Q$ with j the duality map from Q to $Q', q \in Q$, and $(\cdot, \cdot)_Q$ the inner product on Q. For further generalizations, see Kikuchi [1979].

† Actually, the underlying notions of penalty methods are much more general than saddle point techniques based on perturbed Lagrangians, but under the special circumstances described here, these methods are equivalent.

We can generalize these results a step further. Again, let $B : H \rightarrow Q$ be a linear operator in the constraint condition (3.5.1), and let G be any convex, differentiable, nonnegative functional on Q with the property that $G(0) = 0$ and $G(q) > 0$ if $q \neq 0$. Then the functional

$$P(v) = G(Bv - g) \tag{3.5.9}$$

qualifies as a penalty function. The first variation of P at u is computed using the chain rule:

$$\langle \delta P(u), v \rangle_H = [G'(Bu - g), Bv] \tag{3.5.10}$$

Thus, upon comparing (3.5.4) with the first equation in (3.5.5), we see that for $\epsilon > 0$, the penalty method produces an approximation p_ϵ of the Lagrange multiplier p of

$$p_\epsilon = \frac{1}{\epsilon} G'(Bu_\epsilon - g) \tag{3.5.11}$$

For example, consider the Stokes problem for incompressible viscous flow where, if $\mathbf{u}$ is the velocity field, we have the constraint of incompressibility

$$\text{div } \mathbf{u} = 0$$

where $u_i \in H_0^1(\Omega)$. An appropriate penalty functional is

$$P(\mathbf{v}) = \tfrac{1}{2} \int_\Omega (\text{div } \mathbf{v})^2 \, dx \tag{3.5.12}$$

which is of the form $G(B\mathbf{v})$, where $G(w) = \tfrac{1}{2} \int_\Omega w^2 \, dx$ and $B\mathbf{v} = -\text{div } \mathbf{v}$. Since

$$\frac{1}{\epsilon} \langle \delta P(\mathbf{u}_\epsilon), v \rangle_H = \frac{1}{\epsilon} \int_\Omega \text{div } \mathbf{u}_\epsilon \, \text{div } \mathbf{v} \, dx \tag{3.5.13}$$

where $\mathbf{u}_\epsilon$ is a solution of the penalized problem (3.5.4), our approximation of the hydrostatic pressure p for $\epsilon > 0$ is

$$p_\epsilon = -\frac{1}{\epsilon} \text{div } \mathbf{u}_\epsilon \tag{3.5.14}$$

Will the penalty approximation p_ϵ converge to an actual Lagrange multiplier p as ϵ tends to zero? This question is easily resolved if we note the equivalence of the penalty and perturbed Lagrangian methods and retrace the arguments leading up to the Babuška–Brezzi condition (3.2.39). If $\beta > 0$ exists such that

$$\sup_{\substack{v \in H \\ (v \neq 0)}} \frac{|[q, Bv]|}{\|v\|_H} \geq \beta \|q\|_{Q'/\ker B^*} \tag{3.5.15}$$

for all $q \in Q'$, then a sequence $\epsilon \rightarrow 0$ can be found such that p_ϵ converges weakly to p in Q'.

3.5.3 Reduced Integration and Finite Element Penalty Methods

Approximation of a Model Problem: Let us now consider an application of these penalty ideas to a model problem with constraints. Again, Stokes' problem considered in Section 3.2 provides an excellent example. The classical problem is to find a velocity field $\mathbf{u} = [u_1(\mathbf{x}), u_2(\mathbf{x})]$ and a hydrostatic pressure field $p = p(\mathbf{x})$, $\mathbf{x} = (x_1, x_2) \in \Omega$, such that [recall (3.2.29)]

$$\left. \begin{array}{r} -\nu \,\Delta\mathbf{u} + \nabla p = \mathbf{f} \\ \mathrm{div}\,\mathbf{u} = 0 \end{array} \right\} \text{ in } \ \Omega \\ \left. \mathbf{u} = 0 \quad \text{on } \ \partial\Omega \right\} \tag{3.5.16}$$

where ν is the viscosity, a positive constant, $\mathbf{f} = (f_1, f_2)$ is the body force vector, and Ω is a smooth bounded domain in $\mathbb{R}^2$. Recall that this problem is equivalent to the problem of minimizing the functional J of (3.2.15) subject to the constraint $\mathrm{div}\,\mathbf{u} = 0$. In view of (3.2.15) and (3.5.12), an acceptable penalty functional for a penalty formulation of this problem is

$$\left. \begin{array}{l} J_\epsilon : H \longrightarrow \mathbb{R}; \qquad H = H_0^1(\Omega) \times H_0^1(\Omega) \\[2mm] J_\epsilon(\mathbf{v}) = J(\mathbf{v}) + \dfrac{1}{\epsilon}\, P(\mathbf{v}) \\[3mm] \qquad = \dfrac{\nu}{2} \displaystyle\int_\Omega |\nabla\mathbf{v}|^2 \, dx - \int_\Omega \mathbf{f} \cdot \mathbf{v} \, dx + \dfrac{1}{2\epsilon} \int_\Omega (\mathrm{div}\,\mathbf{v})^2 \, dx \end{array} \right\} \tag{3.5.17}$$

where ϵ is a positive penalty parameter.

Thus, for given ϵ and in accordance with (3.5.4), the minimizer $\mathbf{u}_\epsilon$ of J_ϵ is a solution of the variational boundary-value problem,

$$\nu \int_\Omega \nabla\mathbf{u}_\epsilon : \nabla\mathbf{v} \, dx + \epsilon^{-1} \int_\Omega \mathrm{div}\,\mathbf{u}_\epsilon \, \mathrm{div}\,\mathbf{v} \, dx \\ = \int_\Omega \mathbf{f} \cdot \mathbf{v} \, dx \qquad \forall \mathbf{v} \in H \tag{3.5.18}$$

To construct a finite element approximation $\mathbf{u}_h^\epsilon$ of $\mathbf{u}_\epsilon$, we partition Ω into finite elements and develop a conforming C^0-set of basis functions $\{\phi_i\}_{i=1}^N$ which provide the basis for a finite-dimensional subspace H^h of $H_0^1(\Omega) \times H_0^1(\Omega)$. Then, following the usual procedure, we attempt to solve for a $\mathbf{u}_h^\epsilon$ such that

$$\nu \int_\Omega \nabla \mathbf{u}_h^\epsilon : \nabla \mathbf{v}_h \, dx + \epsilon^{-1} \int_\Omega \operatorname{div} \mathbf{u}_h^\epsilon \operatorname{div} \mathbf{v}_h \, dx$$
$$= \int_\Omega \mathbf{f} \cdot \mathbf{v}_h \, dx \qquad \forall \mathbf{v}_h \in H^h \tag{3.5.19}$$

Globally, this leads to a system of algebraic equations for the vector $\mathbf{U}_\epsilon$ of nodal values of $\mathbf{u}_h^\epsilon$ of the form

$$(\mathbf{K} + \epsilon^{-1}\mathbf{K}_P)\mathbf{U}_\epsilon = \mathbf{F} \tag{3.5.20}$$

where $\mathbf{K}$ is the usual stiffness matrix for the unconstrained problem, $\mathbf{F}$ is the load vector, and $\mathbf{K}_P$ is the contribution to the stiffness due to the penalty functional P; for example,

$$\mathbf{U}_\epsilon^T \mathbf{K}_P \mathbf{U}_\epsilon = \int_\Omega (\operatorname{div} \mathbf{u}_h^\epsilon)^2 \, dx \tag{3.5.21}$$

Reduced Integration: Unfortunately, for very small values of ϵ and reasonably fine meshes *the finite element scheme (3.5.19) generally does not work*. The problem arises from the fact that, although the functional P of (3.5.12) is a perfectly legitimate penalty functional for the "continuous" problem (3.5.18), it is not necessarily a penalty functional for the discrete problem (3.5.19) for a fixed mesh size h. To qualify as a penalty functional, recall from (3.5.2) that P must be *positive semidefinite* [in particular, $P(\mathbf{v})$ must be strictly positive if $B\mathbf{v} - g \neq 0$ but it must vanish if $\mathbf{v}$ satisfies the constraint]. In constructing (3.5.19), we have not identified any discrete approximation of the constraint, and the result, in general, is that the discrete approximation of the penalty functional $P_h(\mathbf{v}_h) = \frac{1}{2}\int_\Omega (\operatorname{div} \mathbf{v}_h)^2 \, dx$ or, equivalently, the penalty stiffness matrix $\mathbf{K}_P$ of (3.5.21) will be positive definite. Thus, although (3.5.20) is solvable, we will have $\mathbf{U}_\epsilon \to 0$ as $\epsilon \to 0$. The discrete problem (3.5.19) is thus *overconstrained* or *"locked."*

It should be noted that if, for a given fixed mesh, ϵ in (3.5.19) is large enough, the discrete problem is certainly solvable and the solution $\mathbf{u}_h^\epsilon$ would then correspond to the velocity of a fluid with small compressibility. Of course, the condition $\operatorname{div} \mathbf{u}_h^\epsilon = 0$ would not be satisfied, and the solution would deteriorate (approach $\mathbf{u}_h \equiv 0$) as ϵ approaches zero for a fixed mesh size h. Numerical experiments suggest that for rather fine meshes (i.e., small h), reasonable nonzero solutions $\mathbf{u}_h^\epsilon$ are obtained only for values of ϵ so large that the constraint $\operatorname{div} \mathbf{u} = 0$ is not satisfactorily approximated. This clearly illustrates that for the scheme (3.5.19) it is necessary to choose ϵ as a function of h.

It is well known that the integrals appearing in finite element models such as (3.5.19) are virtually always evaluated using an appropriate numerical integration rule, and when the integrands are polynomials, integration formulas can be selected which produce exact values of these integrals. It has been discovered that if the penalty terms in (3.5.19) are integrated using an integration rule of an order *less* than that required to integrate these terms exactly, reasonable finite element approximations of (3.5.18) can sometimes be obtained. This practice is called *reduced integration* or *selective reduced integration*,* since not all terms in (3.5.19) are integrated inexactly. By employing such underintegrated penalty terms, the discrete penalty functional is positive semidefinite and the "locking" of the solution mentioned earlier does not occur.

To illustrate such reduced-integration penalty methods, suppose that $I(g)$ denotes a quadrature rule for approximating the integral of a function g defined over the E finite elements comprising the mesh. Then

$$I(g) = \sum_{e=1}^{E} I_e(g)$$

where

$$I_e(g) = \sum_{l=1}^{L} w_l^e g(\xi_l^e), \qquad 1 \leq e \leq E \qquad (3.5.22)$$

so that

$$I_e(g) \simeq \int_{\Omega_e} g \, dx$$

Here w_l^e are the quadrature weights (assumed to be positive numbers here) and ξ_l^e are the quadrature points within each element. If g is a polynomial in x_1, x_2, it is possible to choose the order L of the rule I_e such that $I_e(g) = \int_{\Omega_e} g \, dx$ (i.e., the integration is exact).

A reduced-integration penalty approximation of problem (3.5.19) consists of seeking $\mathbf{u}_h^\epsilon \in H^h$ such that

$$\nu \int_\Omega \nabla \mathbf{u}_h^\epsilon : \nabla \mathbf{v}_h \, dx + \epsilon^{-1} I(\operatorname{div} \mathbf{u}_h^\epsilon \operatorname{div} \mathbf{v}_h) = \int_\Omega \mathbf{f} \cdot \mathbf{v}_h \, dx \qquad (3.5.23)$$

$$\forall \mathbf{v}_h \in H^h$$

---

* Selective reduced integration for the analysis of certain plate and shell problems was first suggested by Zienkiewicz et al. [1971] and related ideas were discussed by Fried [1973, 1974], Malkus [1975], Hughes et al. [1976, 1977], Malkus and Hughes [1978], Reddy [1978, 1979a,b], Bercovier [1978], and others. The first analysis of these methods and the establishment of error estimates and convergence and stability criteria was due to Oden et al. [1980]; see also Oden and Kikuchi [1982], Oden et al. [1982], and Carey and Krishnan [1982].

where the integration rule $I(\cdot)$ is selected to integrate the penalty term inexactly in such a way that the resulting matrix $\mathbf{K}_P$ becomes positive semi-definite (and is therefore singular) for a given fixed mesh.

Once an integration rule $I(\cdot)$ is selected, we have effectively also selected a discrete approximation of the constraint div $\mathbf{u} = 0$. This also means that for each choice of a space H^h for approximating the velocities and for each rule $I(\cdot)$, there is intrinsic to the discrete formulation (3.5.23) a finite-dimensional subspace Q^h of the space $Q = L^2(\Omega)$ of Lagrange multipliers. We shall assume that the space Q^h is such that the following conditions hold:

1. For $q_h \in Q^h$,

$$\int_\Omega q_h \operatorname{div} \mathbf{v}_h \, dx = I(q_h \operatorname{div} \mathbf{v}_h) \tag{3.5.24}$$

   for all $\mathbf{v}_h \in H^h$.

2. There is a unique $p_h^\epsilon \in Q^h$ such that

$$I(p_h^\epsilon q_h) = -\frac{1}{\epsilon} I(\operatorname{div} \mathbf{u}_h^\epsilon q_h), \qquad q_h \in Q^h \tag{3.5.25}$$

Thus, whereas the rule $I(\cdot)$ is of insufficient order to integrate the penalty term $\int_\Omega (\operatorname{div} \mathbf{v}_h)^2 \, dx$ exactly, the space Q^h is such that $I(\cdot)$ integrates exactly the product $q_h \operatorname{div} \mathbf{v}_h$ for arbitrary velocities $\mathbf{v}_h \in H^h$. We observe that when (3.5.24) holds, there is generally no loss in the rate of convergence due to inexact integration. Condition (3.5.25) indicates that, instead of (3.5.14), we now have

$$p_h^\epsilon(\xi_l^e) = -\frac{1}{\epsilon} \operatorname{div} \mathbf{u}_h^\epsilon(\xi_l^e); \qquad 1 \le e \le E, \quad 1 \le l \le L \tag{3.5.26}$$

In other words, the values of the approximate pressure p_h^ϵ at the quadrature points ξ_l^e are equal to the values of $-\epsilon^{-1} \operatorname{div} \mathbf{u}_h^\epsilon$ at these integration points. Since p_h^ϵ is uniquely determined by (3.5.26), *the functions $q_h \in Q^h$ cannot be continuous across interelement boundaries.*

We may make another important observation: Suppose that $q_h^0 \in \ker B_h^*$; that is,

$$I(q_h^0 \operatorname{div} \mathbf{v}_h) = \langle B_h^* q_h^0, \mathbf{v}_h \rangle = 0 \qquad \forall \mathbf{v}_h \in H^h$$

Then, from (3.5.25),

$$I(p_h^\epsilon q_h^0) = 0 \qquad \forall q_h^0 \in \ker B_h^* \tag{3.5.27}$$

In other words, *the approximate pressure p_h^ϵ is orthogonal to* ker B_h^*, *this orthogonality being with respect to the discrete inner product* $(f, g)_Q = I(fg)$. In particular, when (3.5.2a) holds or (3.5.2b) with Range $(B_h) \subset Q^h$, we have p_h^ϵ orthogonal to ker B_h^* with respect to the $L^2(\Omega)$-inner product.

Some examples of spaces of pressure approximations Q^h for various choices of H^h and $I(\cdot)$ are given in Fig. 3.6. In examples 1, 6, and 10, exact

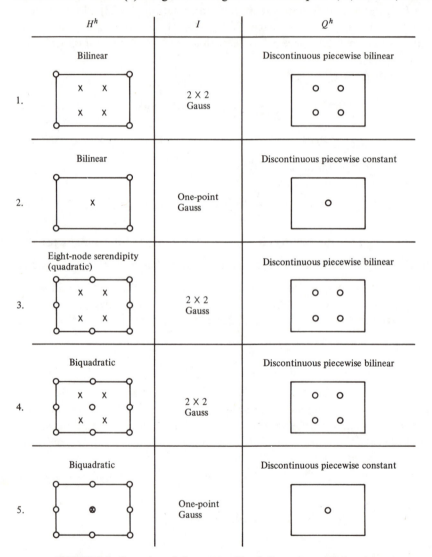

FIGURE 3.6 *Examples of the spaces Q^h of discontinuous finite element approximations of pressure corresponding to various choices of elements for approximating the velocities and various numerical quadrature rules $I(\cdot)$.*

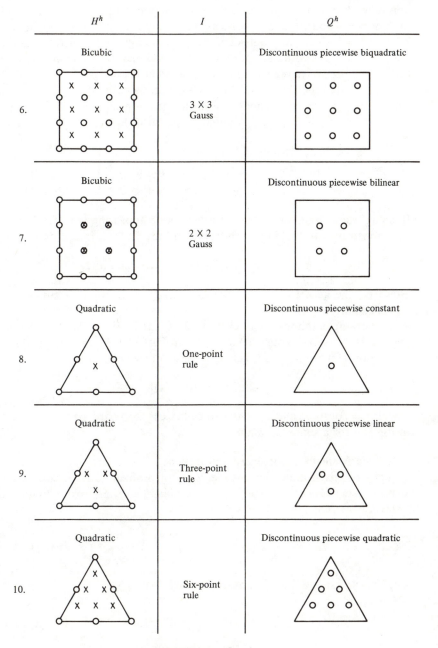

	H^h	I	Q^h
6.	Bicubic	3 × 3 Gauss	Discontinuous piecewise biquadratic
7.	Bicubic	2 × 2 Gauss	Discontinuous piecewise bilinear
8.	Quadratic	One-point rule	Discontinuous piecewise constant
9.	Quadratic	Three-point rule	Discontinuous piecewise linear
10.	Quadratic	Six-point rule	Discontinuous piecewise quadratic

FIGURE 3.6 (*Cont.*)

integration is used and the resulting method locks as described earlier. Condition (3.5.24) holds for examples 2, 3, 4, and 9 whereas a definite underintegration is used in 5 and 8 yielding stable but suboptimally accurate methods. In examples 9 and 10, a three- or six-point quadrature rule need not actually be concocted since the equivalent perturbed Lagrangian method can be used in these cases with discontinuous linear and quadratic pressure approximations respectively. We elaborate on this point below.

We note that the choice of H^h and $I(\cdot)$ (and, therefore, Q^h) also effectively determines discrete approximations of the constraint operator $B = -$divergence, $B^* = $ gradient (plus boundary conditions). Indeed, we may set

$$B_h: H^h \longrightarrow Q^h; \qquad B_h^*: Q^{h'} \longrightarrow H^{h'}$$
$$[q_h, B_h v_h] = \langle B_h^* q_h, v_h \rangle \equiv -I(q_h \operatorname{div} v_h) \qquad\qquad (3.5.28)$$
$$\forall\, q_h \in Q^h, \quad \forall\, v_h \in H^h$$

It is important to note that the exact pressure p appearing in (3.5.16) can be determined only to within an arbitrary constant. Indeed, recall that

$$\ker B^* = \{\text{space of constants defined on } \Omega\}$$

However, the kernel of the approximation B_h^* of B^* may be much more complicated.

For example, in the case of example 2 in Fig. 3.6 (i.e., when v_h is bilinear and q_h is piecewise constant) and when Dirichlet boundary conditions ($u_h = 0$ on $\partial\Omega$) are imposed, $\ker B_h^*$ consists of two elements,

$$\ker B_h^* = \{c, c_b(x)\}$$

where $c = $ constant and c_b is the so-called checkerboard pattern indicated in Fig. 3.7 for a square domain.

Mixed Methods via Reduced Integration: It is interesting to note that a family of mixed finite element methods for problem (3.5.16) can be constructed using *discontinuous pressure approximations* by approximating saddle points of the functional

$$L(v_h, q_h) = \frac{\nu}{2} \int_\Omega |\nabla v_h|^2 \, dx - I(q_h \operatorname{div} v_h) - \int_\Omega f \cdot v_h \, dx \qquad (3.5.29)$$

That is, $(u_h, p_h) \in H^h \times Q^h$ are sought such that

$$\nu \int_\Omega \nabla u_h : \nabla v_h \, dx - I(p_h \operatorname{div} v_h) = \int_\Omega f \cdot v_h \, dx \qquad \forall\, v_h \in H^h$$
$$I(q_h \operatorname{div} u_h) = 0 \qquad\qquad\qquad \forall\, q_h \in Q^h$$

$$(3.5.30)$$

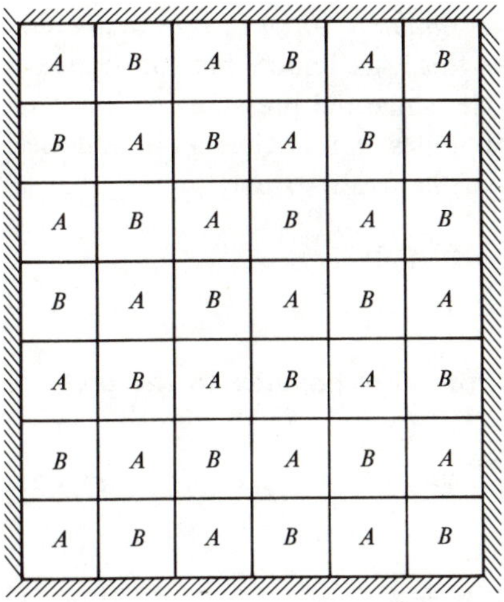

FIGURE 3.7 *Checkerboard function contained in ker B_h^* for the case of Dirichlet boundary conditions in which the elements in example 2 of Fig. 3.6 are used.*

Unfortunately, (3.5.30) is not uniquely solvable; for the boundary conditions in problem (3.5.16), the corresponding global stiffness matrix will be singular. This is because ker $B_h^* \neq$ {constants} and every solution of (3.5.30) for p_h will be of the form

$$p_h = \bar{p}_h + p_h^0, \qquad p_h^0 \in \text{ker } B_h^*$$

This problem may not arise if a perturbed Lagrangian formulation (or, equivalently, a penalty formulation) is used. In particular, if, instead of (3.5.29) we use

$$L_\epsilon(\mathbf{v}_h, q_h) = L(\mathbf{v}_h, q_h) - \frac{\epsilon}{2} I[(\text{div } \mathbf{v}_h)^2]$$

we obtain a system of equations of the form (3.5.30) for $(\mathbf{u}_h^\epsilon, p_h^\epsilon)$ except that the second member has the additional term

$$+\epsilon I(p_h^\epsilon \text{ div } \mathbf{v}_h)$$

Thus, (3.5.25) holds and we can then always solve for p_h^ϵ. Indeed, this approach gives us precisely the relation (3.5.26). In this case,

$$\langle B_h^* p_h^\epsilon, \mathbf{v}_h \rangle = -I(p_h^\epsilon \text{ div } \mathbf{v}_h) \neq 0$$

for any $\mathbf{v}_h$; hence, *the penalty method (or, equivalently, the perturbed Lagrangian method with reduced integration) will always yield a pressure approximation p_h^ϵ with a component outside the kernel ker B_h^* of the approximate gradient.* Thus, the penalty term serves to *regularize* the approximation scheme.

There is another important point that must be made here: Since the perturbed-Lagrangian methods we have described here involve *discontinuous* pressure approximations, *it is possible to eliminate the pressures at the element level* (recall from Section 3.2 that this was impossible for standard conforming mixed methods). For example, since the pressures within each element are not coupled through common nodes on interelement boundaries, the local element approximation will be of the form

$$\mathbf{k}^e \mathbf{u}_\epsilon^e - \mathbf{c}^e \mathbf{p}_\epsilon^e = \mathbf{f}^e + \bar{\boldsymbol{\sigma}}^e$$

$$\mathbf{c}^{e^T} \mathbf{u}_\epsilon^e + \epsilon \mathbf{m}^e \mathbf{p}_\epsilon^e = \mathbf{0}$$

where $\mathbf{m}^e$ is the Gram (or "mass") matrix corresponding to the pressure approximation. Then

$$\mathbf{p}_\epsilon^e = -\epsilon^{-1} \mathbf{m}^{e^{-1}} \mathbf{c}^{e^T} \mathbf{u}_\epsilon^e \tag{3.5.31}$$

and we have

$$\mathbf{k}^e \mathbf{u}_\epsilon^e + \epsilon^{-1} \mathbf{c}^e \mathbf{m}^{e^{-1}} \mathbf{c}^{e^T} \mathbf{u}_\epsilon^e = \mathbf{f}^e + \bar{\boldsymbol{\sigma}}^e \tag{3.5.32}$$

The local contribution to the penalty stiffness matrix $\mathbf{K}_P$ is clearly

$$\mathbf{k}_P^e = \mathbf{c}^e \mathbf{m}^{e^{-1}} \mathbf{c}^{e^T}$$

Note that this reduced-integration penalty formulation (3.5.32) can actually be obtained without introducing an integration rule $I(\cdot)$; we need only introduce a discontinuous local pressure approximation. This is precisely the technique used in example 9 of Fig. 3.6. It is interesting to note that (3.5.31) can, therefore, be used as a device for developing nonconventional integration rules for finite elements.

3.5.4 Stability and Convergence of Reduced-Integration Penalty Methods

Unfortunately, not all of the finite element methods listed in Fig. 3.6 are numerically stable and convergent. As is the case with all the methods of the general type covered in this chapter, stability is governed by a discrete Babuška–Brezzi condition. For the reduced-integration penalty methods, this condition [in analogy with (3.2.39)] takes the following form:

$$\left. \begin{array}{l} \text{There exists a } \beta_h > 0 \text{ such that} \\[2mm] \beta_h \| q_h \|_{Q^{h'}/\ker B_h^*} \le \sup_{\substack{\mathbf{v}_h \in H^h \\ (\mathbf{v}_h \ne 0)}} \dfrac{|I(q_h \operatorname{div} \mathbf{v}_h)|}{\| \mathbf{v}_h \|_1} \\[2mm] \text{for all } q_h \in Q^{h'}. \end{array} \right\} \tag{3.5.33}$$

Here $\| \cdot \|_1$ is the norm defined in (3.2.18).

Suppose that (3.5.24), (3.5.25), and (3.5.33) hold. Also, suppose that H^h and $Q^{h'}$ are endowed with interpolation properties of the type

$$\inf_{v_h \in H^h} \| u - v_h \|_1 \le Ch^k; \qquad \inf_{q_h \in Q^h / \ker B_h^*} \| p - q_h \|_0 \le Ch^r$$

where u and p are sufficiently smooth, $v_h \in H^h$, $q_h \in Q^h$, and k and r denote the degree of the local polynomial approximations. Then it can be shown (see Oden et al. [1980b]) that the errors $u - u_h^\epsilon$ and $p - p_h^\epsilon$ in the approximation (3.5.23) satisfy asymptotic estimates of the type

$$\| u - u_h^\epsilon \|_1 \le C_1 (1 + \beta_h^{-1})(h^k + h^r) + C_2 \beta_h^{-1} \epsilon$$
$$\| p - p_h^\epsilon \|_0 \le C_3 (1 + \beta_h^{-1} + \beta_h^{-2})(h^r + h^k) + C_4 \beta_h^{-2} \epsilon$$
(3.5.34)

where C_1, C_2, C_3, and C_4 are constants independent of ϵ and h.

The theory outlined earlier [and the estimates (3.5.34)] does not apply to the case in which quadratic approximations of the velocities are used but only one-point integration is employed. Such schemes were also studied by Oden et al. [1980a] and it was found that in such cases $\beta_h = $ constant, independent of h. Thus, these methods are stable. However, because of the severe underintegration, one full order of accuracy in velocities in the H^1-norm is lost. Numerical experiments indicate that such elements, although not highly accurate, lead to robust schemes: they are very stable and continue to behave satisfactorily even in certain cases in which the solution possesses strong singularities.

It is clear that the parameter β_h plays a fundamental role in the success of these methods. If $\beta_h = 0$, or $\beta_h \to 0$ as $h \to 0$, these methods are unstable. For certain choices of boundary conditions [particularly those in our model problem (3.5.16)] it has been shown by Oden et al. [1981] that β_h may indeed depend on h. For instance, for a uniform mesh on a square domain and for some of the examples listed in Fig. 3.6, we have the estimates:

Example 2:   Bilinear/constant — $\beta_h = O(h)$.

Example 4:   Biquadratic/bilinear — $\beta_h = O(h)$.

Example 5:   Biquadratic/constant — $\beta_h = $ constant.

Example 8:   Quadratic/constant — $\beta_h = $ constant.

Numerical results also suggest that $\beta_h = $ constant in example 9.

These unstable methods [$\beta_h = O(h)$] are deceptive. Just as in the case of the mixed methods discussed in Section 3.3.4, these methods may actually perform well on uniform rectangular meshes when the solution u is very smooth, but their inherent instability does exhibit itself in problems with irregularities. The approximate pressures p_h^ϵ, for example, are sensitive to singularities in these methods and exhibit oscillations as h tends to zero.

Filters: If $\| \mathbf{u} - \mathbf{u}_h \|_1 = O(h^2)$ and $\beta_h = O(h)$, the penalty approximations of velocity may converge, but, according to (3.5.34), the pressures may be numerically unstable. A situation such as this apparently arises in example 2, the case of bilinear velocity approximations and piecewise constant pressures. In the presence of smooth solutions, Johnson and Pitkaranta [1980] have shown that [despite the first expression in (3.5.34)], $\| \mathbf{u} - \mathbf{u}_h^\epsilon \|_1 = O(h)$ for ϵ very small. If $\beta_h = O(h)$, the second expression (3.5.34) indicates that the pressures may diverge. However, a so-called pressure filter can be devised so that $\beta_h = $ constant and the pressures converge in $L^2(\Omega)$ at a rate of $O(h + \epsilon)$. The idea is simply to adjust the approximate pressures q_h so that the stability condition (3.5.33) is satisfied with β_h independent of h. For example, in the case of a uniform mesh of rectangular elements on a rectangular domain for approximations of the model problem (3.5.16), if the bilinear/constant method of example 2 in Fig. 3.6 is used, it is meaningful to interpret the mesh alternatively as consisting of composite elements in which the velocity is approximated as a piecewise bilinear function defined over four subrectangles and the pressures are constants over each subrectangle. Then if the quadrants surrounding node i are numbered as shown in Fig. 3.8, it is shown in Oden et al. [1981] that β_h is independent of h if

$$q_i^1 + q_i^3 = q_i^2 + q_i^4 \tag{3.5.35}$$

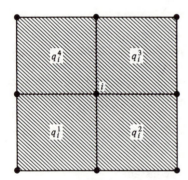

FIGURE 3.8 *"Composite element" over which q_h is piecewise constant, $q_h = q_i^k$, $k = 1, 2, 3, 4$, over quadrant k, and $\mathbf{v}_h$ is piecewise bilinear.*

The relation (3.5.35) defines a "filter" of the pressure that produces a stable and convergent pressure approximation.

Similar filters can be devised for some of the other elements which are unstable in the presence of irregular solutions (see, e.g., Sani et al. [1981]). We mention in particular the element of example 3, Fig. 3.6, in this regard. Numerical experiments indicate that pressure approximations obtained using the method of example 9, Fig. 3.6, are stable and rapidly convergent for many important applications.

Another example of a pressure filter that has proved to be effective is illustrated in Fig. 3.9. Having calculated the piecewise constant pressures

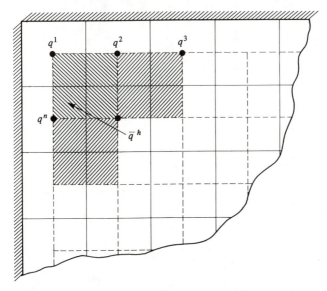

FIGURE 3.9 *Example of a conforming pressure filter: the piecewise constant pressure approximations q^h are interpolated by a conforming bilinear pressure $\bar{q}^h$ defined on a shifted mesh with nodes at the centroids of elements of the initial mesh.*

q_h using the (unstable) scheme 2 of Fig. 3.6, one can construct a "filtered" pressure approximation $\bar{q}_h$ by defining a conforming piecewise-bilinear function on a mesh shifted a half-element length relative to the original mesh. The pressure $\bar{q}_h$ is defined by equating the nodal values (at the centroids of the elements in the original mesh) to the values of the constant pressure for the respective element. This amounts to a projection of the piecewise-constant space Q^h onto a space $\bar{Q}^h$ spanned by conforming basis functions of the same type used in the velocity approximation but defined on a different domain $\bar{\Omega}_h \subset \Omega_h$.

Numerical Experiment: As an example of numerical results obtained using stable and unstable methods, we consider Stokes' problem on the square domain indicated in Fig. 3.10 for the case $\nu = 1000$. An approximation of this problem is obtained using a biquadratic velocity approximation with either 2×2- or one-point Gaussian quadrature, on a 4×4 uniform mesh, as indicated in the figure. The pressure distribution calculated along sections

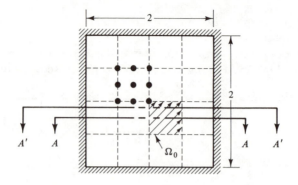

A ——●—— A : Nine-node one-point Gaussian quadrature

A' ——○—— A' : Nine-node 2 × 2-point Gaussian quadrature

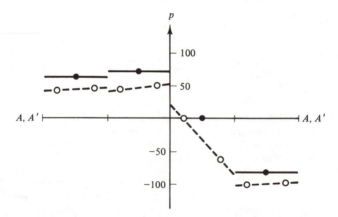

FIGURE 3.10 *Pressure distribution across the sections AA and A'A' of the Example 2 (uniform load).*

AA and A'A' by these two methods for the case of a constant body force

$$\mathbf{f}(\mathbf{x}) = \begin{cases} 200\mathbf{i} + 200\mathbf{j}, & \mathbf{x} \in \Omega_0 \\ 0, & \mathbf{x} \in \Omega_0 \end{cases}$$

Ω_0 being the shaded subdomain shown, are given for $\epsilon = 10^{-5}$. Note that β_h = constant in the nine-node, one-point scheme so that it is stable, but β_h is dependent on h for the nine-node, 2 × 2-point scheme. Remarkably what appear to be reasonable pressure approximations are obtained for this coarse mesh for the unstable scheme. Oscillations in this calculated pressure would be observed if we were to use a nonuniform mesh.

Next we consider the same problem but with a point source, $\mathbf{f} = 200(\delta(x - \bar{x}), \delta(y - \bar{y}))$, as shown in Fig. 3.11. In this case the solution is much less regular, possessing a singularity at the point of application of f. Nevertheless, the stable scheme still produces a reasonable pressure approximation, whereas the nine-node, 2×2-point scheme begins to exhibit erratic oscillations in the pressure calculations. Upon further refinement of the mesh, the nine-node, one-point scheme produces smooth pressures that grow in magnitude near the singularity. The nine-node, 2×2-point scheme,

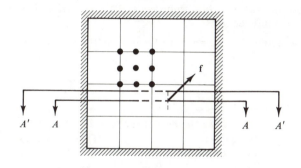

A ——●—— A : Nine-node one-point Gaussian quadrature

A' ––○–– A' : Nine-node 2×2-point Gaussian quadrature

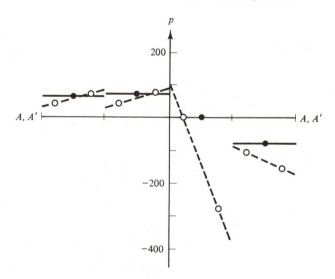

FIGURE 3.11 *Pressure distribution across the sections AA and A'A' of the Example 2 (point load).*

however, yields a pressure approximation which oscillates and does not simulate the correct pressure profile.

EXERCISES

3.5.1 Verify that (3.5.26) is a consequence of (3.5.25).

3.5.2 Discuss possible penalty formulations of the constrained variational problems listed in Exercise 3.2.4.

3.5.3 Furnish additional examples to those listed in Fig. 3.6 for the following cases:

H^h	$I(\cdot)$
(11) trilinear (cube element)	$2 \times 2 \times 2$ Gauss
(12) trilinear	One-point Gauss
(13) triquadratic	$2 \times 2 \times 2$ Gauss
(14) linear (tetrahedron)	One-point Gauss

Which of these elements would you expect to yield pressure approximations that are stable in $L^2(\Omega)$ for the Stokes' problem (3.5.16)?

4

OTHER METHODS

4.1 INTRODUCTION

The finite element methods we have considered to this point have been based
essentially on Galerkin or Rayleigh–Ritz approximations of variational
boundary-value problems. In fact, most finite element computations are made
using these methods. There are, however, other variants of the finite element
method which are effective for analyzing certain classes of problems and also
represent quite interesting extensions of the basic method.

Our purpose in the present chapter is to describe some of these other finite
element methods and to examine some of their approximation properties. We
begin by developing finite element methods based on the concept of collocation and then consider collocation-Galerkin methods in which both the collocation and the Galerkin ideas are combined. Following this, least-squares
and other special finite element schemes are briefly described to provide
additional examples of possible extensions of the method. Some of the more
important features of these schemes lie in the choice of trial and test spaces.
In the final section of the chapter we consider so-called boundary-element
techniques.

Other variants of the finite element method exist which are not dealt with
here. We have selected a sample of the more important extensions of finite
element methods which we feel indicates the broad scope of such methods and
illustrates some of the more important numerical schemes for solving
boundary-value problems.

4.2 COLLOCATION

One of the more straightforward approaches for determining an approximate solution to a boundary-value problem defined on a simple domain such as an interval, rectangle, or cube is the classical method of collocation. The basic idea is to introduce, as usual, an approximate solution that can be uniquely determined by specifying a number of parameters (coefficients). In standard collocation methods, the approximate solution is constructed so that it satisfies specified boundary conditions in advance and the coefficients in the approximation are then determined so that the approximation satisfies the differential equation at a number of distinct points in the domain. We shall show that this classical technique can be used to produce a finite element collocation method.

4.2.1 Two-Point Problem

Global Collocation Example: To introduce the procedure, we begin by describing a classical collocation method for a simple model problem. Assume that we wish to find an approximate solution to the simple linear two-point boundary-value problem

$$-u''(x) + u(x) - x = 0 \quad \text{in} \quad 0 < x < 1$$
$$u(0) = 0, \qquad u(1) = 0 \tag{4.2.1}$$

We shall seek an approximate solution to (4.2.1) which is a linear combination of the global basis functions $\{\phi_i(x)\} = \{x^i(1-x)\}$, $i = 1, 2, \ldots, N$:

$$\tilde{u}(x) = \sum_{j=1}^{N} a_j \phi_j(x) \tag{4.2.2}$$

where the unknown coefficients a_j are to be determined.

The approximation space $H^{(N)}$ is of dimension N and the global basis functions ϕ_i are infinitely differentiable so that $H^{(N)} \subset C_0^\infty[0, 1]$. If, instead of the smooth $f(x) = x$, the forcing function of (4.2.1) were only C^0, the classical solution u of this problem would be in $C^2[0, 1]$. On the other hand, the standard symmetric variational statement of this problem involves seeking solutions only in $H_0^1(0, 1)$. Hence it is clear that a global expansion such as (4.2.2) assumes much more regularity than that required in either the standard variational or the classical statements of this problem.

Given the smoothness of $\tilde{u}$ in (4.2.2), it is possible to substitute in the differential equation to define the residual

$$r(x) = \sum_{j=1}^{N} (-a_j \phi_j''(x) + a_j \phi_j(x)) - x \tag{4.2.3}$$

168

Now, if $\tilde{u}$ were the exact solution of our problem, we would, of course, have $r(x) \equiv 0$ at all x in $(0, 1)$. The idea behind collocation methods is to force this residual to vanish at a large but finite number of points in the domain of the solution. These points are called the collocation points of our approximation and we "collocate" the residual r at N interior collocation points $\{x_i\}$ by setting $r(x_i) = 0$. This leads us to the linear algebraic system for the coefficients $\{a_i\}$,

$$\sum_{j=1}^{N} [-a_j \phi_j''(x_i) + a_j \phi_j(x_i)] - x_i = 0, \qquad i = 1, 2, \ldots, N$$

For example, if $\tilde{u}(x) = a_1 x(1 - x)$, then collocating at $x_1 = \frac{1}{2}$, $2a_1 + \frac{1}{4}a_1 - \frac{1}{2} = 0$ implies that $a_1 = \frac{2}{9}$, which yields a crude approximation to the solution $u(x) = -(\sinh x/\sinh 1) + x$.

Evidently, the approximation $u^{(N)}$ could have been expressed as a Lagrange interpolant $\tilde{u}(x) = \sum_{j=1}^{N+2} u_j \psi_j(x)$, where ψ_j are the Lagrange interpolation polynomials of degree $N + 1$ on $[0, 1]$. Collocation then yields the nodal point values $\{u_i\}$ as the solution of the resulting algebraic system.

Finite Element Collocation: Our next objective is to extend this global collocation technique to include finite element bases. Recall that in a variational formulation of the two-point problem (4.2.1), the admissible functions are in $H_0^1(0, 1)$. This implies that piecewise-continuous functions are adequate for the Galerkin finite element method. In the interior of each element the functions are usually smooth polynomials but across an interface between elements they need only be continuous. That is, the basis functions span a subspace of $C^0(0, 1)$.

In the finite element collocation method, we must require that the approximating functions be locally C^2 (continuous derivatives up to second order) if we are to evaluate terms involving u'' at a given point. This suggests that a finite element collocation scheme be devised in which smooth functions (at least C^2) are used in the interior of each element but which may be less smooth globally. For the previous example the global approximation may be in $C^1(0, 1)$, it being understood that we do not collocate at the element endpoints where the function has only one continuous derivative and u'' is undefined. We shall comment further on the global C^1-continuity requirement later in the section.

For simplicity, we shall describe the method as it applies to the general linear two-point boundary-value problem,

$$-[a(x)u'(x)]' + b(x)u(x) = f(x), \qquad 0 < x < 1 \tag{4.2.4}$$

with mixed boundary conditions of the form

$$\alpha_0 u'(0) + \beta_0 u(0) = \gamma_0, \qquad \alpha_1 u'(1) + \beta_1 u(1) = \gamma_1 \tag{4.2.5}$$

at $x = 0$ and $x = 1$, respectively. The coefficients in the boundary conditions are assumed to be compatible with those of the differential equation in the sense described in Chapter 1. The coefficient a is assumed to be continuous with $a(x) \geq a_0 = $ constant > 0 and b is assumed continuous and nonnegative at all points $x \in [0, 1]$. The data $f = f(x)$ is assumed to be at least piecewise continuous on open subintervals, suffering no worse than simple jump discontinuities at a finite number of points in the domain $\Omega = (0, 1)$. Since f is assumed to be at least piecewise continuous on subintervals, we shall always position an endpoint of an element at points at which f has a simple jump discontinuity. We have at the endpoints x_i between adjacent elements flux continuity,

$$[\![au'(x_i)]\!] = 0 \qquad (4.2.6)$$

where $[\![\cdot]\!]$ designates the jump. Since the coefficient a is continuous in this example, (4.2.6) implies that u' is continuous throughout the domain $\Omega = (0, 1)$. Hence, a C^1-approximation of u is required.

We begin by introducing a partition of N subdomains $\bar{\Omega}_e = [x_{i-1}, x_i]$, over which a space of C^1-finite elements is to be constructed, the C^1-continuity being maintained by using a piecewise-polynomial Hermite basis of the form described in Chapter 2. For example, the Hermite cubic approximation on the master element $\hat{\Omega} = [-1, 1]$ has the form

$$\hat{u}_e(\xi) = \sum_{j=1}^{2} \left[\hat{u}_j^e \hat{\psi}_j^0(\xi) + \frac{d\hat{u}_j^e}{d\xi} \hat{\psi}_j^1(\xi) \right] \qquad (4.2.7)$$

where $\hat{\psi}_j^0$ and $\hat{\psi}_j^1$ are the Hermite cubics on $-1 \leq \xi \leq 1$, and $j = 1, 2$ are local nodal indices. The linear transformation $\xi = [2x - (x_{i-1} + x_i)]/h_e$, with $h_e = x_i - x_{i-1}$, maps $\bar{\Omega}_e$ to $\hat{\Omega}$ and transforms $\hat{u}_e(\xi)$ in (4.2.7) to $u_h^e(x)$ as in our standard Galerkin formulation.

We next transform the governing equation to ξ coordinates, using the chain rule to write, for $x \in \Omega_e$,

$$-\left[\hat{a}(\xi)\left(\frac{2}{h_e}\right)\hat{u}_\xi(\xi) \right]_\xi^e \left(\frac{2}{h_e}\right) + \hat{b}_e(\xi)\hat{u}_e(\xi) = \hat{f}_e(\xi), \qquad \xi \in \hat{\Omega} \qquad (4.2.8)$$

where $(\cdot)_\xi \equiv d(\cdot)/d\xi$. Now we introduce (4.2.7) into (4.2.8) to obtain the element residual,

$$\hat{r}_e(\xi) = -\frac{4}{h_e^2}[\hat{a}(\xi)\hat{u}_\xi(\xi)]_\xi^e + \hat{b}_e(\xi)\hat{u}_e(\xi) - \hat{f}_e(\xi) \qquad (4.2.9)$$

To collocate at a point $\xi_c \in \hat{\Omega}$ means to set $\hat{r}_e(\xi_c)$ equal to zero, and this yields the equation

$$-\frac{h_e^2}{4}[\hat{a}(\xi_c)\hat{u}_\xi(\xi_c)]_\xi^e + \hat{b}_e(\xi_c)\hat{u}_e(\xi_c) - \hat{f}_e(\xi_c) = 0 \qquad (4.2.10)$$

Upon examining the finite element discretization, we observe that there are two degrees of freedom, u_j and du_j/dx, at each node, giving a total of $2(N + 1)$ degrees of freedom. At the boundaries $x = 0$ and $x = 1$ we have conditions (4.2.5), which imply the two equations

$$\left.\begin{array}{llll} \alpha_0 u_1' + \beta_0 u_1 = \gamma_0 & \text{for node 1} & \text{at} & x = 0 \\ \alpha_1 u_{N+1}' + \beta_1 u_{N+1} = \gamma_1 & \text{for node } N+1 & \text{at} & x = 1 \end{array}\right\} \quad (4.2.11)$$

Hence we must add $2N$ collocation equations, two per element, to complete the specification of the finite element collocation system. Setting $c = 1, 2$ in (4.2.10) yields a 2×4-element matrix contribution to the final system.

The following points are noteworthy:

1. Since the approximation is only C^1 globally and the residual r involves second derivatives, the collocation points ξ_c must be in the *interior* of an element.

2. No element quadratures are required, so that element calculations reduce to simply evaluating functions at the collocation points (no quadrature sums).

3. Since the finite element basis functions are C^1-functions, the element complexity and the required smoothness of the approximate solution are increased compared with standard Galerkin finite element methods for this problem.

4. Even though the operator in the governing equation is self-adjoint, the symmetry is lost in the discrete formulation.

5. The assembled system involves the solution derivatives at the nodes as additional degrees of freedom and the system matrix has a block tridiagonal form comprised of 2×2 blocks.

6. C^1-Hermite collocation approximation of higher degree can be constructed by using modified Hermite interpolation, in which additional function values are interpolated in the element interior.

7. Smooth approximations that are globally C^m can be obtained by interpolating higher derivatives $u'', \ldots, u^{(m)}$ at the interface nodes between elements.

8. Results of superior accuracy may be achieved if the collocation points are taken as the roots of appropriate orthogonal polynomials such as the Legendre polynomials. We discuss this point in greater detail under the topic of superconvergence in Chapter 5.

Now that we have seen the basic technique, let us examine some underlying mathematical properties. In particular, we seek to investigate the relationship to the variational and weighted-residual statements. Consider again problem (4.2.4) and, for convenience, let us take $u(0) = u(1) = 0$ as boundary conditions. In the usual variational formulation of this two-point problem, we seek $u \in H_0^1(0, 1)$ such that

$$\int_0^1 (au'v' + buv - fv)\, dx = 0 \qquad (4.2.12)$$

for all test functions v in $H_0^1(0, 1)$. Here trial and test functions have the same regularity. We can integrate by parts in the variational statement to recover the differential equation and obtain, at least symbolically, the weighted residual equation

$$\langle r, v \rangle \equiv \int_0^1 rv\, dx = 0 \qquad \forall v \in H_0^1(0, 1) \qquad (4.2.13)$$

where $\langle \cdot, \cdot \rangle$ denotes duality pairing on $H^{-1}(0, 1) \times H_0^1(0, 1)$, $H^{-1}(0, 1)$ being the dual of $H_0^1(0, 1)$. The standard variational formulation of this problem thus requires that the residual be a linear functional on $H_0^1(0, 1)$ and, therefore, be in $H^{-1}(0, 1)$ and that

$$r = 0 \quad \text{in} \quad H^{-1}(0, 1)$$

The collocation method we have described requires more smoothness of the solution than in the variational method. We can still, however, express collocation in a weighted-residual form similar to (4.2.13) by simply reversing the choice of spaces for r and v. Again let $\langle \cdot, \cdot \rangle$ denote duality pairing on $H^{-1}(0, 1) \times H_0^1(0, 1)$. Then if x_c is a collocation point in $(0, 1)$, the test function v must be such that

$$\langle v, r \rangle = r(x_c) = 0 \qquad (4.2.14)$$

where $v \in H^{-1}(0, 1)$ and $r \in H_0^1(0, 1)$. This is precisely the definition of the Dirac delta at x_c, which we write symbolically as

$$v(x) = \delta(x - x_c) \qquad \text{or} \qquad \int_0^1 vr\, dx = r(x_c) \qquad (4.2.15)$$

Thus, the residual r is now regarded as a smooth (continuous) function in $H_0^1(0, 1)$, whereas v is a linear functional in $H^{-1}(0, 1)$. These observations indicate that the collocation method described above can be regarded as a special weighted-residual method in which the test functions v are Dirac deltas associated with the collocation points. Since the residual involves derivatives of u of at most second order, this implies that $u \in H^3(0, 1)$ $\cap H_0^1(0, 1)$ and hence, as noted earlier, we should have $u_h \in C^2(0, 1)$. Actually, this condition is slightly stronger than we need and we can take u_h to be C^2 in the interior of each element and only C^1 globally provided that we do not try to collocate at the element boundaries. In practice, since most finite elements employ polynomials, the trial function u_h is C^∞ in the interior of each element and C^1 globally.

Although it may appear that C^1 continuity is essential to the method, some qualifying remarks are warranted. The C^1 continuity condition arose from the requirement $[\![\sigma]\!] = [\![au']\!] = 0$, where σ can be identified as the flux. In the variational formulation we saw that this condition holds as a natural boundary condition across any real or artificial interface and is a direct result of the integration-by-parts procedure intrinsic to the standard variational problem. The collocation method, however, does not automatically provide that this condition hold since there is no related integration-by-parts step. Instead, we must enforce the flux balance as an additional condition on the solution. This can be done by embedding the constraint (4.2.6) in the basis as in the present case (since a is continuous) or by alternative means such as the collocation-Galerkin method (Section 4.3).

In the standard Galerkin finite element method, discontinuous coefficients are easily treated by requiring that the element boundary be coincident with a point where the discontinuity occurs. The natural boundary condition then enforces flux continuity at this point. From the preceding discussion, we see then that this implies a restriction on C^1-collocation methods that must be considered in general-purpose finite element collocation programs. Alternative methods can be devised that avoid this difficulty and use the more standard Lagrange bases. For example, the Lagrange cubics can be used as shape functions in the previous example with, instead of (4.2.7),

$$\hat{u}_e(\xi) = \sum_{j=1}^{4} u_j^e \hat{\psi}_j(\xi) \tag{4.2.16}$$

where $\{\hat{\psi}_j(\xi)\}$ are the Lagrange cubic shape functions. For the boundary conditions we obtain, in place of (4.2.11),

$$\left. \begin{aligned} \alpha_0 \sum_{j=1}^{4} u_j^1 (\hat{\psi}_j)_\xi \frac{2}{h_1} + \beta_0 u_1^1 &= \gamma_0 \\ \alpha_1 \sum_{j=1}^{4} u_j^N (\hat{\psi}_j)_\xi \frac{2}{h_N} + \beta_1 u_4^N &= \gamma_1 \end{aligned} \right\} \tag{4.2.17}$$

at $\xi = -1$ and $\xi = +1$ for elements 1 and N, respectively. Two collocation equations such as (4.2.10) again apply in the interior of each element.

Finally, the flux balance at each interface node i between adjacent elements e and $e + 1$ implies that

$$\lim_{x \to x_i^-} a(x)u_e'(x) = \lim_{x \to x_i^+} a(x)u_{e+1}'(x) \tag{4.2.18}$$

which provides an additional $N - 1$ discrete equations to complete the finite element system (Carey and Finlayson [1975]).

4.2.2 Higher Dimensions

The collocation finite element method can be developed for certain classes of two- and three-dimensional problems in a manner similar to that described for one dimension. The main complications that are encountered arise from the flux continuity which is to hold across entire element boundaries and implies restrictions on the shape of the mesh. We shall assume here that the coefficients in the governing equation are smooth, so that the requirement that the flux be continuous reduces once again to the requirement that the admissible functions be globally C^1. Here we sketch only the main formulative steps for a tensor-product Hermite C^1-element in two dimensions.

The natural generalization of the two-point problem (4.2.4)–(4.2.5) is the elliptic boundary-value problem on $\Omega \subset \mathbb{R}^2$:

$$-\nabla \cdot [a(\mathbf{x}) \, \nabla u(\mathbf{x})] + b(\mathbf{x})u(\mathbf{x}) = f(\mathbf{x}) \quad \text{in} \quad \Omega$$

with

$$\alpha(s) \frac{\partial u}{\partial n}(s) + \beta(s)u(s) = \gamma(s) \quad \text{on} \quad \partial\Omega \tag{4.2.19}$$

where the coefficient functions a and b are here assumed to be smooth functions and $a \geq a_0 = \text{const.} > 0$ on $\overline{\Omega}$; $u(s) = u[\mathbf{x}(s)]$ where s parametrically defines $\partial\Omega$. For illustrative purposes we take the domain Ω to be square and $\alpha = 0$, $\beta = 1$, $\gamma = 0$ in (4.2.19), so that we have simply homogeneous Dirichlet data $u = 0$ on $\partial\Omega$.

We described the general form of C^1-families of finite elements in Chapter 2. The simplest C^1-element in two dimensions is the rectangle with tensor-product cubic Hermite basis and we limit the discussion to this case. On the master square defined by $-1 \leq \xi \leq 1$, $-1 \leq \eta \leq 1$, the 16 tensor-product Hermite shape functions are

$$\{\hat{\psi}_{ij}^{kl}(\xi, \eta)\} = \{\hat{\psi}_i^k(\xi)\hat{\psi}_j^l(\eta)\} \tag{4.2.20}$$

where $\hat{\psi}_i^k(\xi)$, $\hat{\psi}_j^l(\eta)$ are Hermite cubics in one dimension for nodes $i = 1, 2$, $j = 1, 2$, and derivative order $k = 0, 1$, $l = 0, 1$. The earlier linear transformation applied to the ξ and η variables determines the shape functions $\{\psi_{ij}^{kl}(x, y)\}^e$ on Ω_e. The associated nodal degrees of freedom are the four values u, u_x, u_y, and u_{xy} at each corner of Ω_e.

Using the chain rule, the element residual may be defined on the master square $\hat{\Omega}$:

$$\hat{r}(\xi, \eta) = -[(\hat{a}\hat{u}_\xi)_\xi \xi_x^2 + (\hat{a}\hat{u}_\eta)_\eta \eta_y^2] + \hat{b}\hat{u} - \hat{f} \tag{4.2.21}$$

For convenience, let us assume a mesh of square elements so that $\xi_x = \eta_y = 2/h$. Collocating at an interior point (ξ_c, η_c) the collocation equations on Ω_e are

$$\hat{r}(\xi_c, \eta_c) = 0$$

or

$$\left\{ -\frac{4}{h^2}[(\hat{a}\hat{u}_\xi)_\xi + (\hat{a}\hat{u}_\eta)_\eta] + \hat{b}\hat{u} - \hat{f} \right\}_{(\xi_c, \eta_c)} = 0 \qquad (4.2.22)$$

where, on element $\hat{\Omega}$ from (4.2.20)

$$\hat{u}(\xi, \eta) = \sum_{i,j=1}^{2} \sum_{k,l=0}^{1} (\hat{u}_{ij}^{kl})^e \hat{\psi}_{ij}^{kl}(\xi, \eta) \qquad (4.2.23)$$

with

$$\hat{u}_{ij}^{00} = (\hat{u})_{ij}, \ \hat{u}_{ij}^{10} = (\hat{u}_\xi)_{ij}, \ u_{ij}^{01} = (\hat{u}_\eta)_{ij}, \ \hat{u}_{ij}^{11} = (\hat{u}_{\xi\eta})_{ij}$$

and $(u_x)_{ij} = (\hat{u}_\xi)_{ij}\xi_x$, etc.

For one dimension we took two interior collocation points per element. In two dimensions the tensor product of the one-dimensional collocation points is employed; that is, four interior collocation points (ξ_c, η_c), $c = 1, 2, 3, 4$, per element. For example, let our square domain Ω contain a mesh of N^2 square elements. There are $(N + 1)^2$ nodes and thus $4(N + 1)^2$ degrees of freedom. Since u is given on $\partial\Omega$, the tangential derivative on the sides and cross derivatives at the corners are known. This provides $8N + 4$ equations and the remaining $4N^2$ equations are obtained by collocation in the interior of each of the N^2 elements.

The resulting finite element system of equations is sparse but not symmetric, even though the given differential operator is self-adjoint. Note that the element matrices for C^1-collocation are not square: for example, the Hermite bicubics in the example above lead to a 4×16 element matrix. In practice the system can still be constructed by combining (assembling) element contributions in a manner similar to that for Galerkin finite element methods. Thus the general design of a finite element collocation program is essentially the same as the Galerkin program, as far as overall structure, logic, and sequence of major computations are concerned. The main differences appear in the element routines, the absence of numerical integration, and the choice of an unsymmetric solver. The method has been examined for prototype problems in one and two dimensions but in an exploratory mode. Some innovative mapping strategies and alternating direction solution techniques have been developed by Hayes [1980] for both Galerkin and collocation methods.

4.2.3 Collocation Points and Estimates

We have not yet considered where the collocation points should be located in an element. Perhaps the most logical choice is equispaced collocation; this is certainly convenient to program, but it does not yield results of optimal accuracy. Optimal results (in the usual L^2- and H^1-error norms) are obtained if special points are selected as collocation points. This was first observed for global collocation methods (see e.g., Finlayson [1972] and Villadsen and Michelsen [1977]) and also applies for finite element collocation (see, e.g., Douglas and Dupont [1974]).

The special collocation points correspond to the zeros of an associated family of orthogonal polynomials and, accordingly, the method has been termed "orthogonal collocation." This optimal choice of collocation points can best be explained by introducing a discrete inner product in an equivalent variational formulation. In this way we shall see that the Gauss-point collocation solution corresponds to the solution of the discrete-quadrature Galerkin equations. We can then appeal directly to the standard error estimates for the Galerkin method, to deduce the optimal global estimates for Gauss-point collocation. Next we examine the choice of collocation points for global polynomial collocation, as the notation is simpler. The ideas are then easily generalized to orthogonal collocation on finite elements.

Let us first summarize some important preliminary ideas on numerical quadrature which will lead to the discrete inner product that is of central importance in the argument. We consider the interval of interest to be $[-1, 1]$, since this is the domain of the master element interval in finite element calculations and the usual domain for which Gauss weights and points are tabulated. By the simple change of variable $\xi = [2x - (a + b)]/(b - a)$ we can map an interval $[a, b]$ to $[-1, 1]$.

The Legendre polynomial of degree n on $[-1, 1]$ is defined by Rodrigues' formula as

$$\mathcal{L}_n(\xi) = \frac{1}{2^n(n!)} \frac{d^n}{d\xi^n}[(\xi^2 - 1)^n]; \qquad n = 0, 1, 2, \ldots \qquad (4.2.24)$$

and has n simple roots $-1 < \zeta_1 < \cdots < \zeta_n < 1$. Since the Legendre polynomials are the family of polynomials on $[-1, 1]$ that are orthogonal in an L^2 sense, with

$$\int_{-1}^{1} \mathcal{L}_n \mathcal{L}_s \, d\xi = \begin{cases} 0 & \text{if } n \neq s \\ \dfrac{2}{2n + 1} & \text{if } n = s \end{cases} \qquad (4.2.25)$$

it follows that

$$\int_{-1}^{1} \mathcal{L}_n q \, d\xi = 0 \qquad (4.2.26)$$

176

for all polynomials q of degree less than n. This property is essential to the derivation of the common Gaussian (Gauss–Legendre) quadrature formula.

The weights w_j of the Gaussian quadrature formula are chosen so that the quadrature rule of order k is exact for polynomials of degree $2k - 1$ or less. That is, if $p \in P_{2k-1}[-1, 1]$, where $P_{2k-1}[-1, 1]$ is the class of polynomials on $[-1, 1]$ of degree $2k - 1$ or less, then

$$\int_{-1}^{1} p(\xi)\, d\xi = \sum_{j=1}^{k} w_j p(\zeta_j) \qquad (4.2.27)$$

where ζ_j are the zeros of the Legendre polynomial of degree k on $[-1, 1]$, $-1 < \zeta_1 < \cdots < \zeta_k < 1$. It is the quadrature property (4.2.27) that leads to the selection of the Gauss points ζ_j as collocation points if optimal global accuracy is to be attained. The key concept here and in the related super-convergence results of Chapter 5 is the derivation of a discrete inner product using (4.2.27) and its relationship to the usual L^2 inner product on $[-1, 1]$.

Using (4.2.27) we define the discrete inner product of two continuous functions f and g on the master element $[-1, 1]$ as

$$(f, g)_d = \sum_{j=1}^{k} w_j f(\zeta_j) g(\zeta_j) \qquad (4.2.28)$$

where $w_j > 0$ are the Gauss weights, ζ_j are the Gauss points, and $(\cdot, \cdot)_d$ denotes the discrete inner product. The discrete norm $|\cdot|_d$ follows from (4.2.28) as

$$|f|_d^2 = (f, f)_d \qquad (4.2.29)$$

If the product fg in (4.2.28) is a polynomial in x of degree $2k - 1$ or less, then $(f, g)_d \equiv (f, g)$, where $(\cdot, \cdot)$ denotes the usual L^2 inner product. We can use the formulas for Gaussian quadrature error to consider integrands fg that are not polynomials of degree $2k - 1$ or less. For example, if $F = fg$ is the integrand, then the Gaussian quadrature error is (Ralston [1965])

$$E = \frac{2^{2k}(k!)^4}{[(2k)!]^3} \frac{d^{2k}F(\eta)}{d\xi^{2k}} \int_{-1}^{1} \mathcal{L}_k^2(\xi)\, d\xi, \qquad \eta \in (-1, 1)$$

whence, from (4.2.25),

$$E = \frac{2^{2k+1}(k!)^4}{(2k + 1)[(2k)!]^3} \frac{d^{2k}F(\eta)}{d\xi^{2k}}, \qquad \eta \in (-1, 1) \qquad (4.2.30)$$

We then have a general relationship between L^2 and discrete inner products of the form

$$(f, g) = (f, g)_d + E \qquad (4.2.31)$$

where $(f, g)_d$ is given in (4.2.28) and E in (4.2.30). We emphasize that for a k-point quadrature, this result is the best possible, E being zero for polynomials of degree $2k - 1$ or less and small (of high order) otherwise. This idea is important, as we shall subsequently replace L^2 inner products in the Galerkin formulation by discrete inner products $(\cdot, \cdot)_d$ as part of our argument.

Having described the properties of the Gauss-quadrature formula and applied it to define a discrete inner product, let us now use this to establish the equivalence between collocation and a discrete quadrature Galerkin method. We consider again the standard self-adjoint two-point problem (4.2.4) with homogeneous Dirichlet data and regular coefficients. In the Galerkin method, the variational statement applies to trial and test functions in $H_0^1(\Omega)$ and, as noted earlier, this is equivalent to a duality pairing (weighted residual variational statement) on $H^{-1}(\Omega) \times H_0^1(\Omega)$. Since the data of the problem are regular and the approximation basis is sufficiently smooth, the residual r is in L^2 and the duality pairing can be replaced by the L^2 inner product. That is, the Galerkin finite element solution is also the solution of

$$((-au_h')' + bu_h - f, v_h) = 0 \qquad (4.2.32)$$

for admissible test functions $v_h \in H^h \subset H_0^1(\Omega)$.

Introducing the quadrature formula (4.2.31) and writing r_h for $(-au_h')' + bu_h - f$, we have

$$(r_h, v_h)_d + E = 0 \qquad (4.2.33)$$

Since the functions a, b, and f in r_h are smooth, the quadrature error E in (4.2.33) is small $[O(h^{2k})$ for a finite element formulation]. Accordingly, let us replace (4.2.33) by the corresponding "nearby" expression involving only the discrete inner product using Gaussian quadrature,

$$(r_h, v_h)_d = 0 \qquad \forall v_h \in H^h \qquad (4.2.34)$$

That is,

$$\sum_{j=1}^{k} w_j r_h(\zeta_j) v_h(\zeta_j) = 0 \qquad \forall v_h \in H^h \qquad (4.2.35)$$

is the discrete quadrature approximation to the Galerkin method. If u_h is the solution to the actual Galerkin system (4.2.32) and $\bar{u}_h$ is the solution to the quadrature Galerkin system (4.2.34), the error due to quadrature satisfies* (Douglas and Dupont [1974])

$$\|u_h - \bar{u}_h\|_0 \leq C \|f\|_{k+1} h^{k+1} \qquad (4.2.36)$$

---

* In fact, at the end nodes, $u_h(x_i) - \bar{u}_h(x_i) = O(h^{2k})$, a result essential to the study of superconvergence in Chapter 5.

Since a similar rate holds for the error $\|u - u_h\|_0$, we have

$$\begin{aligned} \|u - \bar{u}_h\|_0 &= \|u - u_h + u_h - \bar{u}_h\|_0 \\ &\leq \|u - u_h\|_0 + \|u_h - \bar{u}_h\|_0 \qquad (4.2.37) \\ &\leq C\|f\|_{k+1} h^{k+1} \end{aligned}$$

Thus, the discrete quadrature Galerkin method exhibits the optimal rate of convergence.

Now using the discrete formula (4.2.35), since this must hold for *all* $v_h \in H^h$, it follows that

$$r_h(\zeta_j) = 0, \qquad j = 1, 2, \ldots, k \qquad (4.2.38)$$

and this is precisely the system for Gauss-point collocation.* Thus we have a direct equivalence between the solutions from the Gauss-point collocation and Gauss-quadrature Galerkin methods. Since the Galerkin method yields optimal L^2 and H^1 rates of convergence, it follows from this equivalence that the Gauss point collocation method has the same property.

We remark that there are, however, other classes of problems and methods for which different orthogonality properties are needed to obtain an appropriate discrete inner product and hence optimal accuracy. For example, the optimal collocation points for the C^0-collocation-Galerkin method in the next section are the Jacobi points—the zeros of the Jacobi polynomials. Furthermore, when orthogonal collocation is applied, we obtain much greater pointwise accuracy of the solution at the nodes. We discuss this and related material under the general subject of *superconvergence* in Chapter 5.

As an example to indicate the improvement in rate of convergence using Gauss-point collocation, we consider the model two-point problem

$$-u''(x) + u(x) = x, \qquad 0 < x < 1$$

with

$$u(0) = 0, \qquad u(1) = 0$$

Numerical results for the error in H^0 and H^1 norms are given in Tables 4.1 and 4.2 and plotted in Figs. 4.1 and 4.2 for meshes with $h = 1, \frac{1}{2}, \ldots, \frac{1}{64}$. The results are for piecewise-cubic C^1-elements and the rates are given by the

---

* As an alternative argument we note that a solution of the collocation system exists and satisfies (4.2.38), so (4.2.35) follows. Uniqueness for the discrete Galerkin problem can be shown under the given assumptions on the data (Douglas and Dupont [1974]) and the equivalence of (4.2.35) and (4.2.38) follows.

TABLE 4.1
Errors for Equispaced Collocation

h	$\|e\|_0$	$\|e\|_1$
1	5.866×10^{-4}	2.776×10^{-3}
$\frac{1}{2}$	3.433×10^{-4}	1.184×10^{-3}
$\frac{1}{4}$	9.551×10^{-5}	3.231×10^{-4}
$\frac{1}{8}$	2.447×10^{-5}	8.243×10^{-5}
$\frac{1}{16}$	6.154×10^{-6}	2.071×10^{-5}
$\frac{1}{32}$	1.540×10^{-6}	5.184×10^{-6}
$\frac{1}{64}$	3.853×10^{-7}	1.296×10^{-6}
	(Rate = 2)	(Rate = 2)

TABLE 4.2
Errors for Gauss-Point Collocation

h	$\|e\|_0$	$\|e\|_1$
1	9.994×10^{-4}	3.578×10^{-3}
$\frac{1}{2}$	6.458×10^{-5}	4.477×10^{-4}
$\frac{1}{4}$	4.062×10^{-6}	5.582×10^{-5}
$\frac{1}{8}$	2.542×10^{-7}	6.972×10^{-6}
$\frac{1}{16}$	1.590×10^{-8}	8.713×10^{-7}
$\frac{1}{32}$	9.936×10^{-10}	1.089×10^{-7}
$\frac{1}{64}$	6.204×10^{-11}	1.361×10^{-8}
	(Rate = 4)	(Rate = 3)

slopes in Figs. 4.1 and 4.2. We note from the experiment that for equispaced collocation,

$$\|e\|_0 = O(h^2) \quad \text{and} \quad \|e\|_1 = O(h^2)$$

whereas for Gauss-point collocation in agreement with (4.2.37),

$$\|e\|_0 = O(h^4) \quad \text{and} \quad \|e\|_1 = O(h^3)$$

In two dimensions the collocation points may be chosen at the tensor-product Gauss points (ξ_i, η_j), for $i, j = 1, 2, \ldots, N$ on the master square, where ξ_i and η_j are the one-dimensional Gauss points in each of the coordinate directions. Error estimates for C^1-collocation of the second-order equation (4.2.19) with homogeneous Dirichlet data on a square domain have been

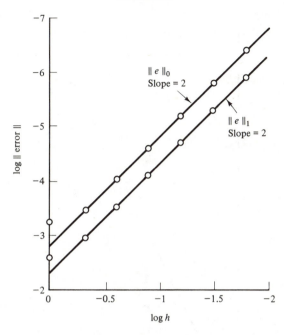

FIGURE 4.1 *Log-log plots of* $\|e\|_0$ *and* $\|e\|_1$ *against mesh size h for equispaced collocation.*

derived by Prenter and Russell [1976] and under less stringent assumptions by Percell and Wheeler [1981] and are of the following form: if $u \in H^{k+2}(\Omega)$, $k \geq 3$, then for a and b in $C^\infty(\bar{\Omega})$ with a strictly positive on Ω, the error estimates for $e = u - u_h$ are

$$\|\nabla e\|_0 \leq Ch^k \|u\|_{k+2} \tag{4.2.39}$$

and if $u \in H^{k+3}(\Omega)$,

$$\|e\|_0 \leq Ch^{k+1} \|u\|_{k+3} \tag{4.2.40}$$

where $\|\cdot\|_0$ is the L^2-norm, h is the mesh size, and k is the polynomial degree of the element shape functions. For example, if tensor-product cubics are used for a uniform mesh on a square as described in Section 4.2.2, the error estimates in the L^2-norm are $O(h^3)$ and $O(h^4)$ for derivative and solution, respectively.

The estimates above have been obtained for the stated class of problems and a square domain. The method is limited in the extent to which it can be applied to less regular domains because of restrictions on the global continuity of the map. This, together with the complexity of the C^1-elements and the smoothness requirement on the coefficient a, does not favor general use of finite element collocation in two or three dimensions. In the next section we

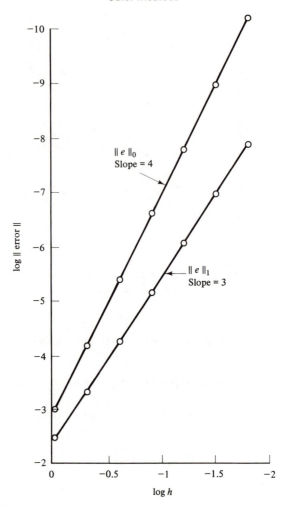

FIGURE 4.2 *Log-log plots of $\|e\|_0$ and $\|e\|_1$ against mesh size h for Gauss-point collocation.*

describe a method that combines the features of the collocation and Galerkin methods to circumvent some of these problems.

EXERCISES

4.2.1 Use quadratic and cubic Lagrange interpolation and global collocation to determine approximations to the solution of equation (4.2.1). Vary the choice of collocation points using random, equispaced, and Gauss-point locations. Graph the error on [0, 1] in each instance and discuss the results.

4.2.2   Set up the finite element collocation system of equations (4.2.7)–(4.2.9) for the simple example of equation (4.2.1). Use two elements ($h = \frac{1}{2}$) and form the linear system. Solve and compare your results with those of Exercise 4.2.1.

4.2.3   Sketch the block-tridiagonal form of the finite element system for C^1-collocation in one dimension using Hermite cubics.

4.2.4   Develop a modified Hermite element for C^1-collocation in one dimension using element polynomials of degree $k \geq 3$. Sketch the structure of the resulting sparse matrix problem. Carry out a similar formulation for an element basis interpolating m nodal derivatives $u^{(j)}, j = 0, 1, \ldots, m - 1$.

4.2.5   Graph the cubic B-splines: $\{s_i(x)\} = \{\alpha(x)(x - x_{i-2}), \quad x \in \Omega_{i-1};$ $h^3\beta(x)(x - x_{i-1})/h, x \in \Omega_i; h^3\beta(x)(x_{i+1} - x)/h, x \in \Omega_{i+1}; \alpha(x)(x_{i+2} - x),$ $x \in \Omega_{i+2}, 0$, otherwise; for $2 \leq i \leq N - 1$ where $\alpha(x) = x^3, \beta(x) = 1 + 3x + 3x^2 - 3x^3$. Extend the partition outside $[0, 1]$ to include $i = -1, N + 2$, obtaining a B-spline basis for approximation. Modify the basis to: $\bar{s}_0(x) = s_0(x) - 4s_{-1}(x); \bar{s}_1(x) = s_1(x) - s_{-1}(x); \bar{s}_i(x) = s_i(x), 2 \leq i \leq N + 1; \bar{s}_N(x) = s_N(x) - s_{N+2}(x); \bar{s}_{N+1}(x) = s_{N+1}(x) - 4s_N(x)$. Use these cubic splines in each of the Galerkin and collocation methods and determine the form of the finite element systems.

4.2.6   Use Lagrange cubics and the limit condition (4.2.18) to determine an alternative form of finite element collocation for the introductory example of equation (4.2.1). Write down the element matrix for this example. If the coefficient of u'' is 1 on $0 < x < \frac{1}{2}$ and $\frac{1}{2}$ on $\frac{1}{2} < x < 1$, write down the flux balance equation (4.2.18) for interface nodes in $0 < x < \frac{1}{2}, \frac{1}{2} < x < 1$ and at $x = \frac{1}{2}$.

4.2.7   The Gauss points for the master square and Hermite bicubics are located at $(\pm 1/\sqrt{3}, \pm 1/\sqrt{3})$. Use these collocation points to determine the collocation equations for an approximation of Laplace's equation.

4.2.8   Describe how the collocation procedure can be applied for Laplace's equation on an "L-shaped" region with prescribed Dirichlet data.

4.3   *C⁰-COLLOCATION-GALERKIN METHODS*

In Section 4.2 we dealt primarily with C^1-collocation methods. There it was observed that the C^1-condition is a natural requirement stemming from the conservation principle and the condition that the (constitutive) coefficient function a be continuous. In more general applications, it is continuity of the flux $\sigma = -au'$ that must be enforced across the interfaces between elements. We saw one approach for including this condition using C^0-Lagrange elements in (4.2.18). In this section we shall discuss another variant of the collocation method in which the flux continuity at the interface is treated in a fashion akin to Galerkin methods and with C^0-Lagrange elements. For this reason these schemes have been termed C^0-collocation-Galerkin methods (see, e.g., Wheeler [1977] and Diaz [1977]).

4.3.1 Two-Point Problem

Let us begin by developing the method for the class of two-point problems in (4.2.4)–(4.2.5). For simplicity, and without loss of generality, let us assume homogeneous Dirichlet data so that the boundary-value problem becomes

$$-[a(x)u'(x)]' + b(x)u(x) = f(x) \quad \text{in} \quad 0 < x < 1 \left.\right\}$$

with

$$u(0) = u(1) = 0 \qquad (4.3.1)$$

The standard variational statement of the problem is to seek $u \in H_0^1(0, 1)$ such that

$$\langle r, v \rangle = \int_0^1 (au'v' + buv - fv) \, dx = 0 \qquad (4.3.2)$$

for all $v \in H_0^1(0, 1)$.

Next we define a discretization of the domain consisting of N elements and construct a finite element space H^h of test functions using a C^0-Lagrange basis. The finite element approximation $u_h \in H^h \subset H_0^1(0, 1)$ is of the usual form,

$$u_h(x) = \sum_{j=1}^{M} u_j \phi_j(x) \qquad (4.3.3)$$

where $\{\phi_i(x)\}$, $i = 1, 2, \ldots, M$, are the global basis functions. Now let $\{x_s\}$ correspond to the $N - 1$ interface nodes between adjacent elements in the discretization and let $\{\bar{\phi}_s\}$ denote the familiar piecewise-linear C^0 basis functions. The test functions in the variational statement (4.3.2) are in $H_0^1(0, 1)$. For the discrete problem with u_h given by (4.3.3), we select $v = \bar{\phi}_s$ for each s to obtain a set of $N - 1$ equations associated with the $(N - 1)$ interface nodes,

$$\sum_{j=1}^{M} \left[\int_0^1 (a\phi_j' \bar{\phi}_s' + b\phi_j \bar{\phi}_s) \, dx \right] u_j = \int_0^1 f\bar{\phi}_s \, dx \qquad (4.3.4)$$

for all $s = 1, \ldots, N - 1$. The restrictions of ϕ_j and $\bar{\phi}_s$ to Ω_e are the corresponding shape functions, here denoted ψ_j^e and $\bar{\psi}_s^e$. Since $\bar{\psi}_s^e$ is linear, the contributions to (4.3.4) from element Ω_e and interface node s are

$$\sum_{j=1}^{N_e} \left\{ \int_{\Omega_e} [a(\psi_j^e)'(\bar{\psi}_s^e)' + b\psi_j^e \bar{\psi}_s^e] \, dx \right\} u_j^e \left.\right\}$$

and

$$\int_{\Omega_e} f\bar{\psi}_s^e \, dx \qquad (4.3.5)$$

for the coefficient matrix and right-hand side, respectively. Here $j = 1, 2, \ldots,$ N_e are local node numbers for element Ω_e and $s \equiv 1, N_e$ identify the end nodes.

We deduce the remaining equations for collocation from (4.3.2) as follows. Let $D_N(0, 1)$ denote the subspace of the space $H_0^1(0, 1)$ of test functions v such that

$$v \in C^\infty(0, 1), \qquad v(0) = v(1) = 0,$$

$$v(x_s) = 0, \qquad \text{interface node } s$$

Then the residual r satisfies formally,

$$\langle r, v \rangle = \int_0^1 [-(au')' + bu - f]v \, dx$$

$$= \sum_{e=1}^N \int_{\Omega_e} [-(au')' + bu - f] v \, dx = 0 \qquad \forall v \in D_N(0, 1)$$

As in the collocation methods discussed earlier, we assume that r is a smooth function and that v is a linear functional on r. We choose v to be Dirac deltas, as in (4.2.14) and (4.2.15), and collocate the approximate residual $r_h = -(au_h')' + bu_h - f$ at $N_e - 2$ points x_c in the interior of each element. The resulting collocation equations for a typical element Ω_e are simply

$$\sum_{j=1}^{N_e} \{[a(x_c)\psi_j^{e'}(x_c)]' + b(x_c)\psi_j^e(x_c)\}u_j^e = f(x_c) \qquad (4.3.6)$$

for $c = 1, 2, \ldots, N_e - 2$, where local element numbering is again implied.

Equations (4.3.4) and (4.3.6), together with the boundary data, constitute the final algebraic system. The element contributions to these equations can be developed directly as in the Galerkin method. This yields an element matrix of size $N_e \times N_e$ and an element vector of size $N_e \times 1$ that can be assembled to the system in the usual manner. For a mesh of N elements of degree $N_e - 1$, there are $N(N_e - 1) + (N + 1)$ nodes. The boundary conditions supply two equations, the interelement projection (4.3.4) an additional $N - 1$ equations, and collocation at $N_e - 2$ points in the interior of each element provides the final $N(N_e - 2)$ equations completing the system.

In C^1-collocation, the element polynomials are of degree three or greater. The C^0-collocation-Galerkin method can be applied using element polynomials of at least quadratic degree. Clearly, for piecewise linears, this method reverts to the standard Galerkin finite element method.

4.3.2 Higher Dimensions

We consider again the elliptic boundary-value problem of (4.2.19) on a square domain Ω,

$$-\nabla \cdot [a(\mathbf{x}) \nabla u(\mathbf{x})] + b(\mathbf{x})u(\mathbf{x}) = f(\mathbf{x}) \quad \text{in} \quad \Omega \qquad (4.3.7)$$

(with $\mathbf{x} = (x, y)$) and homogeneous Dirichlet data

$$u = 0 \quad \text{on} \quad \partial\Omega \tag{4.3.8}$$

Consider a uniform mesh of N^2 square elements with a tensor product C^0-Lagrange approximation of the form

$$u_h(x, y) = \sum_{i,j=1}^{M} u_{ij}\phi_{ij}(x, y) \tag{4.3.9}$$

where $\{u_{ij} = u(x_i, y_j)\}$, $i = 1, 2, \ldots, M$, $j = 1, 2, \ldots, M$, are the nodal degrees of freedom.

The restriction of (4.3.9) to a typical element Ω_e is

$$u_e(x, y) = \sum_{i,j=1}^{N_e} u_{ij}^e \psi_{ij}^e(x, y) \tag{4.3.10}$$

where the shape functions ψ_{ij}^e on Ω_e are tensor products of polynomials of degree $N_e - 1$ in x and y, respectively.

We collocate at an array of $(N_e - 1)^2$ points in the interior of each element,

$$[-\nabla \cdot (a \nabla u) + bu - f]|_{(x_r^e, y_s^e)} = 0 \tag{4.3.11}$$

for $r = 1, 2, \ldots, N_e - 1$ and $s = 1, 2, \ldots, N_e - 1$.

Consider piecewise-bilinear test functions $\bar{\phi}_{mn}$ associated with each of the corner nodes (x_m, y_n). The variational statement implies that

$$\sum_{i,j=1}^{M} \left[\int_{\Omega} (a \nabla\phi_{ij} \cdot \nabla\bar{\phi}_{mn} + b\phi_{ij}\bar{\phi}_{mn}) \, dx \right] u_{ij} = \int_{\Omega} f\bar{\phi}_{mn} \, dx \tag{4.3.12}$$

for each corner node (m, n).

Note that the piecewise bilinears $\bar{\phi}_{mn}$ correspond to the tensor product of the piecewise-linear test functions $\bar{\phi}_m(x)$ and $\bar{\phi}_n(y)$ introduced for the interface nodal equation in one dimension. Now, however, each interface is a line, so that we now have the additional equations obtained by integrating selectively by parts along the sides in each of the x and y directions. For example, if Ω is the unit square $[0, 1] \times [0, 1]$, we have

$$\int_0^1 a(x, y_n) \frac{\partial u_h}{\partial x}(x, y_n)\bar{\phi}_m'(x) \, dx$$

$$- \int_0^1 \frac{\partial}{\partial y}\left[a(x, y_n) \frac{\partial u_h}{\partial y}(x, y_n) \right]\bar{\phi}_m(x) \, dx \tag{4.3.13}$$

$$+ \int_0^1 b(x, y_n)u_h(x, y_n)\bar{\phi}_m(x) \, dx = \int_0^1 f(x, y_n)\bar{\phi}_m(x) \, dx$$

and

$$-\int_0^1 \frac{\partial}{\partial x}\left[a(x_m, y)\frac{\partial u_h}{\partial x}(x_m, y)\right]\bar{\phi}_n(y)\, dy$$

$$+\int_0^1 a(x_m, y)\frac{\partial u_h}{\partial y}(x_m, y)\phi_n'(y)\, dy \qquad (4.3.14)$$

$$+\int_0^1 b(x_m, y)u_h(x_m, y)\bar{\phi}_n(y)\, dy = \int_0^1 f(x_m, y)\bar{\phi}_n(y)\, dy$$

As an example, let us take piecewise biquadratics. The resulting element has nine nodes: the centroid, four corners, and four midsides. In this case we proceed as follows:

1. Collocation at the centroid with (4.3.11)

2. Galerkin contributions at the corners using bilinear test functions (4.3.12)

3. One-dimensional interface equations on each of the four sides according to (4.3.13) and (4.3.14)

4.3.3 Collocation Points and Estimates

As in the collocation method, there is a special choice of collocation points in the C^0-collocation-Galerkin method that will produce optimal global accuracy in the L^2- and H^1-norms. The reason is again an equivalence between the discrete inner product defined by the collocation method in question and the corresponding weighted L^2-inner product. In the case of the C^1-collocation method, we saw that the corresponding discrete inner product is related to Gaussian quadrature and hence to the Legendre polynomials; the quadrature points are then zeros of the Legendre polynomials and are the collocation points. Similarly, in the C^0-collocation-Galerkin method a discrete inner product can be defined but now it is based on the Jacobi points and Jacobi (or Gauss–Jacobi) quadrature. The relationships between the Jacobi quadrature and L^2-inner product are discussed by Diaz [1977].

The Jacobi polynomials of degree r are the polynomials on $[-1, 1]$ given by the weighted Rodrigues formula (see, e.g., Pearson [1974])

$$J_r^{\alpha, \beta}(x) = \frac{(-1)^r}{2^r r!}\frac{1}{(1-x)^{\alpha}(1+x)^{\beta}}\frac{d^r}{dx^r}[(1-x)^{\alpha+r}(1+x)^{\beta+r}], \qquad (4.3.15)$$

$$r = 0, 1, \cdots$$

For each pair (α, β) with $\alpha > -1$, $\beta > -1$ the Jacobi polynomials form a set that satisfy the weighted orthogonality condition

$$\int_{-1}^{1} (1 - x)^{\alpha}(1 + x)^{\beta} J_r^{\alpha, \beta}(x) x^k \, dx = 0, \qquad 0 \le k \le r - 2 \qquad (4.3.16)$$

For the collocation-Galerkin method, the family with $(\alpha, \beta) = (1, 1)$ determines the appropriate Gauss–Jacobi quadrature and discrete inner product. Then we can express $J_r^{1, 1}$ as

$$J_r^{1, 1}(x) = \prod_{j=1}^{r} (x - \zeta_j), \qquad x \in [-1, 1]$$

and the weighted quadrature formula is

$$\int_{-1}^{1} (1 - x^2) p(x) \, dx = \sum_{j=1}^{r} w_j p(\zeta_j), \qquad p \in P_{2r-3}[-1, 1]$$

where $P_{2r-3}[-1, 1]$ is the set of polynomials of degree $2r - 1$ or less on $[-1, 1]$. The points $\{\zeta_j\}$ are the Jacobi points. By defining a discrete inner product based on this Jacobi quadrature and then an equivalent variational formulation to C^0-collocation-Galerkin, optimal convergence rates can be shown to hold for the specific choice of Jacobi collocation points. For example, on the master element $[-1, 1]$ with cubic approximation $r = 3$, we collocate at the Jacobi points ± 0.447213595499958 and for quartics $r = 4$ we collocate at 0 and ± 0.654653670707977144.*

As a numerical example, consider the singular problem

$$u''(x) + \frac{2}{x} u'(x) = \frac{\sinh 2x}{11 x \sinh 2}, \qquad 0 < x < 1, \quad u'(0) = 0, \quad u(1) = 1$$

The exact solution is

$$u(x) = \frac{1}{11} + \frac{10}{11} \frac{\sinh 2x}{x \sinh 2}$$

Numerical results were computed using C^0-cubics and Jacobi-point collocation. The error at a representative interelement node (at $x = \frac{1}{2}$) for uniform meshes of 2, 4, 8, and 16 elements is 3.88×10^{-6}, 8.9×10^{-8}, 7.9×10^{-9}, and 4×10^{-10}, respectively. We see that the C^0-collocation-Galerkin solution is essentially exact at the nodes, the slight variation in computed accuracy arising from inexact numerical quadrature. Additional numerical experiments for linear and nonlinear problems are described by Carey et al. [1981]. Global error estimates have been derived by Diaz [1977] and by Wheeler

---

* See, for example, Stroud and Secrest [1966] for tabulation of Gauss and Jacobi points.

[1977] for the two-point problem. In the former case, it is shown that if $u \in H^{k+1}(0, 1)$, then for elements of degree k

$$\|e\|_0 + h\|e\|_1 \leq Ch^{k+1}\|u\|_{k+1} \tag{4.3.17}$$

Under more general conditions Wheeler [1977] shows that if $u \in H^{r_e}(\Omega_e)$, $3 \leq r_e$, and element size h_e is sufficiently small, there exists a constant $C > 0$ such that the L^2-error estimate is*

$$\left(\sum_{e=1}^{N} \|e\|_{m,\Omega_e}^2\right)^{1/2} \leq C\left[\sum_{e=1}^{N} (h_e^{\mu_e - m}\|u\|_{\mu_e,\Omega_e})^2\right]^{1/2} \tag{4.3.18}$$

and the L^∞ estimate is

$$\|e\|_{m,\infty,\Omega_e} \leq Ch_e^{\mu_e - m} \max_e \|u\|_{\mu_e,\infty,\Omega_e} \tag{4.3.19}$$

where $\mu_e = \min(k + 1, r_e)$ and $m = 0, 1, \ldots, r$ for $r = \min_e r_e$ and an element of degree k.

On a uniform mesh of size h with $u \in H^r(\Omega)$, $3 \leq k + 1 < r$, the estimate (4.3.18) becomes

$$\|e\|_{m,\Omega} \leq Ch^{k+1-m}\|u\|_{k+1,\Omega} \tag{4.3.20}$$

and, under similar regularity assumptions, the L^∞ estimate (4.3.19) is

$$\|e\|_{m,\infty,\Omega} \leq Ch^{k+1-m}\|u\|_{k+1,\infty,\Omega} \tag{4.3.21}$$

Little analysis of the method has been undertaken for problems in higher dimensions. The following estimate has been derived by Diaz [1979] for Laplace's equation on a square using the C^0-collocation-Galerkin formulation (4.3.11)–(4.3.14): Let $u \in H^s(\Omega)$, $1 \leq s \leq k + 1$; then there exists a constant C independent of h and u such that for sufficiently small h,

$$\|e\|_0 + h\|e\|_1 \leq Ch^s\|u\|_s, \qquad 1 \leq s \leq k + 1$$

The restrictions indicated earlier concerning irregular meshes and discontinuous coefficients can be relaxed with the C^0-collocation method due to the Galerkin type of treatment at the interfaces. For these reasons the method appears more versatile than C^1-collocation. There are, however, several questions related to the implementation of the method that merit attention and are currently being studied. In the next section we consider some other variants of the method. These are the least-squares and H^{-1}-finite element methods.

---

* In fact, Wheeler proves more general Sobolev space estimates, but we have restricted our treatment here to the standard L^2-error estimates.

EXERCISES

4.3.1 Using the map to the master element $\hat{\Omega}$, derive the element matrix and vector contributions on $\hat{\Omega}$ for the C^0-collocation-Galerkin scheme for (4.3.1).

4.3.2 Determine an approximate solution to (4.2.1) by applying the C^0-collocation-Galerkin formulation on a partition of two elements. Use the Jacobi points as collocation points. Also solve $u''(x) = 1$ with $u(0) = u(1) = 0$ in this manner and examine the error.

4.3.3 Develop a C^0-collocation-Galerkin scheme for Laplace's equation in two dimensions with biquadratic approximation on the master square. Consider, in particular, a mesh of four squares for a domain Ω with Dirichlet data prescribed on the boundary.

4.4 H^{-1} AND LEAST-SQUARES METHODS

4.4.1 H^{-1} Methods

In the collocation method described earlier, we begin with a weighted residual statement of the problem and select the trial functions to be smooth [locally C^2 for our second-order problem (4.2.4)] and the test functions to be Dirac deltas. On the other hand, the standard Galerkin method is based on a variational statement of the problem in which we have formally integrated once by parts the weighted-residual statement. That is, one of the derivatives on the trial function u is transferred by an integration by parts to the test function v. The result is that both u and v are in H^1 and functions in both the trial and the test spaces have the same regularity. The idea in the so-called H^{-1} method is to continue this integration-by-parts process one step further, transferring another derivative to the test function v. This implies that v must then be locally C^2 but the trial functions may now be discontinuous. Thus in this sequence of integrations we have progressively eased the smoothness requirements on functions in the trial space while concurrently increasing that required of those in the test space.

To illustrate the procedure indicated above, let us return to the example of the two-point problem in (4.2.4),

$$-[a(x)u'(x)]' + b(x)u(x) = f(x) \qquad 0 < x < 1 \qquad (4.4.1)$$

with Dirichlet data

$$u(0) = u(1) = 0 \qquad (4.4.2)$$

As a first step, we write the usual weighted-residual statement

$$\int_0^1 [-(au')' + bu - f]v\, dx = 0 \qquad (4.4.3)$$

where v is the test function and u is assumed to be sufficiently smooth. We next integrate by parts once with $v(0) = v(1) = 0$ to obtain the problem of finding u satisfying for all such v

$$\int_0^1 (au'v' + buv)\, dx = \int_0^1 fv\, dx \qquad (4.4.4)$$

If we work with (4.4.4), we need only seek a solution $u \in H_0^1(0, 1)$ for all test functions $v \in H_0^1(0, 1)$, this being our standard variational statement.

Continuing, however, we integrate (4.4.4) by parts once more and obtain

$$\int_0^1 [-u(av')' + buv]\, dx = \int_0^1 fv\, dx \qquad (4.4.5)$$

provided that $v' = 0$ at $x = 0$ and $x = 1$. Notice that were we given (4.4.5) initially, we would require only that $u \in L^2(0, 1)$, but then $v \in H_0^2(0, 1)$. We have thus reduced the integrability requirements on the trial space while making those on the test space more stringent.

The finite element approximation of problem (4.4.5) need not be globally continuous. If discontinuous approximations are used, the first derivative is now in H^{-1}, and for this reason the method has been termed the H^{-1} Galerkin finite element method (Rachford and Wheeler [1974]). We then seek an approximate solution $u_h \in H^h \subset L^2(0, 1)$ such that

$$\int_0^1 [-u_h(av_h')' + bu_hv_h]\, dx = \int_0^1 fv_h\, dx \qquad (4.4.6)$$

for test functions $v_h \in T^h \subset H_0^2(0, 1)$. Note that the approximation u_h is not required to satisfy the boundary conditions. If, instead, we specify $u_h = 0$ at the boundary points, the condition on the test space can be relaxed to include functions v having $v' \neq 0$ at the boundary; that is, $v \in H^2(0, 1) \cap H_0^1(0, 1)$.

As noted above, since the trial functions need only be in $L^2(0, 1)$, we can use piecewise polynomials that are discontinuous at the interface nodes between adjacent elements in the finite element discretization. On the other hand, the test functions now must be in $H^2(0, 1)$, so a smoother class of test functions is required than in the standard Galerkin finite element method. For example, we can use piecewise linears that are discontinuous at the end nodes of each element for u_h and C^1-cubic Hermites for v_h in (4.4.6) to define a finite element system.

The choice of discontinuous trial functions is perhaps the most interesting aspect of these methods and, from a practical standpoint, would be appropriate for problems in which the solution exhibits boundary-layer behavior; that is, for cases in which the exact solution is "almost discontinuous."

In Figs. 4.3 through 4.6 we show representative numerical results from Rachford and Wheeler [1974] for the problem (4.2.4) with homogeneous data $u(0) = u(1) = 0$ and where the coefficients are chosen to be

$$a(x) = \frac{1}{\alpha} + \alpha(x - \bar{x})^2, \qquad b(x) = 0$$

and

$$f(x) = 2[1 + \alpha(x - \bar{x})\{\tan^{-1} \alpha(x - \bar{x}) + \tan^{-1} \alpha\bar{x}\}]$$

The analytic solution for this choice of data is

$$u(x) = (1 - x)[\tan^{-1} \alpha(x - \bar{x}) + \tan^{-1} \alpha\bar{x}] \qquad (4.4.7)$$

The parameters α and $\bar{x}$ determine the qualitative character of the solution: for certain values of α the solution has an interior layer (large gradient) in the vicinity of $\bar{x}$.

The rate of convergence for a "smooth" problem ($\alpha = 5$, $\bar{x} = 0.2$) can

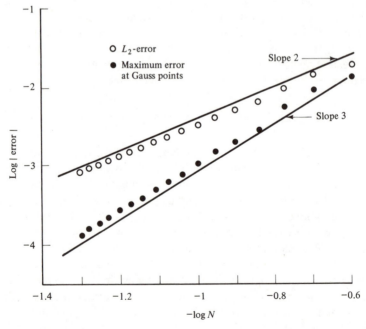

FIGURE 4.3 L^2-norm of error and local Gauss-point superconvergence with mesh refinement for H^{-1}-method and smooth problem; $\alpha = 5$, $\bar{x} = 0.2$. (After Rachford and Wheeler [1974].)

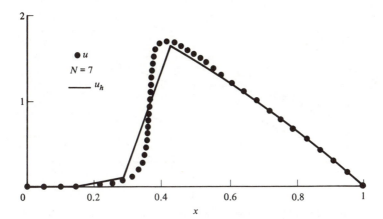

FIGURE 4.4 *Approximate solution u_h and exact solution u for mesh of seven elements. The layer point $\bar{x}$ is near the center of element 3. (After Rachford and Wheeler [1974].)*

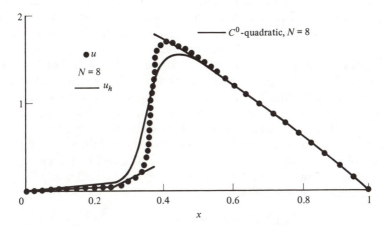

FIGURE 4.5 *Approximate solution u_h for mesh of eight elements. The layer point $\bar{x}$ is near an interelement node. The C^0-quadratic Galerkin approximation is also shown. (After Rachford and Wheeler [1974].)*

be obtained from a log-log plot of the L^2-norm of the error $\|e\|_0$ against mesh size h. For linear discontinuous trial functions a rate of 2 is obtained, so $\|e\|_0 = O(h^2)$, as indicated in Fig. 4.3. A better (local) rate can be obtained for the value of the solution at the Gauss points corresponding to $\xi = \pm 1/\sqrt{3}$ in the interior of the master element. At these points x_g we have the local superconvergence result

$$|e(x_g)| = O(h^3)$$

as indicated by the slope 3 of the corresponding curve in Fig. 4.3.

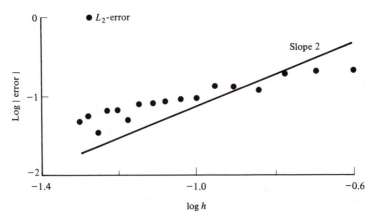

FIGURE 4.6 *L^2-norm of error for rough problem ($\alpha = 100$, $\bar{x} = 0.36388$) with mesh refinement. (After Rachford and Wheeler [1974].)*

The solution for the "interior layer" problem corresponding to $\alpha = 100$, $\bar{x} = 0.36388$, using discontinuous approximations for uniform meshes of seven and eight elements are shown in Figs. 4.4 and 4.5. When the interface between elements occurs near the layer point $\bar{x}$, a significant jump in the solution is evident (Fig. 4.5). In the case when the point $\bar{x}$ is in the middle part of an element, the jumps at the ends are small (Fig. 4.4). The L^2 error is plotted against mesh size in Fig. 4.6 for this "rough" problem and we infer that even though the qualitative behavior of the solution is well characterized, as implied by Figs. 4.4 and 4.5, the asymptotic rate of 2 has not been achieved at the finest discretization of 20 elements considered here.

Error estimates have been derived for the method. In the case of discontinuous piecewise polynomials u_h of degree k on Ω_e and C^1 test functions v_h of degree $k + 2$ on Ω_e, the following estimate holds for sufficiently smooth u (Rachford and Wheeler [1974]):

$$\| e \|_{0,\Omega_e} \leq C \inf_{w_h \in H^h} \| u - w_h \|_{0,\Omega_e} + O(h^{2k+2}) \qquad (4.4.8)$$

where k is the polynomial degree of u_h on element Ω_e and $\| \cdot \|_{0,\Omega_e}$ is the $L^2(\Omega_e)$-norm. The standard interpolation estimate can be introduced on the right to give the rate $O(h^{k+1})$. Superconvergent approximations of u of order $O(h^{2k+2})$ at the nodes are obtained in computations (Dupont [1976]). These estimates may be derived by techniques similar to those discussed in Chapter 5.

4.4.2 Least-Squares Method

Yet another method for determining an approximate solution to a boundary-value problem is the method of least squares. The scheme is based on minimizing the residual function in a least-squares sense. We have established that a measure of *distance* between functions p and q is $\|p - q\|_0$, where $\|\cdot\|_0$ is the L^2-norm. If A denotes the governing differential operator for the boundary-value problem, as defined in Chapter 1, the residual for a trial function w is

$$r = Aw - f \quad \text{in} \quad \Omega \tag{4.4.9}$$

Let us suppose that $f \in L^2(\Omega)$ and $A : H \to L^2(\Omega)$ for some appropriate trial space H of functions satisfying the boundary conditions. If we pick an arbitrary trial function $w \in H$, then in general $r = Aw - f \neq 0$ in $L^2(\Omega)$ with equality if $w = u$, the solution. The squared distance between Aw and f will be nonnegative:

$$\|r\|_0^2 = \int_\Omega (Aw - f)^2 \, dx \geq 0 \tag{4.4.10}$$

A solution u to the problem $Au = f$ can thus be interpreted as a member of H that minimizes the distance squared between Aw and f:

$$0 = \|r(u)\|_0^2 \leq \|r(w)\|_0^2 \qquad \forall w \in H$$

The least-squares method consists of seeking minimizers of the squared distance $\|Aw - f\|_0^2$ in finite-dimensional subspaces H^h of H. This approach leads to finite element methods which are quite different from the standard Galerkin scheme. For example, if A is a second-order operator, the admissible function w must lie in $H^2(\Omega)$ if Aw is to be in $L^2(\Omega)$. This implies that, as in the collocation methods, the approximation must be more regular than in the standard Galerkin finite element method. For example, if the coefficients in the differential equation are continuous, we need to ensure that the trial functions are continuous across element interfaces to an appropriate order by using, for instance, C^1-elements.

Following the notation introduced in Chapter 3, we write the quadratic functional in (4.4.10) as

$$J(w) = \int_\Omega (Aw - f)^2 \, dx = \|r(w)\|_0^2 \tag{4.4.11}$$

For the present analysis, let us again take $u = 0$ on the boundary $\partial\Omega$ and assume that A is a second-order linear elliptic operator. A necessary condition that $u \in H = H^2(\Omega) \cap H_0^1(\Omega)$ be a minimizer of the functional J in (4.4.11) is that its first variation (Gâteaux derivative) vanish at u for all admissible w. That is,

$$\langle \delta J(u), w \rangle = 2 \int_\Omega (Au - f)Aw \, dx = 0 \qquad \forall w \in H \tag{4.4.12}$$

where $\langle \cdot, \cdot \rangle$ denotes the duality paring on $H' \times H$ and

$$\langle \delta J(u), w \rangle \equiv \lim_{\epsilon \to 0} \frac{\partial}{\partial \epsilon}[J(u + \epsilon w)]$$

Thus, the least-squares method leads us to a variational boundary-value problem in which the test functions v are images of the trial functions under the operation A; that is, $v = Aw, w \in H$, and

$$\int_\Omega AuAw \, dx = \int_\Omega fAw \, dx \qquad \forall w \in H \qquad (4.4.13)$$

An alternative interpretation and formulation of the least-squares problem is possible whenever the trial functions and data are sufficiently smooth. In particular, let $(\cdot, \cdot)$ denote the usual L^2-inner product and let A be a linear operator mapping H into $L^2(\Omega)$. Suppose that u and v are sufficiently smooth functions. Then successive integrations by parts (or successive applications of the divergence theorem) yield a Green's formula of the type

$$(v, Au) = (u, A^*v) + \Gamma(u, v) \qquad (4.4.14)$$

where A^* is an operator with range in L^2 called the *adjoint* of A and $\Gamma(u, v)$ is the so-called *bilinear concomitant*: the form that contains boundary terms produced by the integrations by parts.

With this Green's formula in hand, let us return to (4.4.13) and construct the corresponding Euler equation. We obtain

$$\langle A^*Au - A^*f, w \rangle + \Gamma(Au - f, w) = 0 \qquad (4.4.15)$$

for all admissible $w \in H$, whence we have the Euler equation (in a distributional sense)

$$A^*Au = A^*f \quad \text{in} \quad \Omega \qquad (4.4.16)$$

The natural boundary condition can be obtained from the boundary term $\Gamma(Au - f, w) = 0$ for all admissible w. Hence the least-squares problem is formally equivalent to the higher-order problem with differential equation given in (4.4.16). That is, the functional $J = \| r \|_0^2$ is the classical variational functional corresponding to the higher-order Euler equation (4.4.16).

For example, if A is the Laplacian Δ, the functional J in (4.4.11) is

$$J(w) = \int_\Omega (\Delta w - f)^2 \, dx \qquad (4.4.17)$$

The Laplacian is formally self-adjoint, so $A = A^* = \Delta$, and the Euler equation corresponding to the variational functional J in (4.4.17) is, from (4.4.16), the biharmonic equation

$$\Delta^2 u = \Delta f \tag{4.4.18}$$

Observe that the operator A^*A in (4.4.15) is formally self-adjoint. For the given problem with Dirichlet data prescribed on $\partial\Omega$, it follows that $\Gamma(Au-f, w) = 0$ and the resulting finite element system is symmetric, even though A itself may not be self-adjoint. The computational advantages associated with symmetry are offset by the increase in complexity of the basis functions. Moreover, the condition number is approximately the square of that for the standard Galerkin system.

Let us return now to the least-squares statement in (4.4.13) and, as an example, consider the element derivations for the linear two-point problem (4.2.4) with homogeneous boundary conditions. In this case we have

$$\| r \|_0^2 = \int_0^1 [-(aw')' + bw - f]^2 \, dx \tag{4.4.19}$$

and the admissible functions w are required to be in $H^2(0, 1)$. Hence, a C^1-basis for finite element approximations is appropriate. Introducing the piecewise-polynomial Hermite cubic basis $\{\phi_i^m\}$ for nodes $i = 1, 2, \ldots, N+1$ defining N elements Ω_e of the mesh with $m = 0, 1$ corresponding to the order of derivative interpolated, we construct trial functions for the approximation of the form

$$u_h(x) = \sum_{i=1}^{N+1} [u_i \phi_i^0(x) + u_i' \phi_i^1(x)] \tag{4.4.20}$$

From (4.4.13), the minimization condition implies that u satisfies

$$\int_0^1 \{-(au')' + bu - f\}\{-(av')' + bv\} \, dx = 0 \tag{4.4.21}$$
$$\forall v \in H^2(0, 1) \cap H_0^1(0, 1)$$

For the approximate problem, we seek $u_h \in H^h \subset H^2(0, 1) \cap H_0^1(0, 1)$ such that

$$\int_0^1 [-(au_h')' + bu_h - f][-(av_h')' + bv_h] \, dx = 0 \qquad \forall v_h \in H^h \tag{4.4.22}$$

Substituting for u_h from (4.4.20) and setting $v_h = \{\phi_i^m\}$ in (4.4.22), we obtain the least-squares finite element system. Writing (4.4.20) as $u_h =$

$\sum_{j=1}^{N+1} \sum_{m=0}^{1} u_j^m \phi_j^m$, where $u_j^0 \equiv u$ and $u_j^1 \equiv du/dx$ at $x = x_j$, we have

$$\int_0^1 \left[-\left(a \sum_{j=1}^{N+1} \sum_{m=0}^{1} u_j^m (\phi_j^m)' \right)' + b \sum_{j=1}^{N+1} \sum_{m=0}^{1} u_j^m \phi_j^m - f \right] [-(a(\phi_i^s)')' + b\phi_i^s] \, dx = 0$$

$$(4.4.23)$$

for $i = 2, 3, \ldots, N$ with $s = 0$ and $i = 1, 2, \ldots, N+1$ with $s = 1$.

The shape functions $(\psi_j^m)^e$, $m = 0, 1$, are the restrictions of the C^1-piecewise-polynomial basis functions to Ω_e. Setting $(\psi_j^m)^e$ for ϕ_j^m in (4.4.23) determines the element matrix and vector contributions to the least-squares system. For the C^1-Hermite cubic basis we obtain a 4×4 element matrix and 4×1 element vector as contributions to the system. The element matrix is symmetric and contributions are assembled to the global system in the standard manner.

4.4.3 A Boundary Least-Squares Method

The least-squares approach can be extended to include approximation of the boundary data. One form is obtained by adding a least-squares penalty term to the variational functional for the Ritz method. The basic concepts are precisely those encountered in Chapter 3, where penalty methods were introduced to treat linear constraints. For the elliptic problem (4.3.7) with $u = g$ on $\partial\Omega$, the penalized functional is

$$J_\epsilon(v) = \int_\Omega \left(\frac{1}{2} a \, \nabla v \cdot \nabla v + \frac{1}{2} b v^2 - fv \right) dx + \frac{1}{2\epsilon} \int_{\partial\Omega} (v - g)^2 \, ds \quad (4.4.24)$$

where ϵ is the penalty parameter. On examining the first variation of $J_\epsilon(v)$, we recover the governing equation

$$-\nabla \cdot (a \, \nabla u) + bu = f \quad \text{in} \quad \Omega \quad (4.4.25)$$

as the Euler equation. The natural boundary condition becomes

$$\epsilon a \frac{\partial u}{\partial n} + u = g \quad \text{on} \quad \partial\Omega \quad (4.4.26)$$

so that as $\epsilon \to 0$ this reduces to the given boundary condition $u = g$ on $\partial\Omega$ and u_ϵ converges weakly to the solution u in H^1 (see Section 3.5.1). In practice, a large value of ϵ^{-1} is selected. For the two-point problem this approach is identical to the standard computational practice of "doping" the diagonal of the coefficient matrix with a large number to enforce Dirichlet boundary data.

The least-squares formulation can also be applied to *both* the residual of the differential equation and boundary residual. The penalized functional for the least-squares residual then becomes

$$J_\epsilon(v) = \int_\Omega [-\nabla \cdot (a\,\nabla v) + bv - f]^2\,dx + \frac{1}{2\epsilon} \int_{\partial\Omega} (v - g)^2\,ds \qquad (4.4.27)$$

Bramble and Schatz [1971] have studied a similar form of the least-squares method in which the penalty parameter is replaced by a mesh-dependent parameter. In the case above we get

$$J_h(v) = \int_\Omega [-\nabla \cdot (a\,\nabla v) + bv - f]^2\,dx + \beta \int_{\partial\Omega} (v - g)^2\,ds \qquad (4.4.28)$$

where $\beta = \beta(h)$ is chosen to balance the accuracy in approximating the solution u in the interior and on the boundary. For a, b, and f sufficiently regular, the following estimate is obtained for the choice $\beta(h) = h^{-3}$:

$$\|e\|_0 \le Ch^k(\|f\|_{k-2} + |g|_{k-1/2})$$

where $\|\cdot\|$ and $|\cdot|$ denote the L^2 norm on Ω and $\partial\Omega$, respectively, and $k \ge 4$.

4.4.4   Discrete Least-Squares Procedures

In the approaches discussed above we applied the least-squares procedure to the residual of the differential equation. An alternative is to use a discrete least-squares formulation in conjunction with another finite element procedure. We next describe briefly the main steps for such a least-squares finite element collocation method to conclude this section.

Consider again the C^1-collocation procedure for our model two-point problem (4.2.4). We proceed exactly as in Section 4.2 to determine the residual equation on an arbitrary element. For a piecewise cubic Hermite approximation we earlier collocated at two interior Gauss points in each element. Instead, we now collocate at a single interior point to obtain, after assembling, an underdetermined system of N equations in $2N$ unknowns:

$$\mathbf{Au = b} \qquad (4.4.29)$$

In a least-squares approximation to this algebraic problem, we solve the normal system

$$\mathbf{A}^T\mathbf{Au} = \mathbf{A}^T\mathbf{b} \qquad (4.4.30)$$

Now $\mathbf{A}^T\mathbf{A}$ is symmetric, positive definite, and square. Observe that the bandwidth of this normal system is larger than that of $\mathbf{A}$.

Since this least-squares approach automatically yields a symmetric, positive system, this strategy may have some merit in problems where the differential equation is of mixed type—elliptic in part of the domain and hyperbolic in another part of the domain. Such problems are encountered in compressible flows and are discussed in Volume VI.

EXERCISES

4.4.1 Use piecewise-linear C^0 approximation and piecewise-linear discontinuous approximation in the H^{-1} formulation of equation (4.4.3). Describe the resulting finite element systems.

4.4.2 Consider the example problem $-u''(x) = f(x)$ in $0 < x < 1$ with $u(0) = u(1) = 0$ and f in $L^2(0, 1)$. Construct the element contributions on the master element $\hat{\Omega}$ for the least-squares scheme in (4.4.22)–(4.4.23) using a C^1-finite element basis of Hermite cubics.

4.4.3 Determine the adjoint operator A^* corresponding to $Au = a_0 u'' + a_1 u' + a_2 u$, where a_0, a_1, and a_2 are smooth functions of x. Show that the operator A in our standard example $Au = (-au')' + bu = f$ is formally self-adjoint. For a general operator A that is not necessarily self-adjoint, verify that the higher-order operator A^*A is formally self-adjoint.

4.4.4 Derive and compare the systems of finite element equations resulting for Laplace's equation with Dirichlet data on a smooth bounded domain if:
(a) A least-squares method is applied to the differential equation as in (4.4.11).
(b) A least-squares method is applied to the algebraic system arising from a Galerkin finite element formulation using C^0-Lagrange elements.

4.5 THE BOUNDARY ELEMENT METHOD

Linear elliptic boundary-value problems of the form considered here can be recast as integral equations involving traces of the solution and its derivatives defined only on the boundary. The formulation of the problem as a boundary integral equation is based on the use of an appropriate Green's formula and fundamental solution to relate integrals over the interior of the domain to integrals on the boundary. By introducing the finite element approximation on a discretization of the boundary domain, an approximate solution of the boundary integral equation can be determined. This procedure has been termed the *boundary element method* (see, e.g., Cruse and Rizzo [1975], and Brebbia [1978]).*

---

* Numerous applications of the method are described in *Proc. 2nd Conf. Innovative Numerical Methods*, Shaw et al. (eds.), Montreal, Canada, 1980.

In this section we describe the principal features of the method. Toward this end, we must first record some classical results of potential theory to establish how boundary-value problems are formulated as boundary integral equations. From this theory we then show how finite elements can be introduced to yield boundary element methods for computing approximate solutions to the boundary integral equation.

4.5.1   Boundary Integral Equation

To begin our study let us consider again the typical self-adjoint elliptic problem with Dirichlet boundary data,

$$-\mathbf{V} \cdot [a(\mathbf{x}) \, \mathbf{V}u(\mathbf{x})] + b(\mathbf{x})u(\mathbf{x}) = f(\mathbf{x}), \qquad \mathbf{x} \in \Omega \left.\right\} \tag{4.5.1}$$
$$u(s) = \hat{u}(s), \qquad s \in \partial\Omega$$

where, as before, the coefficient functions a and b are smooth [e.g., in $C^2(\bar{\Omega})$], f and u are smooth functions, Ω is a smooth bounded domain in $\mathbb{R}^2$, $\mathbf{x} = (x_1, x_2) \equiv (x, y) \in \Omega$, and $u(s) = u[\mathbf{x}(s)]$, where $\mathbf{x} = \mathbf{x}(s)$ defines parametrically the boundary curve.

We shall need to introduce the idea of a fundamental solution and an appropriate Green's formula to recast problem (4.5.1) as a boundary integral equation. Let us first consider the *fundamental solution* (or *free-space Green's function*) for the operator $A(\cdot) = -\mathbf{V} \cdot [a \, \mathbf{V}(\cdot)] + b(\cdot)$ appearing in (4.5.1). The fundamental solution $\chi(\mathbf{x}; \boldsymbol{\xi})$ is defined as the solution of the governing differential equation in $\mathbb{R}^2$ when the data f is a unit point source,

$$-\mathbf{V} \cdot [a(\mathbf{x}) \, \mathbf{V}\chi(\mathbf{x}; \boldsymbol{\xi})] + b(\mathbf{x})\chi(\mathbf{x}; \boldsymbol{\xi}) = \delta(\mathbf{x} - \boldsymbol{\xi}) \tag{4.5.2}$$

where $\delta(\mathbf{x} - \boldsymbol{\xi})$ is the Dirac delta concentrated at point $\boldsymbol{\xi}$. Some basic properties of the fundamental solution should be noted:

1. The fundamental solution χ is defined on the entire plane $\mathbb{R}^2$; it is not required to satisfy the boundary conditions on $\partial\Omega$. If we also demand that the solution of (4.5.2) satisfy the boundary condition $\chi = 0$ on $\partial\Omega$, we obtain the Green's function $G(\mathbf{x}; \boldsymbol{\xi})$ for problem (4.5.1).

2. The Dirac delta, we recall, is defined only by its action on smooth test functions. Thus, if $C_0^\infty(\mathbb{R}^2)$ denotes the space of infinitely differentiable functions ϕ such that $\phi(\mathbf{x}) = 0$ for $|\mathbf{x}|$ sufficiently large, then

$$\delta(\mathbf{x} - \boldsymbol{\xi})\phi(\mathbf{x}) = \phi(\boldsymbol{\xi}) \qquad \forall \phi \in C_0^\infty(\mathbb{R}^2) \tag{4.5.3}$$

We note, however, that it is standard practice to use the integral symbol $\int$ for this operation and to write

$$\int_{\mathbb{R}^2} \delta(\mathbf{x} - \xi)\phi(\mathbf{x}) \, dx = \phi(\xi)$$

for all $\phi \in C_0^\infty(\mathbb{R}^2)$. Thus, it is convenient to write (4.5.2) *symbolically* as

$$\int_{\mathbb{R}^2} (-\nabla \cdot [a \, \nabla\chi(\mathbf{x}; \xi)] + b\chi(\mathbf{x}; \xi))\phi \, dx = \phi(\xi) \tag{4.5.4}$$
$$\forall \phi \in C_0^\infty(\mathbb{R}^2)$$

3. Being independent of boundary conditions, fundamental solutions of many different operators can be calculated and tabulated for future reference. The fundamental solution $\chi(\mathbf{x}; \xi)$ is *singular* at the point $\mathbf{x} = \xi$ (i.e., either χ or its derivatives are unbounded as $\mathbf{x} \rightarrow \xi$). For example, if $a = 1$, $b = 0$, the operator in (4.5.1) reduces to the Laplacian, so that

$$-\Delta\chi(\mathbf{x}; \xi) = \delta(\mathbf{x} - \xi) \tag{4.5.5}$$

and, for two-dimensional domains,

$$\chi(\mathbf{x}; \xi) = \frac{-1}{2\pi} \ell \text{n} \, r$$

where

$$r = [(x_1 - \xi_1)^2 + (x_2 - \xi_2)^2]^{1/2} \tag{4.5.6}$$

The fundamental solution here can be interpreted physically as the solution due to a point load such as a point force on a membrane in elasticity, a sink in potential flow, or a point charge in electrostatics. Some other examples of fundamental solutions are given in Tables 4.3 and 4.4 (Brebbia and Wrobel [1980]). For conciseness we have used indicial notation in the tables (summation is implied on repeated indices and differentiation implied by the comma subscript). Coefficients $\lambda, a_1, a_2, a_3, c, \mu, b, v, G$ are constants. We do not elaborate here upon the nature and origin of the classes of problems summarized as these are well known in the related literature.

4. For the particular self-adjoint problem (4.5.1), χ is symmetric; that is, $\chi(\mathbf{x}; \xi) = \chi(\xi; \mathbf{x})$.

5. The fundamental solution χ is related to the Green's function by

$$G(\mathbf{x}; \xi) = \chi(\mathbf{x}; \xi) + H(\mathbf{x}; \xi) \tag{4.5.7}$$

TABLE 4.3

Common Fundamental Solutions: Two Dimensions; $r = [(x_1 - \xi_1)^2 + (x_2 - \xi_2)^2]^{1/2}$

Problem	Equation	Fundamental Solution
Laplace equation	$-\Delta \chi(\mathbf{x}; \boldsymbol{\xi}) = \delta(\mathbf{x} - \boldsymbol{\xi})$	$\chi(\mathbf{x}; \boldsymbol{\xi}) = \dfrac{-1}{2\pi} \ell n\, r$
Helmholtz equation	$-\Delta \chi(\mathbf{x}; \boldsymbol{\xi}) - \lambda^2 \chi(\mathbf{x}; \boldsymbol{\xi}) = \delta(\mathbf{x} - \boldsymbol{\xi})$	$\chi(\mathbf{x}; \boldsymbol{\xi}) = \dfrac{1}{4i} B_0^{(2)}(\lambda r)$ where $B_0^{(2)}$ is a Hankel function
Darcy flow equation (steady, orthotropic)	$-\left[a_1 \dfrac{\partial^2 \chi}{\partial x_1^2}(\mathbf{x}; \boldsymbol{\xi}) + a_2 \dfrac{\partial^2 \chi}{\partial x_2^2}(\mathbf{x}; \boldsymbol{\xi}) \right] = \delta(\mathbf{x} - \boldsymbol{\xi})$	$\chi(\mathbf{x}; \boldsymbol{\xi}) = \dfrac{-1}{(a_1 a_2)^{1/2}} \left(\dfrac{1}{2\pi}\, \ell n\, \bar{r} \right)$ where $\bar{r} = \left[\dfrac{(x_1 - \xi_1)^2}{a_1} + \dfrac{(x_2 - \xi_2)^2}{a_2} \right]^{1/2}$
Wave equation	$\dfrac{\partial^2 \chi}{\partial t^2}(\mathbf{x}, t; \boldsymbol{\xi}, \tau) - c^2\, \Delta \chi(\mathbf{x}, t; \boldsymbol{\xi}, \tau)$ $= \delta(\mathbf{x} - \boldsymbol{\xi})\, \delta(t - \tau)$	$\chi(\mathbf{x}, t; \boldsymbol{\xi}, \tau) = \dfrac{-H[c(t - \tau) - r]}{2\pi c[c^2(t - \tau)^2 - r^2]}$ where H is the Heaviside unit step function
Plate equation	$\dfrac{\partial^2 \chi}{\partial t^2}(\mathbf{x}, t; \boldsymbol{\xi}, \tau) - \mu^2\, \Delta^2 \chi(\mathbf{x}, t; \boldsymbol{\xi}, \tau)$ $= \delta(\mathbf{x} - \boldsymbol{\xi})\, \delta(t - \tau)$	$\chi(\mathbf{x}, t; \boldsymbol{\xi}, \tau) = \dfrac{-H(t - \tau)}{4\pi \mu} \text{Si} \dfrac{r}{4\mu(t - \tau)}$ where Si is the integral sine function $\text{Si}(w) = \displaystyle\int_w^\infty \dfrac{\sin \theta}{\theta}\, d\theta$
Reduced plate equation	$-\Delta^2 \chi(\mathbf{x}; \boldsymbol{\xi}) + b\chi(\mathbf{x}; \boldsymbol{\xi}) = \delta(\mathbf{x} - \boldsymbol{\xi})$ $(b = \sqrt{\omega/\mu})$	$\chi(\mathbf{x}; \boldsymbol{\xi}) = \dfrac{-1}{8i\sqrt{b}} \left[B_0^{(2)} \sqrt{b}\, r - \dfrac{2i}{\pi} K_0(\sqrt{b}\, r) \right]$ where $B_0^{(2)}$ and K_0 are Hankel functions
Navier's equation (Kelvin solution)	$\dfrac{-\partial \sigma_{jk}}{\partial x_j}(\mathbf{x}; \boldsymbol{\xi}) = \nu_k^{(l)} \delta(\mathbf{x} - \boldsymbol{\xi})$ (component of unit vector in direction l)	Displacement in direction k: $\chi_k(\mathbf{x}; \boldsymbol{\xi}) = U_{kl} n_l = \dfrac{-[(3 - 4\nu)\, \ell n\, r\, \delta_{kl} - r_{,k} r_{,l}] n_l}{8\pi G(1 - \nu)}$ Traction in direction k: $T_k(\mathbf{x}; \boldsymbol{\xi}) = T_{kl} n_l = -\dfrac{1}{r} \left\{ \dfrac{dr}{dn} [(1 - 2\nu)\, \delta_{lk} + 2 r_{,l} r_{,k}] \right.$ $\left. - (1 - 2\nu)(n_l r_{,k} - n_k r_{,l}) \right\} \dfrac{n_l}{4\pi(1 - \nu)}$

TABLE 4.4

Common Fundamental Solutions: Three Dimensions; $r = [(x_1 - \xi_1)^2 + (x_2 - \xi_2)^2 + (x_3 - \xi_3)^2]^{1/2}$

Problem	Equation	Fundamental Solution
Laplace equation	$-\Delta\chi(\mathbf{x};\boldsymbol{\xi}) = \delta(\mathbf{x}-\boldsymbol{\xi})$	$\chi(\mathbf{x};\boldsymbol{\xi}) = \dfrac{1}{4\pi r}$
Helmholtz equation	$-\Delta\chi(\mathbf{x};\boldsymbol{\xi}) - \lambda^2\chi(\mathbf{x};\boldsymbol{\xi}) = \delta(\mathbf{x}-\boldsymbol{\xi})$	$\chi(\mathbf{x};\boldsymbol{\xi}) = \dfrac{1}{4\pi r}\,e^{-i\lambda r}$
Darcy flow equation	$-\left[a_1\dfrac{\partial^2\chi}{\partial x_1^2}(\mathbf{x};\boldsymbol{\xi}) + a_2\dfrac{\partial^2\chi}{\partial x_2^2}(\mathbf{x};\boldsymbol{\xi}) \right.$ $\left. + a_3\dfrac{\partial^2\chi}{\partial x_3^2}(\mathbf{x};\boldsymbol{\xi})\right] = \delta(\mathbf{x}-\boldsymbol{\xi})$	$\chi(\mathbf{x};\boldsymbol{\xi}) = \dfrac{1}{(a_1 a_2 a_3)^{1/2}}\dfrac{1}{4\pi\bar{r}}$ where $\bar{r} = \left[\dfrac{(x_1-\xi_1)^2}{a_1} + \dfrac{(x_2-\xi_2)^2}{a_2} + \dfrac{(x_3-\xi_3)^2}{a_3}\right]^{1/2}$
Wave equation	$\dfrac{\partial^2\chi}{\partial t^2}(\mathbf{x},t;\boldsymbol{\xi},\tau) - c^2\,\Delta\chi(\mathbf{x},t;\boldsymbol{\xi},\tau)$ $= \delta(\mathbf{x}-\boldsymbol{\xi})\,\delta(t-\tau)$	$\chi(\mathbf{x},t;\boldsymbol{\xi},\tau) = -\dfrac{1}{4\pi r}\,\delta\!\left(t-\tau-\dfrac{r}{c}\right)$
Navier's equation (isotropic, homogeneous, Kelvin solution)	$-\dfrac{\partial\sigma_{jk}}{\partial x_j} = \nu_k^{(l)}\,\delta(\mathbf{x}-\boldsymbol{\xi})$ (component of unit vector in direction l)	Displacement in direction k: $\chi_k(\mathbf{x};\boldsymbol{\xi}) = U_{kl}n_l$ $U_{kl} = \dfrac{1}{16\pi G(1-\nu)}\left(\dfrac{3-4\nu}{r}\delta_{lk} + r,_l\,r,_k\right)$ Traction in direction k: $T_k(\mathbf{x};\boldsymbol{\xi}) = T_{kl}n_l$ $= \dfrac{-1}{8\pi(1-\nu^2)r^2}\left[\dfrac{\partial r}{\partial n}\big[(1-2\nu)\,\delta_{lk} + 3r,_l\,r,_k\big| \right.$ $\left. + (1-2\nu)(n_l r,_k - n_k r,_l)\right]n_l$

204

where $H(\mathbf{x}; \xi)$ is a regular solution of the homogeneous equation

$$-\nabla \cdot [a(\mathbf{x}) \nabla H(\mathbf{x}; \xi)] + b(\mathbf{x})H(\mathbf{x}; \xi) = 0 \qquad (4.5.8)$$

Let us now assume that we know the fundamental solution χ for the operator $A(\cdot)$ appearing in problem (4.5.1). Multiplying the first equation in (4.5.1) by χ and (4.5.2) by the solution u of (4.5.1), subtracting, and changing variables yields (since χ is symmetric),

$$\{-\nabla \cdot [a(\xi) \nabla u(\xi)] + b(\xi)u(\xi) - f(\xi)\}\chi(\mathbf{x}; \xi)$$
$$-\{-\nabla \cdot [a(\xi) \nabla\chi(\mathbf{x}; \xi)] + b(\xi)\chi(\mathbf{x}; \xi) - \delta(\mathbf{x} - \xi)\}u(\xi) = 0 \qquad (4.5.9)$$

Next, we integrate this result over Ω and make use of the Green's formula*

$$\int_\Omega [v\nabla \cdot (a \nabla u) - u\nabla \cdot (a \nabla v)] \, dx = \oint_{\partial\Omega} \left(av\frac{\partial u}{\partial n} - au\frac{\partial v}{\partial n}\right) ds \qquad (4.5.10)$$

to obtain

$$\int_\Omega f(\xi)\chi(\mathbf{x}; \xi) \, dx + \oint_{\partial\Omega}\left[\chi(\mathbf{x}; s)a(s)\frac{\partial u(s)}{\partial n}\right.$$
$$\left. - a(s)u(s)\frac{\partial\chi(\mathbf{x}; s)}{\partial n}\right] ds - \int_\Omega \delta(\mathbf{x} - \xi)u(\xi) \, d\xi = 0 \qquad (4.5.11)$$

where, as usual, if $\mathbf{x} = \mathbf{x}(s)$ on $\partial\Omega$, we denote $u[\mathbf{x}(s)] = u(s)$, $a[\mathbf{x}(s)] = a(s)$, and so on. Note that $u(s) = \hat{u}(s)$ is known on $\partial\Omega$.

Now χ is defined on the entire Euclidean plane $\mathbb{R}^2$. Hence, we can choose the point ξ to be interior to $\bar\Omega$ or completely outside $\bar\Omega$. Thus, if $\mathbf{x}$ is not on the boundary $\partial\Omega$ we have immediately

$$\int_\Omega \delta(\mathbf{x} - \xi)u(\xi) \, d\xi = \begin{cases} u(\mathbf{x}) & \text{if } \mathbf{x} \in \Omega \\ 0 & \text{if } \mathbf{x} \notin \bar\Omega \end{cases} \qquad (4.5.12)$$

Choosing $\mathbf{x} \in \Omega$, we obtain the following *integral representation* for the solution u of (4.5.1):

$$u(\mathbf{x}) - \oint_{\partial\Omega} a(s)\frac{\partial u(s)}{\partial n}\chi(\mathbf{x}; s) \, ds = -\oint_{\partial\Omega} a(s)\hat{u}(s)\frac{\partial\chi(\mathbf{x}; s)}{\partial n} \, ds$$
$$+ \int_\Omega f(\xi)\chi(\mathbf{x}; \xi) \, d\xi \qquad (4.5.13)$$

---

* This formula is easily derived for sufficiently regular functions u and v by a direct application of the Gauss divergence theorem: $\int_\Omega \nabla \cdot \mathbf{q} \, dx = \oint_{\partial\Omega} \mathbf{q} \cdot \mathbf{n} \, ds$, $\mathbf{q}$ being a vector field and $\mathbf{n}$ a unit vector outward and normal to $\partial\Omega$.

at any point $\mathbf{x}$ in the interior of Ω. Since the fundamental solution χ is presumed known and functions a, $\hat{u}$, and f are prescribed as data, the integrals on the right in (4.5.13) can be evaluated. Observe that in this integral equation the unknown solution derivative $\partial u / \partial n$ enters in the first boundary integral and as a point value $u(\mathbf{x})$ for $\mathbf{x}$ an interior point.

Our intent is to pose the problem as an integral equation involving values of the solution and its derivatives on the boundary. Hence we now let the point $\mathbf{x}$ approach the boundary $\partial \Omega$. Returning to the integral expression on the left in (4.5.12), let $\mathbf{x}$ now lie on the boundary $\partial \Omega$ and delete from Ω a small region Ω_ϵ of radius ϵ centered at $\mathbf{x}$. Next apply Green's second identity (4.5.10) to the region $\Omega - \Omega_\epsilon$ (that part of Ω exterior to the ϵ-neighborhood of $\mathbf{x}$) and take the limit as $\epsilon \longrightarrow 0$ to obtain, for a general form corresponding to (4.5.12) (Exercise 4.5.4),

$$\int_\Omega \delta(\mathbf{x} - \xi) u(\xi) \, d\xi = \begin{cases} \dfrac{\theta}{2\pi} u(\mathbf{x}) & \text{if } \mathbf{x} \in \bar{\Omega} \\ 0 & \text{if } \mathbf{x} \notin \bar{\Omega} \end{cases} \tag{4.5.14}$$

where θ is the interior angle at $\mathbf{x}$. If $\mathbf{x}$ is in the interior of Ω, then $\theta = 2\pi$ and we recover (4.5.12). If $\mathbf{x}$ is a point on $\partial \Omega$ where there is a continuous tangent, then $\theta = \pi$ and we obtain $(\theta/2\pi) u(\mathbf{x}) = 1/2 \, u(\mathbf{x})$ in (4.5.14). If $\mathbf{x}$ is located at a corner point on the boundary, the corner angle θ determines the weighting factor $\theta/2\pi$. We denote by α the factor $\theta/2\pi$ and rewrite (4.5.13) accordingly as

$$\alpha(\mathbf{x}) u(\mathbf{x}) - \oint_{\partial \Omega} a(s) \frac{\partial u(s)}{\partial n} \chi(\mathbf{x}; s) \, ds$$
$$= -\oint_{\partial \Omega} a(s) \hat{u}(s) \frac{\partial \chi}{\partial n}(\mathbf{x}; s) \, ds + \int_\Omega f(\xi) \chi(\mathbf{x}; \xi) \, d\xi \tag{4.5.15}$$

where $\alpha(\mathbf{x}) = 1, \frac{1}{2}$, or 0 for $\mathbf{x}$ in the interior, at a point on the boundary where there is a continuous tangent, or exterior to Ω, respectively. In particular, if $\mathbf{x}$ is restricted to the boundary $\partial \Omega$, equation (4.5.15) is an integral equation in which $\alpha(\mathbf{x}) u(\mathbf{x}) = \alpha(s) \hat{u}(s)$ is given as data and $\partial u(s)/\partial n$ is unknown.

Now that the general procedure has been established, the construction of boundary integral equations for other types of boundary conditions is straightforward. For example, if, instead of Dirichlet data, the flux

$$\sigma(s) \equiv -a(s) \frac{\partial u}{\partial n}(s) = \hat{\sigma}(s) \tag{4.5.16}$$

is prescribed as a Neumann boundary condition on $\partial \Omega$, we can substitute in

(4.5.11) to obtain a boundary integral equation analogous to (4.5.15),

$$\alpha(s)u(s) + \oint_{\partial\Omega} a(s)u(s)\frac{\partial\chi}{\partial n}(\mathbf{x};s)\,ds = -\oint_{\partial\Omega} \hat{\sigma}(s)\chi(\mathbf{x};s)\,ds$$
$$+ \int_{\Omega} f(\xi)\chi(\mathbf{x};\xi)\,d\xi \tag{4.5.17}$$

where $u(s)$ is the unknown.

Similarly, consider the problem with mixed boundary conditions

$$u(s) = \hat{u}(s) \quad \text{on} \quad \partial\Omega_1$$

and

$$-a(s)\frac{\partial u}{\partial n}(s) = \hat{\sigma}(s) \quad \text{on} \quad \partial\Omega_2 \tag{4.5.18}$$

where $\partial\Omega = \overline{\partial\Omega_1 \cup \partial\Omega_2}$. From (4.5.11) we now have

$$\alpha(s)u(s) - \int_{\partial\Omega_1} a(s)\frac{\partial u}{\partial n}(s)\chi(\mathbf{x};s)\,ds + \int_{\partial\Omega_2} a(s)u(s)\frac{\partial\chi}{\partial n}(\mathbf{x};s)\,ds$$
$$= \int_{\Omega} f(\xi)\chi(\mathbf{x};\xi)\,d\xi - \int_{\partial\Omega_1} a(s)\hat{u}(s)\frac{\partial\chi}{\partial n}(\mathbf{x};s)\,ds \tag{4.5.19}$$
$$- \int_{\partial\Omega_2} \hat{\sigma}(s)\chi(\mathbf{x};s)\,ds$$

where $u(s)$ is unknown on $\partial\Omega_2$ and $\sigma(s)$ is unknown on $\partial\Omega_1$. The boundary integral equation (4.5.19) is an appropriate statement from which we can develop the boundary element method.

4.5.2 Boundary Elements

We shall now describe how integral representations of the type discussed in the preceding section can be used as a basis for the construction of special finite element methods. Let us consider the problem with mixed boundary conditions (4.5.18), as both the Dirichlet and Neumann cases can be deduced from this form. The corresponding boundary integral equation is given in (4.5.19).

To obtain a finite element approximation of (4.5.19), we first introduce a finite element discretization of the boundary domain $\partial\Omega$. In Figs. 4.7 and 4.8, the boundary element discretization is indicated for a smooth boundary contour. Observe that, in approximating the boundary by $\partial\Omega_h$, we may no longer have a continuously turning tangent at some nodal points, so that the factor $\alpha(s) = \theta/2\pi$ in (4.5.19) may differ from $\frac{1}{2}$ in the approximate problem.

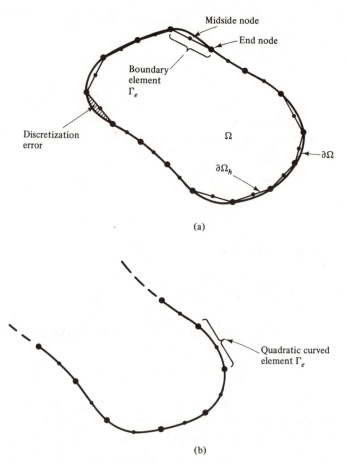

FIGURE 4.7 *Discretization of $\partial\Omega$ as boundary elements. In (a) straight elements with quadratic approximation are used; in (b) the elements are also quadratically curved.*

Let $\partial\Omega_{1h}$ and $\partial\Omega_{2h}$ denote the finite element approximations to $\partial\Omega_1$ and $\partial\Omega_2$ with

$$\partial\Omega_h = \partial\Omega_{1h} \cup \partial\Omega_{2h}$$

and

$$\partial\Omega_{1h} = \bigcup_{e=1}^{n_1} \Gamma_e^1, \qquad \partial\Omega_{2h} = \bigcup_{e=1}^{n_2} \Gamma_e^2 \tag{4.5.20}$$

Here Γ_e^1 and Γ_e^2 denote boundary elements on $\partial\Omega_{1h}$ and $\partial\Omega_{2h}$, respectively.

Using this notation in (4.5.19) the corresponding boundary integral equation on $\partial\Omega_h$ can be obtained by summing contributions from each element

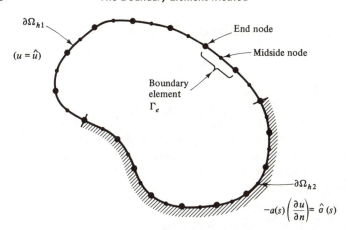

FIGURE 4.8 *Boundary problem with mixed data $u = \hat{u}$ on $\partial\Omega_{1h}$ and $-a\,\partial u/\partial n = \hat{\sigma}$ on $\partial\Omega_{2h}$.*

in the usual way:

$$\alpha(s)u(s) + \sum_{e=1}^{n_1} \int_{\Gamma_e^1} \sigma_e(s)\chi(\mathbf{x};s)\,ds + \sum_{e=1}^{n_2} \int_{\Gamma_e^2} a(s)u_e(s)\frac{\partial\chi}{\partial n}(\mathbf{x};s)\,ds$$

$$= \int_{\Omega_h} f(\xi)\chi(\mathbf{x};\xi)\,d\xi - \sum_{e=1}^{n_1} \int_{\Gamma_e^1} a(s)\hat{u}_e(s)\frac{\partial\chi}{\partial n}(\mathbf{x};s)\,ds \qquad (4.5.21)$$

$$- \sum_{e=1}^{n_2} \int_{\Gamma_e^2} \hat{\sigma}_e(s)\chi(\mathbf{x};s)\,ds$$

The unknown solution enters as u on $\partial\Omega_{2h}$ and σ on $\partial\Omega_{1h}$.

We next introduce finite element approximations to u on elements Γ_e^2 and σ on elements Γ_e^1 of the form

$$u_e(s) = \sum_{j=1}^{N_e} u_j^e \psi_j^e(s)$$

$$\sigma_e(s) = \sum_{j=1}^{M_e} \sigma_j^e \tilde{\psi}_j^e(s) \qquad (4.5.22)$$

where $\psi_j^e(s)$ and $\tilde{\psi}_j^e(s)$ are the element shape functions for $u_e(s)$ and $\sigma_e(s)$, respectively; N_e and M_e are the number of degrees of freedom for elements Γ_e^2 and Γ_e^1.

If $\mathbf{x}$ is on $\partial\Omega_{1h}$, then $u(s) = \hat{u}(s)$ is prescribed as data and in (4.5.21) $\alpha(s)u(s) = \alpha(s)\hat{u}(s)$ can be transposed to the right. On the other hand, if $\mathbf{x}$ is on $\partial\Omega_{2h}$, then $u(s)$ is not known and by (4.5.22),

$$\alpha(s)u_e(\mathbf{x}) = \alpha(s)\sum_{j=1}^{N_e} u_j^e \psi_j^e[s(\mathbf{x})]$$

In particular, if $\mathbf{x}$ is positioned at node i on $\partial\Omega_{1h}$,

$$\alpha[s(\mathbf{x}_i)]u_h(\mathbf{x}_i) = \alpha_i\hat{u}_i \qquad (4.5.23)$$

whereas on $\partial\Omega_{2h}$,

$$\alpha[s(\mathbf{x}_i)]u_h(\mathbf{x}_i) = \alpha_i u_i \qquad (4.5.24)$$

Note that we shall have $\alpha_i = \frac{1}{2}$ at nodes i interior to an element and, in general, $\alpha_i \neq \frac{1}{2}$ at interface (corner) nodes between adjacent elements. Substituting the element expansions (4.5.22) into (4.5.21), we obtain the discrete equation for node i,

$$
\begin{aligned}
\alpha_i u_h(\mathbf{x}_i) &+ \sum_{e=1}^{n_1} \int_{\Gamma_e^2} \left[\sum_{j=1}^{M_e} \sigma_j^e \tilde{\psi}_j^e(s) \right] \chi(\mathbf{x}_i; s)\, ds \\
&+ \sum_{e=1}^{n_2} \int_{\Gamma_e^2} a(s) \left[\sum_{j=1}^{N_e} u_j^e \psi_j^e(s) \right] \frac{\partial\chi}{\partial n}(\mathbf{x}_i; s)\, ds \\
&= \int_{\Omega_h} f(\xi)\chi(\mathbf{x}_i, \xi)\, d\xi - \sum_{e=1}^{n_1} \int_{\Gamma_e^1} a(s)\hat{u}_e(s) \frac{\partial\chi}{\partial n}(\mathbf{x}_i; s)\, ds \\
&\quad - \sum_{e=1}^{n_2} \int_{\Gamma_e^2} \hat{\sigma}_e(s)\chi(\mathbf{x}_i; s)\, ds \qquad i = 1, 2, \dots, N
\end{aligned}
\qquad (4.5.25)
$$

where $u_h(\mathbf{x}_i)$ is defined in (4.5.23) and (4.5.24) and there are N nodes.

As a point of particular interest, we note that the integral on Ω_h remains to be evaluated as a right-hand-side term. The fundamental solution $\chi(\mathbf{x}; \xi)$ is assumed known and the data f is also prescribed. Analytic evaluation of this integral may be possible for simple domains, but in general numerical integration will be required. For instance, we can discretize the interior as a number of quadrature "cells" Ω_c, so that

$$\int_{\Omega_h} \chi(\mathbf{x}_i; \xi) f(\xi)\, d\xi = \sum_{c=1}^{N_c} \int_{\Omega_c} \chi(\mathbf{x}_i; \xi) f(\xi)\, d\xi \qquad (4.5.26)$$

Employing numerical quadrature on each cell, we have

$$\int_{\Omega_c} \chi(\mathbf{x}_i; \xi) f(\xi)\, d\xi = \sum_{l=1}^{N_l} w_l \chi(\mathbf{x}_i; \xi_l) f(\xi_l) \qquad (4.5.27)$$

where w_l are the quadrature weights and $\{\xi_l\}$ are the quadrature points. This quadrature is repeated for each of the boundary nodes $i = 1, 2, \dots, N$.

Although the problem remains a boundary problem, the need to generate an interior mesh for this quadrature unfortunately reintroduces some of the data generation required in standard finite element calculations. Only in the case of homogeneous problems ($f = 0$) is this calculation avoided.

4.5.3 System Calculations

The boundary element system may now be described. The representative equation for node i in (4.5.25) can be written compactly in indicial form as

$$\alpha_i u_i + \sum_{j=1}^{N_1} a_{ij}\sigma_j + \sum_{j=1}^{N_2} b_{ij}u_j = \tau_i + \mu_i + \nu_i \qquad i = 1, 2, \ldots, N \qquad (4.5.28)$$

where $N = N_1 + N_2$ and N_1, N_2 are the number of nodes on $\partial\Omega_{1h}$ and $\partial\Omega_{2h}$, respectively. In (4.5.28) we have $u_i = \hat{u}_i$ if i is on $\partial\Omega_{1h}$,

$$\tau_i = \int_{\Omega_h} \chi(\mathbf{x}_i; \xi)f(\xi)\,d\xi \qquad (4.5.29)$$

and

$$a_{ij}^e = \int_{\Gamma_e^1} \chi(\mathbf{x}_i; s)\tilde{\psi}_j^e(s)\,ds,$$

$$b_{ij}^e = \int_{\Gamma_e^2} a(s)\frac{\partial\chi}{\partial n}(\mathbf{x}_i; s)\psi_j^e(s)\,ds$$

$$(4.5.30)$$

$$\mu_i^e = -\int_{\Gamma_e^1} a(s)\hat{u}_e(s)\frac{\partial\chi}{\partial n}(\mathbf{x}_i; s)\,ds,$$

$$\nu_i^e = -\int_{\Gamma_e^2} \hat{\sigma}_e(s)\chi(\mathbf{x}_i; s)\,ds$$

The assembled matrices containing the above element contributions are

$$\mathbf{A} = [a_{ij}] = \sum_{e=1}^{n_1} (a_{ij}^e), \qquad \mathbf{B} = [b_{ij}] = \sum_{e=1}^{n_2} (b_{ij}^e)$$

$$(4.5.31)$$

$$\boldsymbol{\mu} = \{\mu_i\} = \sum_{e=1}^{n_1} (\mu_i^e), \qquad \mathbf{v} = \{\nu_i\} = \sum_{e=1}^{n_2} (\nu_i^e)$$

Writing (4.5.28) in matrix form, we have

$$\mathbf{Cu} + \mathbf{A}\boldsymbol{\sigma} + \mathbf{Bu} = \boldsymbol{\tau} + \boldsymbol{\mu} + \mathbf{v} \qquad (4.5.32)$$

where $\mathbf{C}$ is a diagonal matrix with $C_{ii} = \alpha_i$. Setting $\mathbf{v}^T = [\mathbf{u}^T \quad \boldsymbol{\sigma}^T]$ as the vector of unknowns and transposing $\alpha_i u_i = \alpha_i \hat{u}_i$ for nodes i on $\partial\Omega_{1h}$, we may write the final boundary system in the form

$$\mathbf{Mv} = \mathbf{g} \qquad (4.5.33)$$

where $\mathbf{M}$ is composed of block submatrices derived from $\mathbf{A}$, $\mathbf{B}$, and $\mathbf{C}$ and the right-side vector is similarly obtained from $\mathbf{Cu}$, $\boldsymbol{\tau}$, $\boldsymbol{\mu}$, and $\mathbf{v}$.

For example, if nodes on $\partial\Omega_{1h}$ are numbered first, then $\nu_i = \sigma_i$, $i =$

$1, 2, \ldots, N_1$, and $v_i = u_i$, $i = N_1 + 1, N_1 + 2, \ldots, N_1 + N_2$. The system (4.5.32) or (4.5.33) can be expressed in block matrix form as

$$
\begin{matrix}
& \begin{matrix} N_1 & \;\; N_2 \\ \text{columns} & \text{columns} \end{matrix} & & \\
\begin{matrix} N_1 \text{ rows} \\ N_2 \text{ rows} \end{matrix} &
\begin{bmatrix} \mathbf{B}^{(1)} & \mathbf{A}^{(1)} \\ \mathbf{C}^{(2)} + \mathbf{B}^{(2)} & \mathbf{A}^{(2)} \end{bmatrix}
\begin{bmatrix} \mathbf{u} \\ \boldsymbol{\sigma} \end{bmatrix} =
\begin{bmatrix} \boldsymbol{\tau}^{(1)} + \boldsymbol{\mu}^{(1)} + \mathbf{v}^{(1)} - \mathbf{C}^{(1)}\hat{\mathbf{u}} \\ \boldsymbol{\tau}^{(2)} + \boldsymbol{\mu}^{(2)} + \mathbf{v}^{(2)} \end{bmatrix}
\end{matrix}
\tag{4.5.34}
$$

where the superscripts 1 and 2 are included to indicate that the terms are computed for $\mathbf{x}_i$ positioned on $\partial\Omega_{1h}$ and $\partial\Omega_{2h}$, respectively. The system is full and of size $(N_1 + N_2) \times (N_1 + N_2)$.

Once the solution u on $\partial\Omega_2$ and normal flux $-a\, \partial u/\partial n = \sigma$ on $\partial\Omega_1$ have been determined by the boundary element method, we can use (4.5.13) to calculate $u(\mathbf{x})$ at any interior point by numerical quadrature.

When dealing with a Dirichlet problem, $\partial\Omega = \partial\Omega_1$ and equation (4.5.25) becomes $(n = n_1)$

$$
\alpha_i \hat{u}_i + \sum_{e=1}^{n} \int_{\Gamma_e^1} \left[\sum_{j=1}^{M_e} \sigma_j^e \tilde{\psi}_j^e(s) \right] \chi(\mathbf{x}_i; s)\, ds
$$
$$
= \int_{\Omega_h} f(\xi) \chi(\mathbf{x}_i; \xi)\, d\xi - \sum_{e=1}^{n} \int_{\Gamma_e^1} a(s) \hat{u}(s) \frac{\partial \chi}{\partial n}(\mathbf{x}_i; s)\, ds
\tag{4.5.35}
$$

so that the system (4.5.34) simplifies to

$$
\mathbf{A}^{(1)}\boldsymbol{\sigma} = \boldsymbol{\tau}^{(1)} + \boldsymbol{\mu}^{(1)} - \mathbf{C}^{(1)}\hat{\mathbf{u}}
\tag{4.5.36}
$$

Similarly, for the Neumann problem (4.5.17), we have $n = n_2$ and

$$
\alpha_i u_i + \sum_{e=1}^{n} \int_{\Gamma_e^2} a(s) \left[\sum_{j=1}^{N_e} u_j^e \psi_j^e(s) \right] \frac{\partial \chi}{\partial n}(\mathbf{x}_i; s)\, ds
$$
$$
= \int_{\Omega_h} f(\xi) \chi(\mathbf{x}_i; \xi)\, d\xi - \sum_{e=1}^{n} \int_{\Gamma_e^2} \hat{\sigma}_e(s) \chi(\mathbf{x}_i; s)\, ds
\tag{4.5.37}
$$

and (4.5.34) becomes

$$
(\mathbf{C}^{(2)} + \mathbf{B}^{(2)})\mathbf{u} = \boldsymbol{\tau}^{(2)} + \mathbf{v}^{(2)}
\tag{4.5.38}
$$

As an illustrative example we consider Laplace's equation in the domain exterior to the circle of unit radius with zero flux specified on the circle. The far field asymptotic solution is given and corresponds to unit horizontal flux. Here we use a boundary discretization of 16 quadratic curved isoparametric boundary elements. In Fig. 4.9 the potential on the cylinder surface $(\pi/2 \le \theta \le 0)$ is plotted against the x-axis ordinate $0 < x < 1$.

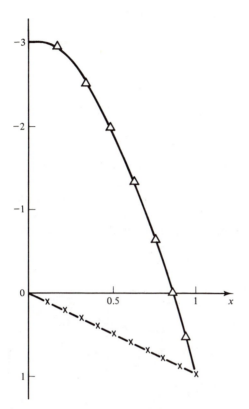

FIGURE 4.9 *Flow around a circular cylinder: x, solution for potential; Δ, solution for pressure coefficient. Solid lines denote exact solution.*

Two additional comments can be made:

1. The calculation of $\alpha(\mathbf{x})$ is a point that merits attention. We noted earlier that $\alpha = 1$ for $\mathbf{x}$ in the interior of Ω and $\alpha = \frac{1}{2}$ if $\mathbf{x}$ is at a point on the boundary where there is a continuously turning tangent. If, instead, $\mathbf{x}$ is at an interface node i between two elements, the interior angle determines the magnitude of α_i. A simple alternative scheme for determining the value of α at a node and that avoids prescribing α_i directly as data may be developed as follows: Consider the subsidiary problem, $-\Delta\phi = 0$ in Ω_h with $\phi = \phi_0$, constant on the boundary. The solution is $\phi = \phi_0$ in Ω so that $\partial\phi_0/\partial n = 0$ on $\partial\Omega_h$. Setting this trivial example as a Neumann boundary problem in (4.5.37)–(4.5.38), we have $\boldsymbol{\tau}^{(2)} = \mathbf{0}$ since $f = 0$ and $\mathbf{v}^{(2)} = \mathbf{0}$ for $\partial\phi_0/\partial n = 0$. This

implies that in (4.5.38),

$$(\mathbf{C}^{(2)} + \mathbf{B}^{(2)})\boldsymbol{\phi}_0 = \mathbf{0} \tag{4.5.39}$$

whence

$$c_{ii}^{(2)} = \alpha_i = -\sum_{j=1}^{N} b_{ij}^{(2)} \tag{4.5.40}$$

determines the diagonal entries α_i.

2.　The boundary element method may be of greatest value in solving three-dimensional problems. In Fig. 4.10 the discretization of part of a boundary surface by plane triangles and curved triangles is indicated. The reduction in data preparation and simplicity of the discretization relative to that of standard finite element methods are evident. Again,

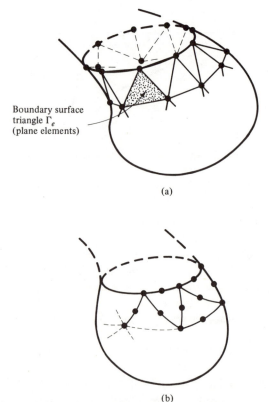

Boundary surface
triangle Γ_e
(plane elements)

(a)

(b)

FIGURE 4.10 *Discretization of a part of a boundary surface as (a) plane 3-node triangles and (b) quadratic curved triangles.*

however, the main limitations stem from the requirement that we have an appropriate Green's formula and fundamental solution.

EXERCISES

4.5.1 Construct boundary integral equations for Laplace's equation

$$-\Delta u = 0 \quad \text{in} \quad \Omega$$

with mixed boundary conditions

$$u = \hat{u} \quad \text{on} \quad \partial\Omega_1, \qquad -\frac{\partial u}{\partial n} = \hat{\sigma} \quad \text{on} \quad \partial\Omega_2$$

where $\partial\Omega = \overline{\partial\Omega_1} \cup \overline{\partial\Omega_2}$. Consider both the two- and three-dimensional cases. Develop the boundary element system in a general form.

4.5.2 Develop the element matrix contributions in integral form for the potential problems in Exercise 4.5.1 using piecewise-linear approximation on line segments and triangles, respectively.

4.5.3 For the Laplacian $A(\cdot) \equiv \Delta(\cdot)$ in three dimensions, set $\chi_\epsilon(\mathbf{x}; \xi) = 1/4\pi(r + \epsilon)$, $0 < \epsilon \ll 1$, for v in Green's formula (4.5.10) to evaluate $\Delta\chi_\epsilon$. Take the limit as $\epsilon \longrightarrow 0$ in the resulting expression $-2\epsilon/4(r + \epsilon)^3$ to obtain the integral equation

$$u(\mathbf{x}) = \int_{\partial\Omega} \left[\chi(\mathbf{x}; s) \frac{\partial u}{\partial n}(s) - u(s) \frac{\partial \chi}{\partial n}(\mathbf{x}; s) \right] ds - \int_\Omega \chi(\mathbf{x}; \xi) \, \Delta u(\xi) \, d\xi$$

where $\chi(\mathbf{x}; \xi) = \lim_{\epsilon \to 0} \chi_\epsilon(\mathbf{x}; \xi) = 1/4\pi r$ and we have taken the point $\mathbf{x}$ to lie within the domain Ω.

4.5.4 Consider the cases in two dimensions in which the location $\mathbf{x}$ of the point source is on the boundary at a point where
(a) there is a continuously turning tangent so $\theta = \pi$, and
(b) there is a corner of angle θ. As indicated in the text, delete from Ω a small region of radius ϵ, centered at $\mathbf{x}$, and by means of Green's formula (4.5.10) on $\Omega - \Omega_\epsilon$, show that the fundamental solution becomes $\chi = (1/\alpha)[(1/2\pi) \log r]$, where $\alpha = \theta/2\pi$.

4.5.5 Introduce a quadratic map from the master element $\hat{\Gamma} = [-1, 1]$ to a quadratically curved element Γ_e in the (x, y)-plane. Transform the integrals to integrals on $\hat{\Gamma}$. Use a low-order quadrature rule to evaluate a representative contribution to each of a_{ij}^e, b_{ij}^e, μ_i^e, and v_i^e for linear elements in (4.5.30).

4.5.6 For standard C^0 linear triangle and bilinear quadrilateral surface elements write down the form of the element contributions for the Poisson equation in three dimensions.

5

OTHER TOPICS

5.1 INTRODUCTION

In this chapter we discuss several special topics in the theory and application of finite element methods to elliptic problems. Many of these topics concern situations in which the previous assumptions concerning either the data of the boundary-value problem or the form of the finite element approximation do not hold.

We begin by examining the local behavior of the approximate solution for problems where the data are smooth to determine if there are distinguished points at which superior rates of convergence are achieved. This leads to the idea of local superconvergence and to special techniques for improving accuracy of computed results. Next, we study a very different question associated with local behavior: the treatment of singularities and the development of special "singular elements."

In Section 5.4 we give a brief account of certain consistency problems associated with the boundary conditions for some fourth-order boundary-value problems. The effects of relaxing the standard assumption of conformity of the finite element approximation are examined in Section 5.5. There we give examples of nonconforming elements and also criteria for convergence of these methods. The final section deals with finite element methods for eigenvalue problems. In our treatment of eigenvalue problems the approximate formulation, properties, and accuracy of approximations of solutions to linear, second-order, self-adjoint eigenvalue problems are examined.

5.2 SUPERCONVERGENCE

Numerical experiments with finite element methods reveal that for some classes of problems there are special points within the elements at which the approximation or its derivatives are much more accurate than at other points in the mesh. Moreover, calculations show that in many cases the asymptotic rate of convergence of values of the approximation or derivatives at these points may be higher than the maximum possible global rates of convergence in the energy or L^2-norms. This *superconvergence*, as it is termed, is a local property of such approximations that one would like to exploit in finite element computations. To do so we must be able to ascertain the location of the superconvergence points and to estimate the higher asymptotic rates of convergence at these points. These goals can be realized in some classes of problems for both the standard Galerkin finite element method and some of the special finite element schemes discussed in Chapter 4.

The existence of special points at which greater accuracy and higher rates of convergence are observed has been observed in practice and is well known in connection with global polynomial methods. In recent years it has been the subject of several analyses for finite element methods. For example, De Boor and Swartz [1973] found that for a particular collocation method the error at the interelement nodes of certain spline approximations was of higher order in mesh size h than the global estimates. Douglas and Dupont [1973, 1974] studied superconvergence for Galerkin finite element methods and Dupont [1976] has described a general theory for Galerkin finite element methods for one-dimensional problems. Similar results have been demonstrated by Diaz [1977] and Wheeler [1977] for C^0-collocation-Galerkin finite element methods.

In this section we examine the superconvergence property for standard finite element methods and summarize similar results for the collocation and collocation-Galerkin techniques. We also show how related ideas can be employed to compute the flux to greater accuracy. Representative numerical results are included to illustrate the superconvergence behavior.

5.2.1 Galerkin Method for a Two-Point Problem

The self-adjoint ordinary differential equation in (4.2.4) is again considered:

$$-[a(x)u'(x)]' + b(x)u(x) = f(x) \quad \text{in} \quad 0 < x < 1$$
$$u(0) = 0 \quad \text{and} \quad u(1) = 0$$

(5.2.1)

with $a(x) \geq a_0 > 0$ and $b(x) > 0$ for $x \in [0, 1]$.

217

Let u and $u_h \in H^h \subset H_0^1(0, 1)$ be the exact and Galerkin finite element solutions to this two-point problem. Then u and u_h satisfy, respectively,

$$B(u, v) = \int_0^1 fv \, dx \qquad \forall v \in H_0^1(0, 1) \tag{5.2.2}$$

and

$$B(u_h, v_h) = \int_0^1 fv_h \, dx \qquad \forall v_h \in H^h \tag{5.2.3}$$

where $B(\cdot, \cdot)$ is the bilinear form,

$$B(u, v) = \int_0^1 (au'v' + buv) \, dx \tag{5.2.4}$$

As in Section 1.4, the bilinear form (5.2.4) is assumed to be such that constants $\alpha, M > 0$ exist satisfying

$$B(u, v) \leq M \| u \|_1 \| v \|_1 \qquad \text{and} \qquad B(v, v) \geq \alpha \| v \|_1^2$$

for all $u, v \in H_0^1(0, 1)$, where $\| v \|_1^2 = \int_0^1 (v'^2 + v^2) \, dx$.

Since $H^h \subset H_0^1(0, 1)$, we can take $v = v_h$ in (5.2.2) and subtract (5.2.3) to obtain the orthogonality condition [recall (1.3.47)],

$$B(e, v_h) = 0 \qquad \forall v_h \in H^h \tag{5.2.5}$$

where $e = u - u_h$ is the approximation error.

Let us suppose that the finite element space H^h is endowed with the usual interpolation properties of finite elements associated with quasiuniform refinements of the mesh and with piecewise polynomial approximations of degree k. Then, as noted in Chapter 1, if the solution u to problem (5.2.1) is in $H^r(0, 1) \cap H_0^1(0, 1)$ and if the finite element basis functions contain complete piecewise polynomials of degree $\leq k$, then, as h tends to zero, the error e satisfies the estimate

$$\| e \|_s \leq Ch^\mu \| u \|_r, \qquad s = 0, 1$$
$$\mu = \min (k + 1 - s, r - s) \tag{5.2.6}$$

In particular, if $r \geq k + 1$, the errors in $H^1(0, 1)$ and $L^2(0, 1)$ are of order h^k and h^{k+1}, respectively:

$$\| e \|_1 \leq Ch^k \| u \|_{k+1} = O(h^k)$$
$$\| e \|_0 \leq Ch^{k+1} \| u \|_{k+1} = O(h^{k+1}) \tag{5.2.7}$$

These global rates of convergence are optimal; that is, the estimated rates are the best possible in the stated norms.

Local Solution Estimates: What we want to investigate now is the *pointwise error* $|e(\xi)|$ at some specific points $\xi \in (0, 1)$. There are several ways to approach this problem, but perhaps the most direct in the case of one-dimensional problems is to introduce the *Green's function* $G(x; \xi)$ for the boundary-value problem (5.2.1). Recall that the Green's function G is defined as the solution of (5.2.1) for the case in which the given function f is the Dirac delta concentrated at point ξ [symbolically, $\int_0^1 fv \, dx = \int_0^1 \delta(x - \xi)v(x) \, dx = v(\xi)$]. Thus,

$$B[G(\cdot \, ; \xi), v] = v(\xi) \qquad \forall v \in H_0^1(0, 1) \tag{5.2.8}$$

Since $e \in H_0^1(0, 1)$, we can set $v = e$ in (5.2.8) to obtain

$$e(\xi) = B[G(\cdot \, ; \xi), e] \tag{5.2.9}$$

Moreover, in view of the orthogonality condition (5.2.5),

$$
\begin{aligned}
e(\xi) &= B[G(\cdot \, ; \xi), e] - B(w_h, e) \\
&= B[G(\cdot \, ; \xi) - w_h, e] \\
&= \int_0^1 [a(G'(\cdot \, ; \xi) - w_h')e' + b(G(\cdot \, ; \xi) - w_h)e] \, dx
\end{aligned}
\tag{5.2.10}
$$

for any test function $w_h \in H^h$.

It is clear from (5.2.9) and (5.2.10) that an estimate of the pointwise error $e(\xi)$ can be obtained provided that it is possible to obtain an estimate of $G(\cdot \, ; \xi) - w_h$ in an appropriate norm.

In the special case of a one-dimensional problem with smooth coefficients a, b satisfying

$$a \in C^t[0, 1], \qquad b \in C^{t-1}[0, 1] \tag{5.2.11}$$

for $t \geq 1$, the Green's function G is in $H^1(0, 1)$.* Then, from continuity for any point $\xi \in (0, 1)$,

$$
\begin{aligned}
|e(\xi)| &= |B[G(\cdot \, ; \xi) - w_h, e]| \\
&\leq M \|G(\cdot \, ; \xi) - w_h\|_1 \|e\|_1
\end{aligned}
\tag{5.2.12}
$$

for any $w_h \in H^h$.

From (5.2.6), we know that $\|e\|_1 \leq Ch^{\min(k, r-1)} \|u\|_r$. Thus, the problem of estimating the pointwise error $|e(\xi)|$ reduces to one of estimating

---

* This is not true for two- or three-dimensional problems.

$\|G(\cdot\,;\xi) - w_h\|_1$. At first glance, the prognosis is somewhat pessimistic: since $G(\cdot\,;\xi)$ is in $H^1(0, 1)$, we know that if ξ is completely arbitrary, then, according to the usual interpolation results, $\|G(\cdot\,;\xi) - w_h\|_1$ is bounded by $Ch^{\min(k+1-1,1-1)}\|G(\cdot\,;\xi)\|_1 = Ch^0\|G(\cdot\,;\xi)\|_1 = $ constant. Hence, (5.2.12) indicates that $|e(\xi)| \le M(\text{constant})\|e\|_1$; that is, the order of the pointwise error is no better than that of the error in the H^1-norm.

Numerical experiments, however, indicate that exceptional pointwise accuracy in the finite element approximation can be obtained at the inter-element nodes x_i. Indeed, if we choose the point ξ to coincide with an inter-element node x_i instead of an arbitrary point, then $G(\cdot\,;\xi)$ is very regular on the open element interval. This fact, together with the fact that $w_h \in H^h$ need not be differentiable at x_i suggests that a much better estimate of $\|G(\cdot\,;\xi) - w_h\|_1$ for $\xi = x_i$ than ξ an arbitrary point can be anticipated. We shall show, in fact, that if $\xi = x_i$, then for some w_h

$$\|G(\cdot\,;x_i) - w_h\|_1 \le Ch^p, \quad p = \min(k, t) \tag{5.2.13}$$

where C is a constant independent of h, t is the parameter introduced in (5.2.11) to describe the regularity of the coefficients a and b, and k is the element degree.

To prove (5.2.13), we first note that from (5.2.11) it follows that

$$G(\cdot\,;x_i) \in H^{t+1}(0, x_i) \cap H^{t+1}(x_i, 1) \tag{5.2.14}$$

and hence

$$\|G(\cdot\,;x_i)\|_{t+1,\,(0,\,x_i)} + \|G(\cdot\,;x_i)\|_{t+1,\,(x_i,\,1)} \le C \tag{5.2.15}$$

where C is a constant.

Next, we construct a local projection g_e of $G(\cdot\,;x_i)$ into the polynomial space $P_k(\Omega_e)$, where $\Omega_e = (x_{j-1}, x_j)$ is a typical element of length h_e with end nodes located at x_{j-1} and x_j. The projection g_e is required to agree with $G(\cdot\,;x_i)$ at the end nodes x_{j-1}, x_j and to satisfy

$$\int_{\Omega_e} [g_e - G(\cdot\,;x_i)]' v' \, dx = 0 \tag{5.2.16}$$

for all $v \in P_k(\Omega_e)$ having $v(x_{j-1}) = v(x_j) = 0$. Thus, in this auxiliary problem, g_e is the H_0^1-projection of $G(\cdot\,;x_i)$ into $P_k(\Omega_e)$ and

$$\|[G(\cdot\,;x_i) - g_e]'\|_{0,\Omega_e} \le Ch_e^p \|G^{t+1}(\cdot\,;x_i)\|_{0,\Omega_e} \tag{5.2.17}$$

with $p = \min(k, t)$. Finally, let g_e^0 be the extension of g_e outside Ω_e with

$$g_e^0 = \begin{cases} g_e(x), & x \in \Omega_e \\ 0, & x \notin \Omega_e \end{cases} \tag{5.2.18}$$

and define w_h as

$$w_h = \sum_{e=1}^{E} g_e^0 \tag{5.2.19}$$

for a partition of E elements. With w_h so defined, the bound (5.2.13) follows immediately from (5.2.17) with

$$\| G(\,\cdot\,; x_i) - w_h \|_1 \le Ch^p, \qquad h = \max_e h_e \tag{5.2.20}$$

Combining this result with (5.2.12) and the global error estimate for $\| e \|_1$ given in (5.2.6), we obtain the following estimate for $| e(x_i) |$:

$$| e(x_i) | \le Ch^{\mu+p} \| u \|_r, \tag{5.2.21}$$

with $\mu = \min(k, r - 1)$, $p = \min(k, t)$. Again, we are reminded that k is the degree of the element polynomial basis and r denotes the regularity of the solution u; that is, we have assumed that f is of such regularity that $u \in H^r(0, 1) \cap H_0^1(0, 1)$ with $1 \le r \le t + 1$. The best rate is attained when $r \ge k + 1$, $t \ge k$ and we then have

$$| e(x_i) | \le Ch^{2k} \| u \|_{k+1} \tag{5.2.22}$$

Douglas and Dupont [1974] have constructed an example for which $| e(x_i) | \ge Cx_i h^{2k}$, $0 \le x_i \le \frac{1}{2}$, thereby demonstrating that the estimate (5.2.22) is indeed optimal.

As a special case, if $a(x) = 1$ and $b(x) = 0$, the Green's function is

$$G(x; \xi) = \begin{cases} x(1 - \xi), & x \in [0, \xi] \\ (1 - x)\xi, & x \in [\xi, 1] \end{cases}$$

Thus $G(\,\cdot\,; x_i)$ is piecewise linear on $[0, 1]$ and as x_i is located at an interelement node, we see that $G(\,\cdot\,; x_i)$ is in the approximation space H^h. Since $G(\,\cdot\,; x_i) \in H^h$ and w_h in (5.2.12) is an arbitrary element of H^h, then $e(x_i) = 0$. That is, *the approximation u_h interpolates the exact solution u at the interelement nodes x_i.*

To demonstrate the results of numerical experiments on superconvergence for a simple model problem, let us consider once again the case $a = b = 1$ in (5.2.1):

$$-u''(x) + u(x) = x, \qquad 0 < x < 1$$
$$u(0) = 0, \quad u(1) = 0$$

The local error at a fixed interior node $x = \frac{1}{3}$ and at a fixed interelement node $x = \frac{1}{2}$ is examined for an approximate solution obtained using C^0-cubic

Lagrange elements and mesh sizes $h = \frac{1}{2}, \frac{1}{4}$, and $\frac{1}{8}$. The calculated rate curves are given for each of the two points in Fig. 5.1. We observe that the local rate of convergence at the representative interior point is 4, which is the same rate as the global error in the L^2-norm. The calculated pointwise rate of conver-

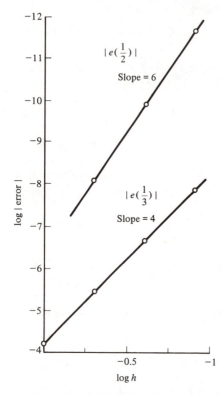

FIGURE 5.1 *Local superconvergence for piecewise cubics at interelement node* $(x = \frac{1}{2})$ *and standard rate at interior node* $(x = \frac{1}{3})$.

gence at the interelement node is 6, which is in perfect agreement with the predicted rate of $O(h^{2k})$.

Flux Computation: In many applications the flux or stress at the boundary or at specific points in the interior of the domain is of particular interest. The superconvergence properties and special quadrature formulas can again be combined to yield a highly accurate approximation to the flux or derivative.

Assume that we wish to compute the boundary flux at $x = 1$ from a finite element solution to the two-point problem (5.2.3). The most direct approach is to differentiate the approximate solution u_h on the last element and then evaluate the flux at $x = 1$. That is, on element N, compute $-a(1)u_h'(1)$

directly. Using this result, the flux or stress error is $O(h^k)$. By means of the following quadrature technique we are able to compute the flux accurate to $O(h^{2k})$.

The essential argument is as follows. We begin with the weighted residual statement and integrate by parts as in the usual variational construction (5.2.1)–(5.2.2). Now, however, let us select specifically a linear test function to obtain the desired quadrature result for the approximation σ_h^* of the flux $-au'$ at $x = 1$.

The weighted-residual statement is, for arbitrary $v \in H^1(0, 1)$,

$$\int_0^1 rv \, dx = \int_0^1 [(-au')'v + buv - fv] \, dx = 0 \qquad (5.2.23)$$

Integrating by parts yields

$$\int_0^1 (au'v' + buv - fv) \, dx = au'v \Big|_0^1 \qquad (5.2.24)$$

If we set $v(0) = 0$ and $v(1) = 1$, then (5.2.24) yields as an expression for the flux $\sigma = -au'$ at $x = 1$,

$$\sigma(1) = -a(1)u'(1) = -\int_0^1 (au'v' + buv - fv) \, dx \qquad (5.2.25)$$

This result suggests that an approximation to the flux may be calculated by quadrature according to

$$\sigma_h^* = -\int_0^1 (au_h'v' + bu_hv - fv) \, dx \qquad (5.2.26)$$

where u_h is the finite element solution to (5.2.3) and $v \in H^1(0, 1)$, $v(0) = 0$, $v(1) = 1$. In particular we can select v to be linear on the last element Ω_N and zero elsewhere:

$$v(x) = \begin{cases} \dfrac{x - x_N}{h_N}, & x \in \Omega_N \\ 0, & \text{otherwise} \end{cases} \qquad (5.2.27)$$

The integrals in (5.2.26) are nonzero only on the last element, so that

$$\sigma_h^* = -\int_{\Omega_N} \left[au_h'\left(\frac{1}{h_N}\right) + \frac{bu_h(x - x_N)}{h_N} - \frac{f(x - x_N)}{h_N} \right] dx \qquad (5.2.28)$$

As in the case of the corresponding result (5.2.22) for the solution, this quadrature formula yields a local error for the flux that is $O(h^{2k})$ if u is sufficiently smooth. Numerical results are summarized in Fig. 5.2 for the

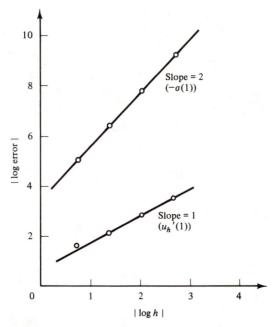

FIGURE 5.2 *Log-log plot of error in derivatives* $-\sigma(1)$
and $u_h'(1)$ *for linear elements giving rates of convergence as
respective slopes.* (*From Carey* [1982].)

model problem $-u'' + u = x$, $u(0) = u(1) = 0$ (Carey [1982]). In the figure
we see that the rate for the improved flux approximation $\sigma_h^*(1)$ is 2 for linear
elements, whereas $u_h'(1)$ converges with the standard rate 1. The same procedure can be used to calculate the flux at an interior point. In Fig. 5.3 the rates
are obtained for linear elements (rate 2) and quadratic elements (rate 4) for
the flux at $x = 0.5$ and compared with rates for u_h' (0.5).

To summarize the foregoing discussion, we have seen that the Galerkin
finite element solution is superconvergent at any interelement node and the
error here is $O(h^{2k})$ for elements of degree k. In special cases the finite element
solution may be exact at these points. By means of simple quadratures,
superconvergent approximations for u and u' can be achieved at any point $\bar{x}$
in the interval. Using the computed values of u_h and u_h' at the superconvergent
points and the differential equation, we can obtain a quadrature expression
yielding a superconvergent approximation to u''. Similarly, by differentiating
equation (5.2.1) repeatedly we can develop formulas for calculating accurate
estimates of higher derivatives at any point $\bar{x}$ in [0, 1]. Analogous superconvergence results have been obtained for other methods, such as C^1-collocation
and C^0-collocation-Galerkin finite element techniques (see, e.g., Wheeler

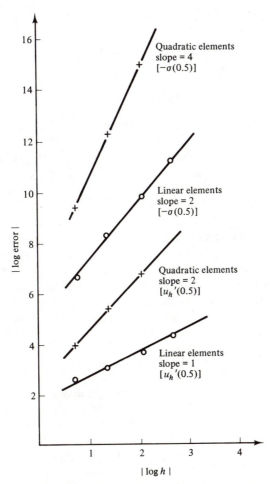

FIGURE 5.3 *Log-log plot of error in derivatives* $-\sigma(0.5)$ *and* $u_h'(0.5)$ *giving rates of convergence as respective slopes for linear and quadratic elements.*

[1977], Diaz [1977], Dupont [1976]). The main results for these finite element methods are summarized below.

5.2.2 Other Methods

As noted early in this section, superconvergent finite element rates were observed by De Boor and Swartz [1973] for C^1-collocation finite element methods. Both the solution and derivative values exhibit higher rates of convergence at the interelement nodes: If u is sufficiently smooth, then for orthogonal collocation*

$$|e(x_i)| \le Ch^{2k-2} \|u\|_{2k} \tag{5.2.29}$$

and

$$|e'(x_i)| \le Ch^{2k-2} \|u\|_{2k} \tag{5.2.30}$$

at the interelement nodes x_i. Douglas and Dupont [1974] prove these estimates by construction of a quasi-interpolant associated with the differential operator and then apply this approach to parabolic problems. [Note that the estimate here is $O(h^{2k-2})$ as compared with $O(h^{2k})$ for the Galerkin finite element method.]

In the C^0-collocation-Galerkin method we have global L^2 and H^1 estimates that are $O(h^{k+1})$ and $O(h^k)$ as in the corresponding Galerkin method (provided that the solution u is sufficiently smooth and the Jacobi points are selected as collocation points in the interior of each element). Again the Green's function and its approximation in the finite element space are employed in the superconvergence arguments. The derivations are similar to those of De Boor and Swartz for C^1 collocation and are given in Diaz [1977] and in Wheeler [1977]. The main result is that the approximation is again super-convergent at the interelement nodes, with

$$|e(x_i)| \le Ch^{2k} \|u\|_{2k} \tag{5.2.31}$$

where u is assumed sufficiently smooth and u_h is determined using collocation at the Jacobi points. The techniques developed earlier in connection with the Galerkin method can be extended to C^0-collocation-Galerkin to yield quadrature formulas for the solution and derivatives that are $O(h^{2k})$ at any point in the interval. (See Carey et al., [1981] for related numerical studies.)

EXERCISE

5.2.1 Use a simple program (such as CODE1 in Volume I) with linear elements to verify the interpolation property for the example $-u''(x) = x$, on $0 < x < 1$ with $u(0) = u(1) = 0$. Examine the derivatives at the midpoints. Compare these results with those for $-u''(x) + u(x) = x$. Plot $|e(x_i)|$ at an inter-element node against mesh size h for a sequence of uniform refinements (use a log-log graph) to check the superconvergent rate for the latter problem and compare your results with those for the cubic element given in Fig. 5.1.

---

* Recall that the choice of the Gauss points as collocation points gave us optimal global estimates in Chapter 4.

Derive formulas for superconvergent approximations to u'' and u''' using the results in the text for u and u'.

5.3 SINGULAR PROBLEMS

When values of a solution of a boundary-value problem or its partial derivatives approach infinity at points, lines, or surfaces in the domain, the solution is said to possess *singularities* at these places. The approximation of functions with singularities presents some serious numerical difficulties, as we shall see in this section. Nevertheless, calculation of solutions with singularities is extremely important; such problems arise in fracture mechanics, various flow phenomena, heat conduction problems, and, in fact, in any boundary-value problem in which strong irregularities occur in the data: that is, in the geometry of the domain, the coefficients in the governing differential equation, or in the prescribed functions f, $\hat{u}$, $\hat{\sigma}$, and so on.

Various finite element methods for singular problems have been studied by a number of authors. Surveys of finite element methods for linear fracture mechanics problems have been written by Gallagher [1971] and Pian [1975] and mathematical aspects of approximations of certain types of singular problems have been discussed by Babuška [1970], Babuška and Rosenweig [1972], Schatz and Wahlbin [1979], Barnhill and Whiteman [1975], Strang and Fix [1973], and others. Our aim in this section is to discuss the finite element approximation of a typical class of problems with solutions exhibiting singularities and to describe several families of singular elements that have been used successfully in a particular class of problems: stress singularities in linear fracture mechanics. Many of the ideas are, however, quite broadly applicable to general problems with singularities.

5.3.1 Effect of Irregularities of the Domain on the Rate of Convergence

As a model elliptic boundary-value problem, consider once again the task of determining a function u such that

$$\left.\begin{array}{rl} -\nabla \cdot [a(\mathbf{x}) \nabla u(\mathbf{x})] + b(\mathbf{x})u(\mathbf{x}) = f(\mathbf{x}) & \text{in} \quad \Omega \\ u(\mathbf{x}) = 0 & \text{on} \quad \partial\Omega \end{array}\right\} \qquad (5.3.1)$$

the variational form of which consists of finding $u \in H_0^1(\Omega)$ such that

$$\int_\Omega (a \nabla u \cdot \nabla v + buv)\, dx = \int_\Omega fv\, dx \qquad \forall v \in H_0^1(\Omega) \qquad (5.3.2)$$

If the data (a, b, f) are smooth and if the boundary $\partial\Omega$ of Ω is regular, the solution to (5.3.2) is regular and, as we have noted earlier, coincides with the classical solution of problem (5.3.1). In fact, for nice domains, if $f \in H^r(\Omega)$ for some $r \geq 0$, and if $a, b \in C^\infty(\bar{\Omega})$, $a(\mathbf{x}) \geq a_0 > 0$, $b(\mathbf{x}) > 0$, then for this second-order problem, u has $r + 2$ derivatives in $H^0(\Omega)$ [i.e., $u \in H^{r+2}(\Omega)$ $\cap H_0^1(\Omega)$]. This follows from the fact that u must be differentiated twice to give the H^r-function f.

Recall that it is possible to solve (5.3.2) approximately using regular finite element methods so that the error in the energy (i.e., the H^1-norm) satisfies an estimate of the type

$$\|u - u_h\|_1 \leq Ch^\mu \|u\|_r$$
$$\mu = \min(k, r - 1) \tag{5.3.3}$$

where k is the degree of the complete polynomials used in the approximation u_h. Thus, if u is regular so that r is very large, we can improve the rate of convergence of our method by increasing k (i.e., by using local polynomial approximations of high degree). In particular, if $r - 1 > k$, the convergence rate depends on k. For piecewise linear elements ($k = 1$), we have

$$\|u - u_h\|_1 \leq Ch \|u\|_r$$

For quadratic elements, we have

$$\|u - u_h\|_1 \leq Ch^2 \|u\|_r$$

For cubics, the H^1-error is bounded by $Ch^3 \|u\|_r$; and so on.

However, if the coefficients (a, b) and the nonhomogeneous term f are very smooth but the boundary of Ω is irregular, the solution to (5.3.2) may be very irregular, with the result that r is so small that $r - 1 \leq k$. Then, no matter how large k is, the error in $H^1(\Omega)$ will be of order h^{r-1} and the rate of convergence will be very small (perhaps zero!).

To assess the effects of irregularities in $\partial\Omega$, let us consider the special case of (5.3.2) corresponding to $a = 1, b = 0$, and $f \in C^\infty(\Omega)$, $\Omega \subset \mathbb{R}^2$. Then (5.3.2) reduces to

$$\int_\Omega \nabla u \cdot \nabla v \, dx = \int_\Omega fv \, dx \qquad \forall v \in H_0^1(\Omega) \tag{5.3.4}$$

which corresponds to the Poisson problem, $-\Delta u = f$ in Ω, $u = 0$ on $\partial\Omega$. Suppose that $\partial\Omega$ is smooth everywhere except at a point P at which there is a corner with interior angle α as shown in Fig. 5.4.

The loss of regularity in u occurs because of its singular behavior at P for certain values of the interior angle α. To study the behavior of u near P, we suppose that u is well behaved outside a conical region Ω_0 of radius r_0

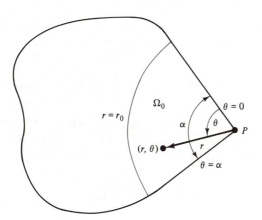

FIGURE 5.4 *Domain in $\mathbb{R}^2$ with a corner at P.*

and with vertex at P. If we establish a polar coordinate system (r, θ) at P, as shown in the figure, the solution u can be expanded in a series of the form

$$u(r, \theta) = \sum_{n=1}^{\infty} u_n(r)\psi_n(\theta)$$

$$\psi_n(\theta) = \sqrt{\frac{2}{\alpha}} \sin \frac{n\pi\theta}{\alpha}$$

(5.3.5)

and $u_n(r)$ is a function of only r for each $n \geq 1$. Representing f by the series

$$f(r, \theta) = \sum_{n=1}^{\infty} \sum_{m=1}^{\infty} f_{mn} r^m \psi_n(\theta)$$

$$\sum_{m=1}^{\infty} f_{mn} r^m = \int_0^{\alpha} f\psi_n(\theta)\, d\theta$$

and setting $v = g(r)\psi_n(\theta)$ in (5.3.4), where g is an arbitrary smooth function vanishing for $r \geq r_0$, we arrive at the series of equations (after a transformation to polar coordinates)

$$\int_0^{r_0} \left[\frac{d}{dr}\left(r\frac{du_n}{dr} \right) - \left(\frac{n\pi}{\alpha} \right)^2 \frac{1}{r} u_n - r \sum_{m=1}^{\infty} f_{mn} r^m \right] g(r)\, dr = 0$$

which holds for all suitable test functions g. The general solution of this equation is

$$u_n(r) = A_n r^{\mu_n} + B_n r^{-\mu_n} + \sum_{m=0}^{\infty} f_{mn}\chi_{mn}(r)$$

(5.3.6)

where A_n and B_n are constants, $\mu_n = n\pi/\alpha$, and

$$\chi_{mn}(r) = \begin{cases} [(m + 2)^2 - \mu_n^2]^{-1} r^{m+2}; & \mu_n \neq m + 2 \\ [2(\mu_n + 1)]^{-1} r^{\mu_n} \ln r; & \mu_n = m + 2 \end{cases}$$

Since the energy must be finite, we set $B_n = 0$; A_n is selected so that (5.3.6) yields the correct Fourier coefficient of u at $r = r_0$. The final solution is of the form

$$u(r, \theta) = \sum_{n=1}^{\infty} A_n r^{\mu_n} \psi_n(\theta) + \sum_{n=1}^{\infty} \sum_{m=0}^{\infty} f_{mn} \chi_{mn}(r) \psi_n(\theta) \qquad (5.3.7)$$

which consists of smooth terms, but also *singular* terms [i.e., terms for which derivatives of u may become unbounded $(\to \pm\infty)$ at $r = 0$]. Indeed, for $\pi/\alpha \neq$ integer, the leading term in the singular part of u is

$$r^{\pi/\alpha} \sin \frac{\pi\theta}{\alpha}$$

Thus, the order of the singularity increases as α increases. If $\alpha/\pi > 1$ (i.e., if α is obtuse), P is referred to as a *re-entrant corner* and the *first derivatives of u are then unbounded* as $r \to 0$. Indeed, if $\alpha \approx 2\pi$, the domain is said to have a *crack* or *slit* at P, and u behaves like $r^{1/2} \sin \theta/2$ near $r = 0$.

If the singular term is of the form

$$s = r^\mu \sin \mu\theta$$

then

$$\left\| \frac{d^j s}{dr^j} \right\|_{0, \Omega_0} = \left[\int_0^\alpha \int_0^{r_0} [\mu(\mu - 1)(\mu - 2) \cdots (\mu + 1 - j) r^{\mu-j} \sin \mu\theta]^2 r \, dr \, d\theta \right]^{1/2}$$

$$= C(j, \alpha, \mu) \frac{r_0^{\mu+1-j}}{(\mu + 1 - j)^{1/2}}$$

where $C(j, \alpha, \mu)$ is a constant depending on j, α, and μ. Thus, s has fewer than $\mu + 1$ derivatives in $L^2(\Omega)$! This means that in the case of re-entrant corners, u lies in $H^1(\Omega)$ but not in $H^2(\Omega)$; u effectively has fractional derivatives of order $1 \leq r \leq 2$, so that r in (5.3.3) is so small that, in general, it governs the rate of convergence of finite element methods in the H^1-norm.

These observations suggest that standard finite element methods may be very inaccurate or, at best, very slowly convergent whenever the solution possesses singularities of the type described above.

5.3.2 Generalizations and Further Comments

We now describe a straightforward generalization of the ideas described above and comment on the actual rates of convergence of conventional finite element methods for model problem (5.3.4). Consider the case of a plane polygonal domain $\Omega \subset \mathbb{R}^2$ with a boundary $\partial\Omega$ consisting of M straight-line segments (or smooth arcs) meeting at vertices $P_j, j = 1, 2, \ldots, M$ having interior angles $0 < \alpha_1 \leq \alpha_2 \leq \cdots \leq \alpha_M < 2\pi$ (Fig. 5.5). Suppose that we wish to solve Poisson's equation on this domain; that is, $-\Delta u = f$ in Ω, $u = 0$ on $\partial\Omega$, or, more specifically, the variational form of it in (5.3.4), with $f \in C^\infty(\Omega)$.

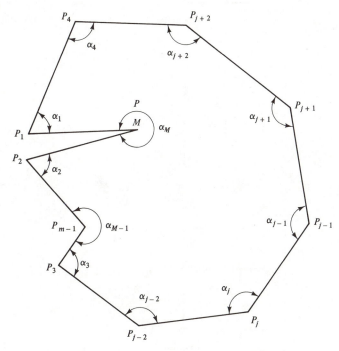

FIGURE 5.5 *Polygonal domain $\Omega \subset \mathbb{R}^2$.*

Then, if

$$\beta_j = \frac{\pi}{\alpha_j} \tag{5.3.8}$$

the solution u of (5.3.4) is just smooth enough to satisfy

$$u \in H^{1+\beta_M-\epsilon}(\Omega) \tag{5.3.9}$$

where ϵ is an arbitrary positive number. Thus, for the domains indicated in Fig. 5.6, we have:

$$\left. \begin{array}{ll} \text{Rectangular:} & u \in H^{3-\epsilon}(\Omega) \cap H^1_0(\Omega) \\ \text{L-shaped:} & u \in H^{5/3-\epsilon}(\Omega) \cap H^1_0(\Omega) \\ \text{Cracked:} & u \in H^{3/2-\epsilon}(\Omega) \cap H^1_0(\Omega) \end{array} \right\} \tag{5.3.10a}$$

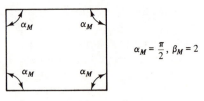

$$\alpha_M = \frac{\pi}{2}, \ \beta_M = 2$$

Rectangular

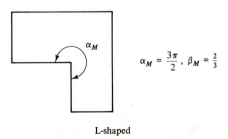

$$\alpha_M = \frac{3\pi}{2}, \ \beta_M = \frac{2}{3}$$

L-shaped

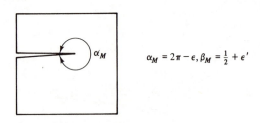

$$\alpha_M = 2\pi - \epsilon, \ \beta_M = \frac{1}{2} + \epsilon'$$

Cracked

FIGURE 5.6 *Some common irregular domains.*

Thus, if polynomials of degree k are employed in the finite element approximation of (5.3.4), the global rates of convergence in $H^1(\Omega)$ for the domains in Fig. 5.6 are:

$$\left. \begin{array}{ll} \text{Rectangular:} & \|u - u_h\|_{1,\Omega} \leq Ch^{\min(k, \, 2-\epsilon)} \|u\|_{3-\epsilon,\Omega} \\ \text{L-shaped:} & \|u - u_h\|_{1,\Omega} \leq Ch^{2/3-\epsilon} \|u\|_{5/3-\epsilon,\Omega} \\ \text{Cracked:} & \|u - u_h\|_{1,\Omega} \leq Ch^{1/2-\epsilon} \|u\|_{3/2-\epsilon,\Omega} \end{array} \right\} \quad (5.3.10b)$$

Thus, in all these cases except the rectangular domain, the rate of convergence in $H^1(\Omega)$ is not influenced at all by the order of the polynomial k, and, even in the rectangular case, only the choice $k = 1$ controls the rate of convergence.

5.3.3 Local Estimates

Estimates such as (5.3.10) reflect the worst behavior of the solution on the entire domain Ω. Locally, the situation may be much better. Let Ω_j denote the intersection of Ω with a collection of open discs centered at each vertex with radii small enough that each disc contains only one vertex and denote $\tilde{\Omega} = \Omega - \bigcup_{j=1}^M \Omega_j$ (see Fig. 5.7). Then locally

$$\| u - u_h \|_{1,\Omega_j} \leq Ch^{\min(k, \beta_j - \epsilon)} \| u \|_{\beta_j + 1 - \epsilon, \Omega_j}$$
$$\| u - u_h \|_{1,\tilde{\Omega}} \leq Ch^k \| u \|_{k+1,\tilde{\Omega}} \tag{5.3.11}$$

These estimates are given in Schatz and Wahlbin [1979].

Schatz and Wahlbin [1979] obtained some of the most important results to date on local interior estimates for polygonal domains. Let Ω_0 and $\hat{\Omega}$ be

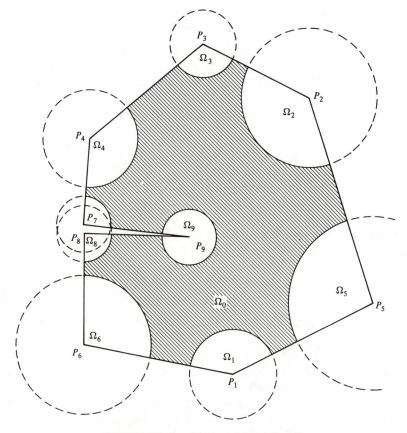

FIGURE 5.7 Localization of corner domains.

233

interior subdomains of Ω such that $\Omega_0 \subset\subset \hat{\Omega} \subset\subset \Omega$; that is, the closure of $\hat{\Omega}$ is strictly interior to Ω and the closure of Ω_0 is strictly interior to $\hat{\Omega}$ [by "strictly interior" we mean that dist $(\hat{\Omega}, \partial\Omega) > 0$, etc.]. Then, it can be shown that for any $v_h \in H^h(\Omega)$,

$$\left. \|u - u_h\|_{\infty, \Omega_0} \leq C\left[\left(\ell n \frac{1}{h}\right)^\gamma \|u - v_h\|_{\infty, \Omega_0} + \||u - u_h|\|_{-s, \hat{\Omega}}\right] \right\}$$

wherein

$$\gamma = \begin{cases} 1 & \text{if } k = 1 \\ 0 & \text{if } k \geq 2 \end{cases}$$

(5.3.12)

and $\|| \cdot \||_{-s, \Omega}$ denotes the norm on the dual space $[H_0^s(\Omega)]'$ for s an arbitrary positive integer. Locally, it can be shown that

$$\inf_{v_h \in H^h} \|u - v_h\|_{\infty, \Omega_0} \leq Ch^{k+1} \|u\|_{k+1, \hat{\Omega}} \tag{5.3.13}$$

The term $\|u - u_h\|_{-s, \hat{\Omega}}$ in (5.3.12) reflects the influence on the local error of irregularities outside $\hat{\Omega}$. Schatz and Wahlbin have shown that for $s > \beta_M - 1$,

$$\||u - u_h|\|_{-s, \hat{\Omega}} \leq C(u)h^{\min(2k, 2\beta_M - \epsilon)} \tag{5.3.14}$$

Hence, upon combining (5.3.12), (5.3.13), and (5.3.14), we find that

$$\|u - u_h\|_{\infty, \Omega_0} \leq C(u)h^{\min(k+1, 2\beta_M - \epsilon)} \tag{5.3.15}$$

Clearly, whenever $\frac{1}{2} < \beta_M < 1$, which is the case in which Ω is a cracked or L-shaped domain, the local interior error in $L^\infty(\Omega_0)$ is of order $h^{2\beta_M - \epsilon}$. For example, for the domains in Fig. 5.6, we have

$$\left. \begin{array}{ll} \text{Rectangular:} & \|u - u_h\|_{\infty, \Omega_0} \leq C(u)h^{\min(k+1, 4-\epsilon)} \\ \text{L-shaped:} & \|u - u_h\|_{\infty, \Omega_0} \leq C(u)h^{(4/3)-\epsilon} \\ \text{Cracked:} & \|u - u_h\|_{\infty, \Omega_0} \leq C(u)h^{1-\epsilon} \end{array} \right\} \tag{5.3.16}$$

The dominating factor here, of course, is the *pollution* of the local behavior due to irregularities in the solution outside the test domain Ω_0. No matter how rich the local interpolation functions (i.e., no matter now high the degree of the polynomials used in the finite element approximation), the local rates of convergence are unchanged for the L-shaped and cracked domains.

Schatz and Wahlbin [1979] have also obtained the rate of convergence at a point **x** a distance d from a corner. Suppose that $\beta_M < 1$ (e.g., cracked or L-shaped domains) and that regular mesh refinements are used. Then

$$|(u - u_h)(\mathbf{x})| \leq \begin{cases} C(u)h^{2\beta_M - \epsilon}d^{-\beta_M}, & h \leq d \leq d_0 \\ C(u)h^{\beta_M - \epsilon}, & d \leq h \end{cases} \quad (5.3.17)$$

In other words, a substantial increase in the rate of convergence is experienced as one goes from the element nearest the vertex (or crack) to adjacent elements a distance $d \leq h$ away.

5.3.4 Other Sources of Singularities

Singularities in the solution of a boundary-value problem can be present even when the boundary is very smooth. Other sources of singularities include:

1. *Singular data.* The nonhomogeneous part f of problem (5.3.1) may be very irregular (e.g., a Dirac delta or worse) and the boundary conditions $u = g$ or $\partial u/\partial n = h$ on $\partial\Omega$ may involve boundary data g, h which are irregular.

2. *Mixed boundary conditions.* If $u = g$ is prescribed on a portion $\partial\Omega_1$ of the boundary and $\partial u/\partial n = h$ is prescribed on $\partial\Omega_2$, the solution u may possess singularities at the junction points $\overline{\partial\Omega_1} \cap \overline{\partial\Omega_2}$.

3. *Discontinuous coefficients.* If the coefficient a in model problem (5.3.1) is discontinuous at points in Ω, or vanishes at certain points, the solution u will generally exhibit singularities at these points.

Finite element methods for the first type of singularity listed above have been studied by Scott [1973]. In practical computations in which f is a point source, numerical difficulties can generally be minimized by positioning a nodal point at the point source. For singularities of type 2 or 3, nodal points should again be positioned to lie on the junction points $\overline{\partial\Omega_1} \cap \overline{\partial\Omega_2}$ or at the interface between jumps in values of the coefficient $a(x)$. In any of these cases, improvement in the accuracy of finite element solutions can generally be obtained if the local behavior of the singularity can be characterized, as in Section 5.3.2 and if special elements are devised using shape functions that contain the same type of singularity. We discuss examples of such singular elements for domains with a crack in Section 5.3.6.

5.3.5 Improvement of Accuracy for Singular Problems

Despite the difficulties just described, finite element methods can be devised that yield excellent results for singular problems. Basically, there are three general ways the problem can be approached:

Nonuniform Meshes: By this, we simply mean that a finer gradation of the mesh (i.e., smaller elements) is used in the neighborhood of singular points in order to capture large changes in the gradients of the solution nearby, and a coarser grid (larger elements) is used in regions far removed from singularities where the solution is much smoother. This is often a straightforward and effective way to handle singularities and it requires no special modification of the code or special elements, but it may be expensive.

Use of Singular Functions: This is, perhaps, the most logical device for "removing" a singularity from the solution. In view of (5.3.7) the actual solution to the model problem (5.3.4) may be expressed in the form

$$u(r, \theta) = \sum_{j=1}^{N} A_j p_j(r, \theta) + w(r, \theta) \qquad (5.3.18)$$

where the A_j are constants, $w(r, \theta)$ is smooth [e.g., $w \in H^2(\Omega)$], and the p_j are singular functions of the form

$$p_j(r, \theta) = \begin{cases} r^{\mu_j} \sin \mu_j \theta, & 0 \le r \le r_0 \\ \gamma_j(r) \sin \mu_j \theta, & r_0 \le r \le r_1 \\ 0, & r_1 \le r \end{cases} \qquad (5.3.19)$$

where the functions $\gamma_j(r)$ are designed so that r^{μ_j} decays smoothly to zero.

If the functions p_j are known in advance, we can use finite element approximations of the type

$$u_h = \sum_{j=1}^{N} A_j^h p_j + w_h \qquad (5.3.20)$$

where w_h is constructed in the conventional way using, for instance, piecewise polynomials of degree k. The finite element subspace H^h of $H_0^1(\Omega)$ is then spanned by $\{p_1, p_2, \ldots, p_N, \phi_1, \phi_2, \ldots, \phi_M\}$, where ϕ_j are the usual global basis functions. One can easily show that the error then satisfies an estimate of the form

$$\| u - u_h \|_1 \le Ch^k \| w \|_k \qquad (5.3.21)$$

That is, the full rate of convergence is obtained.

Unfortunately, the method just described is rarely used in applications. The singular functions are not always known explicitly for general elliptic problems and when they are known, they are difficult to integrate when formulating the stiffness matrix. More important, the convenient band structure of the stiffness matrix is destroyed and special attention must be given to the conditioning of the matrix and to the management of arrays.

Special Singular Elements: By far the most popular technique for handling singularities is to devise special finite elements in which the approximation mimics the singularity in elements neighboring singular points. In a sense, these methods are related to the use of singular functions described in 2 above. However they may employ only crude approximations of the functions p_j, and these approximations to p_j may be present in only a few elements near the singular points.

The use of special singular elements is most prevalent in fracture mechanics applications, where the mission is not so much one of determining an accurate approximation of the solution u but of calculating the coefficient A_1 in the leading term of the expansion of singular functions [as in (5.3.7)]. This term is proportional to the *stress intensity factor*, which determines when a crack in an elastic material will begin to propagate and possibly lead to failure.

5.3.6 Examples of Special Singular Elements

A multitude of special "crack elements" have been proposed in the fracture mechanics literature. These include degenerate isoparametric elements, elements that employ rational polynomials [i.e., shape functions of the form $p_1(x)/p_2(x)$, p_1 and p_2 being polynomials], mixed elements, hybrid elements, and so on. Surveys of elements of this type have been compiled by Gallagher [1971] and Pian [1975] (see also, e.g., Tong et al. [1973]). We will be content here to describe representative examples of some effective singular elements.

Degenerate Isoparametric Elements: Recalling our discussion of element maps, invertibility, and the Jacobian as described in Chapter 5 of Volume I, it is understandable that when two or more nodes of an isoparametric element are mapped into a single point, singularities of some type should arise in the element shape functions. Barsoum [1976] and Henshell [1975] independently exploited this idea to produce special singular elements with a $1/\sqrt{r}$-singularity in the derivatives at the so-called degenerate node. They have used such elements successfully in two- and three-dimensional fracture mechanics problems. The beauty of such degenerate elements is that they can be produced in any existing code that employs isoparametric elements by simply specifying the appropriate data for the node locations.

Consider, for example, the eight-node isoparametric element shown in Fig. 5.8a. The shape functions along line 1–2 ($\eta = -1$) are:

$$\psi_1(\xi) = -\tfrac{1}{2}\xi(1 - \xi), \quad \psi_2(\xi) = \tfrac{1}{2}(1 + \xi), \quad \psi_5(\xi) = (1 - \xi^2)$$

and the transformed x-coordinate there satisfies

$$x = -\tfrac{1}{2}\xi(1 - \xi)x_1 + \tfrac{1}{2}\xi(1 + \xi)x_2 + (1 - \xi^2)x_5$$

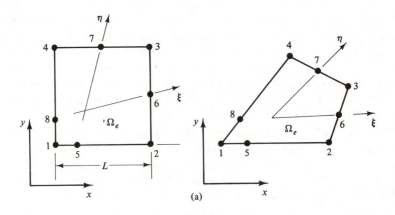

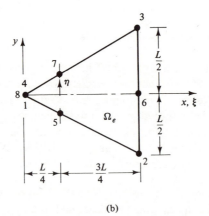

(b)

FIGURE 5.8 *Degenerate isoparametric elements with* $1/\sqrt{r}$ *singularities at node 1 (see Barsoum [1976]).*

x_1, x_2, x_5 being the x-coordinates of nodes along side 1–2. If we set $x_1 = 0$, $x_5 = L/4$, $x_2 = L$, L being the length of side 1–2, then we have

$$\xi = \left(2\sqrt{\frac{x}{L}} - 1\right), \quad \frac{\partial x}{\partial \xi} = \frac{L}{2}(1 + \xi) = \sqrt{xL}$$

The local finite element approximation along this edge of master element $\hat{\Omega}$ is

$$\hat{u}_e(\xi, \eta) = -\tfrac{1}{2}\xi(1 - \xi)u_1^e + \tfrac{1}{2}\xi(1 + \xi)u_2^e + (1 - \xi^2)u_5^e$$

and under the element map becomes

$$u_h^e(x, y) = -\frac{1}{2}\left(2\sqrt{\frac{x}{L}} - 1\right)\left(2 - 2\sqrt{\frac{x}{L}}\right)u_1^e$$

$$+ \frac{1}{2}\left(-1 + 2\sqrt{\frac{x}{L}}\right)2\sqrt{\frac{x}{L}}u_2^e + 4\left(\sqrt{\frac{x}{L}} - \frac{x}{L}\right)u_5^e$$

Thus,

$$\frac{\partial u_h^e}{\partial x} = -\frac{1}{2}\left(\frac{3}{\sqrt{xL}} - \frac{4}{L}\right)u_1^e + \frac{1}{2}\left(-\frac{1}{\sqrt{xL}} + \frac{4}{L}\right)u_2^e + \left(\frac{2}{\sqrt{xL}} - \frac{4}{L}\right)u_5^e$$

In other words, the partial derivative of u_h grows at a rate $1/\sqrt{r}$ as r approaches zero along this side. A similar result holds for side 1–4 and hence we have a $1/\sqrt{r}$ singularity at node 1 in the direction of these sides.

Similarly, when we collapse the entire edge 1–8–4 into a single node, as indicated in Fig. 5.8b, we obtain the six-node triangular element with nodes at quarter points as shown. In this case, a calculation similar to that above shows that on $\eta = 0$, $\xi = (2\sqrt{x/L} - 1)$, as before, so that again a singularity in $\partial u_h/\partial x$ of order $1/\sqrt{r}$ is obtained along this axis.

The method can also be applied to three-dimensional elements. For example, if the four midside nodes of the 20-node cubic isoparametric element in Fig. 5.9 are positioned at the quarter points, as shown, the singularity on faces $\overline{1234}$ and $\overline{5678}$ is, again, of the form $1/\sqrt{r}$.

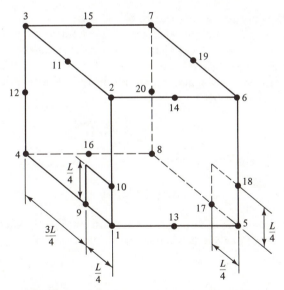

FIGURE 5.9 *Twenty-node degenerate isoparametric element with 1/r singularity (see Barsoum [1976]).*

Numerical experiments performed using elements of this type show excellent agreement with known exact solutions available for a few model problems. Some analysts prefer, however, to use special singular elements that model the angular behavior of the solution about the singularity better than the degenerate isoparametrics. For further discussion of these elements, see Hellen [1977].

Akin's Family of Singular Elements: Akin [1976] has developed a technique for constructing singular elements of any order with an arbitrary growth $O(r^{-\lambda})$ (i.e., like $r^{-\lambda}$), $0 < \lambda < 1$, at any node of the element. Suppose that the local finite element approximation for a conventional conforming element is of the form

$$\hat{u}_e(\xi, \eta) = \sum_{k=1}^{N} u_k^e \hat{\psi}_k(\xi, \eta) \tag{5.3.22}$$

where (ξ, η) are appropriate local coordinates and the $\hat{\psi}_k(\xi, \eta)$ are element shape functions satisfying the local conditions

$$\hat{\psi}_k(\xi_j, \eta_j) = \delta_{kj}; \quad k, j = 1, 2, \ldots, N$$

$$\sum_{k=1}^{N} \hat{\psi}_k(\xi, \eta) = 1,$$

$$\sum_{k=1}^{N} \frac{\partial}{\partial \xi} \hat{\psi}_k(\xi, \eta) = 0, \quad \sum_{k=1}^{N} \frac{\partial}{\partial \eta} \hat{\psi}_k(\xi, \eta) = 0 \tag{5.3.23}$$

(ξ_j, η_j) being the coordinates of node j.

We now describe a technique whereby the given functions $\hat{\psi}_k$ can be used to define a new system of local shape functions that satisfy requirements (5.3.23) and exhibit an $r^{-\lambda}$-singularity, $0 < \lambda < 1$, at any node j. Let us again choose node 1 to be the singular point. We introduce a function $W(\xi, \eta)$, defined over the element by

$$W(\xi, \eta) = 1 - \hat{\psi}_1(\xi, \eta) \tag{5.3.24}$$

For a given λ we define the function

$$R(\xi, \eta) = (W(\xi, \eta))^{\lambda} \tag{5.3.25}$$

Clearly, R attains a value of unity at all nodes except node 1, where it is zero. The new shape functions for the singular element are then given by $\tilde{\psi}_j, j = 1, 2, \ldots, N$, where

$$\tilde{\psi}_1(\xi, \eta) = 1 - \frac{W(\xi, \eta)}{R(\xi, \eta)}$$

$$\tilde{\psi}_j(\xi, \eta) = \frac{\hat{\psi}_j(\xi, \eta)}{R(\xi, \eta)}, \quad 2 \leq j \leq N \tag{5.3.26}$$

so that instead of (5.3.22) we have

$$\hat{u}_e(\xi, \eta) = \sum_{j=1}^{N} u_j^e \tilde{\psi}_j(\xi, \eta) \tag{5.3.27}$$

We easily verify that

$$\sum_{j=1}^{N} \tilde{\psi}_j(\xi, \eta) = 1$$

Moreover, a compatible (conforming) set of basis functions is obtained if the singular point is surrounded with special elements of the same type and if that collection of elements is, in turn, surrounded with appropriate standard compatible elements.

The functions (5.3.26) are rational functions. For example, for the three-node triangle in Fig. 5.10a, $\hat{u}_e(\xi, \eta)$ takes on the following values on the indicated edges

Edge 1–2: $\hat{u}_e(\xi, \eta) = u_1^e + (u_2^e - u_1^e)\sqrt{\xi}$

Edge 1–3: $\hat{u}_e(\xi, \eta) = u_1^e + (u_3^e - u_1^e)\sqrt{\eta}$

Edge 2–3: $\hat{u}_e(\xi, \eta) = u_3^e + (u_2^e - u_3^e)\xi$

whereas for the four-node quadrilateral in Fig. 5.10b, we have

Edge 1–2: $\hat{u}_e(\xi, \eta) = u_1^e + \sqrt{\xi}\,(u_2^e - u_1^e)$

Edge 1–4: $\hat{u}_e(\xi, \eta) = u_1^e + \sqrt{\eta}\,(u_4^e - u_1^e)$

Edge 2–3: $\hat{u}_e(\xi, \eta) = u_2^e + \eta(u_3^e - u_2^e)$

Edge 3–4: $\hat{u}_e(\xi, \eta) = u_4^e + \xi(u_3^e - u_4^e)$

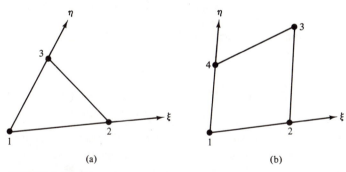

FIGURE 5.10 *Triangular and quadrilateral singular elements of Akin [1976].*

In general, the local finite element approximation for these quadrilateral elements will be of the form

$$\hat{u}_e(\xi, \eta) = u_1^e + (u_2^e - u_1^e)\xi(\xi + \eta)^{-2} + (u_3^e - u_1^e)\eta(\xi + \eta)^{-2} \qquad (5.3.28)$$

Along any radial line, we have $\xi = ar$ and $\eta = br$, $a, b = $ constants, so that (5.3.28) reduces to

$$\hat{u}_e(\xi(r), \eta(r)) = u_1^e + cr^{(1-\lambda)}$$

c being a constant. Hence

$$\frac{du_e}{dr} = (1 - \lambda)cr^{-\lambda}$$

as desired.

Akin [1976, 1979] has also developed quadrature formulas for these elements and has used similar techniques to devise three-dimensional elements with line singularities (i.e., singularities along a curve Γ lying on the boundary of the element).

The Stern Family of Singular Elements: Another ingenious technique for constructing two- and three-dimensional singular elements was developed by Stern [1979]. These elements have the following general properties:

1. The finite element approximation u_h within an element can be enriched with a singular term of any desired order λ, $0 < \lambda < 1$, at any nodal point in the element.

2. The element shape functions contain complete polynomials of degree $k \leq 1$.

3. The elements are conforming; that is, they can be made compatible with standard polynomial elements of any order so as to produce a conforming C^0-mesh of finite elements.

4. Simple quadrature formulas are available for these elements, which allow an exact integration of element stiffness matrices.

We will outline the development for a family of singular triangular elements. First, we construct a more convenient system of coordinates. The transformation

$$\left. \begin{aligned} x &= x_1 + \rho[(x_2 - x_1) + \sigma(x_3 - x_2)] \\ y &= y_1 + \rho[(y_2 - y_1) + \sigma(y_3 - y_2)] \end{aligned} \right\} \tag{5.3.29}$$

where $x_j, y_j, j = 1, 2, 3$, are the Cartesian coordinates of the vertices, maps the unit square in the (ρ, σ)-plane onto a triangular region as shown in Fig. 5.11.

Note that the lines $\sigma =$ constant are rays emanating from node 1. The lines $\rho =$ constant are parallel to side 2–3 of the triangle and ρ is proportional to the distance from node 1 along any ray $\sigma =$ constant. The interior of the triangle is then $0 < \rho < 1, 0 < \sigma < 1$, and (5.3.29) is invertible everywhere except at node 1 where $\rho = 0$. In these coordinates, a differential element of area dA is of the form

$$dA = 2A\rho \, d\rho \, d\sigma$$

where A is the total area of the triangle.

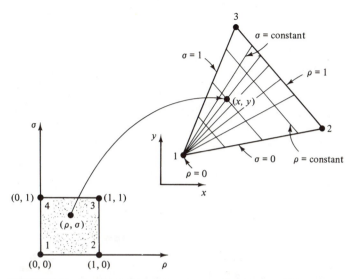

FIGURE 5.11   *Transformation of cartesian coordinates to coordinates* (ρ, σ).

Let (r, θ) denote the usual polar coordinates with origin at node 1. Then we wish to approximate singular functions of the type

$$u(r, \theta) = p(\theta)r^\lambda + q(r, \theta) \tag{5.3.30}$$

where $0 < \lambda < 1$ and $q(r, \theta)$ is well behaved in the sense that it is differentiable near $r = 0$ and its gradient is $o(r^{\lambda-1})$ as $r \to 0$. Our local finite element approximation of (5.3.30) will be of the form

$$u_h^e(\rho, \sigma) = P(\sigma)\rho^\lambda + Q(\rho, \sigma) \tag{5.3.31}$$

where P and Q are polynomials. The basic idea is to make $Q(\rho, \sigma)$ linear in σ and $\rho\sigma$ so as to include polynomials of degree $k \geq 1$.

Each triangular element has $5 + K$ nodes: the vertices, numbered 1, 2, and 3; midside nodes 4 and 5 on edges 1–2 and 1–3; and $K(\geq 0)$ distinct nodes distributed along the edge 2–3. Some examples are shown in Fig. 5.12. The local representation (5.3.31) then assumes the specific form

$$u_h^e(\rho, \sigma) = A_1 + \rho[A_2(1 - \sigma) + A_3\sigma]$$
$$+ \rho^\lambda[A_4(1 - \sigma) + A_5\sigma + \sigma(1 - \sigma)p_K(\sigma)] \tag{5.3.32}$$

where $A_1, \ldots, A_5$ are constants and p_K is a polynomial of degree $K - 1$ in $\sigma(p_K \equiv 0$ if $K = 0)$. Thus, p_K has K independent coefficients which are to be selected along with the five constants A_i. We determine these constants in terms of the nodal values u_k of u_h, $k = 1, 2, \ldots, 5 + K$ in the usual way to

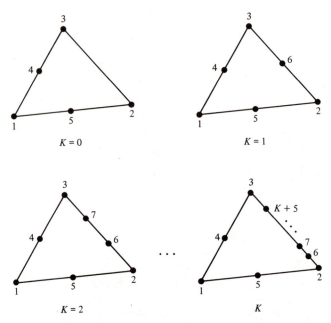

FIGURE 5.12 Families of singular triangular elements studied by Stern [1979].

obtain the final local form,

$$u_h^e(\rho, \sigma) = u_1^e + \frac{\rho}{\beta - 1}\{[\beta u_2^e - 2u_4^e + (2 - \beta)u_1^e](1 - \sigma)$$
$$+ [\beta u_3^e - 2u_5^e + (2 - \beta)u_1^e]\sigma\}$$
$$+ \rho^\lambda\{\frac{1}{\beta - 1}[(2u_4^e - u_2^e - u_1^e)(1 - \sigma) \qquad (5.3.33)$$
$$+ (2u_5^e - u_3^e - u_1^e)\sigma]\}$$
$$+ \sum_{k=1}^{K} [u_{5+k}^e - (1 - s_k)u_2^e - s_k u_3^e]L_k(\sigma)$$

where $\beta = 2^{1-\lambda}$ and $L_k(\sigma)$ are the Lagrange interpolation polynomials corresponding to nodes s_k, $k = 0, 1, 2, \ldots, K + 1$, on side 2–3:

$$L_k(\sigma) = \prod_{\substack{j=0 \\ j \neq k}}^{K+1} \frac{\sigma - s_j}{s_k - s_j}, \qquad k = 0, 1, \ldots, K \qquad (5.3.34)$$

The following properties of these elements can be easily verified:

1. If the local approximation u_h in an element joining the singular ele-

ment on edge 2–3 is linear, then (5.3.33) is also linear along edge 2–3; that is, the singular shape functions are compatible with linear fields.

2. If we interpolate a function of the form $u = f(\theta)r^\lambda$ at the nodes, the resulting local finite element approximation assumes the form $u_h^e = P(\sigma)\rho^\lambda$, $P(\sigma)$ being a polynomial of degree $K + 1$.

3. The basis functions produced by the procedure just described [i.e., those in equation (5.3.33)] are continuous across interelement boundaries of two singular elements of the same type having a common singular vertex on their common edge.

4. The basis functions are continuous across the edge opposite the singular node if the adjoining element has the same boundary nodes and the interpolation function on that edge is a polynomial of degree one less than the number of edge nodes $(K + 2)$.

For an element with $5 + K$ nodes, the shape function corresponding to node j assumes the form

$$\psi_j^e(\rho, \sigma) = \rho^\lambda m_j(\sigma) + n_j(\rho, \sigma), \qquad j = 1, 2, \ldots, 5 + K \quad (5.3.35)$$

in which

$$
\begin{aligned}
n_1 &= 1 + X(2 - \beta)\rho, \quad n_2 = X\beta(1 - \sigma)\rho \\
n_3 &= X\beta\rho, \quad n_4 = -2X(1 - \sigma)\rho, \quad n_5 = -2X\sigma\rho \\
&\quad n_{5+k} \equiv 0, \quad k = 1, 2, \ldots, K \\
m_1 &= -X, \quad m_4 = 2X(1 - \sigma), \quad m_5 = 2X\sigma \\
m_2 &= -X(1 - \sigma) - \sum_{k=1}^{K} (1 - s_k)L_k(\sigma) \\
m_3 &= -X\sigma - \sum_{k=1}^{K} s_k L_k(\sigma); \quad m_{5+k} = L_k(\sigma) \\
&\quad k = 1, 2, \ldots, K
\end{aligned}
\right\} \quad (5.3.36)
$$

where

$$\beta = 2^{\lambda - 1}, \qquad X = (\beta - 1)^{-1} \quad (5.3.37)$$

To extract the coefficients of the dominant growth terms, we note that along the ray $\sigma = 0$,

$$\rho^\lambda P(0) = \rho^\lambda \sum_{j=1}^{5+K} u_j m_j(0) \quad (5.3.38)$$

so that

$$P(0) = X[(x_2 - x_1)^2 + (y_2 - y_1)^2]^{-\lambda/2}(2u_4 - u_1 - u_2) \quad (5.3.39)$$

Stern has also developed convenient quadrature rules for these singular elements by which exact stiffnesses can be calculated. In addition, he has extended the ideas to three-dimensional prismatic elements [in prismatic coordinates (ρ, σ, ζ)] of the type shown in Fig. 5.13 and has developed simple formulas for the corresponding shape functions.

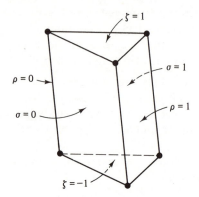

Prismatic coordinates

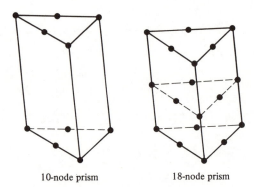

10-node prism   18-node prism

FIGURE 5.13 *Singular prismatic elements developed by Stern [1979].*

EXERCISES

5.3.1 Give asymptotic error estimates in the H^1-norm for quadratic finite element approximations of Laplace's equation on the following domains for the case $f \in C^\infty(\Omega)$:

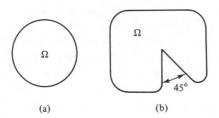

(a)   (b)

5.3.2   Describe L^∞-interior error estimates for the domains Ω given in Exercise 5.3.1.

5.3.3   Using Akin's method, construct formulas for shape functions for a six-node triangle with a singularity with $\lambda = \frac{1}{2}$ at node 1. Show, for example, that along edge 1–4–2,

$$u_e(\xi, \eta) = u_1^e + \sqrt{\frac{\xi}{3 - 2\xi}}(-3u_1^e - u_2^e + 4u_4^e)$$

$$+ \sqrt{\frac{\xi}{3 - 2\xi}}(2u_1^e + 2u_2^e - 4u_4^e), \text{ etc.}$$

5.3.4   Using Stern's method, develop a triangular element with a $\lambda = \frac{1}{2}$ singularity at node 1 which has sufficient degrees of freedom to produce a conforming approximation when embedded in a mesh of nine-node biquadratic elements.

5.3.5   Describe the steps needed to employ a Stern element to model a point singularity with $\lambda = \frac{1}{3}$ in a finite element approximation of a two-dimensional problem in which 16-node bicubic isoparametric elements are to be used.

5.4   FOURTH-ORDER PROBLEMS

5.4.1   Introduction

We have briefly described the standard variational formulation and Galerkin finite element approximation of fourth-order problems in Chapter 6 of Volume I and reviewed some of these ideas in Chapter 1 of the present volume. One can extend many of the special finite element methods described in Chapter 4, such as collocation techniques, to higher-order problems in a straightforward manner as indicated in the exercises of Chapter 4. In addition, we also showed, in the treatment of mixed methods in Chapter 3, that one could replace a fourth-order problem by a second-order system and then construct a mixed variational statement and corresponding finite element method. Hence, many of the formulative aspects of various finite element methods for fourth-order problems have already been discussed and require no further elaboration.

There are, however, situations that arise in finite element analysis of certain fourth-order problems which are not encountered in second-order problems. Perhaps the most intriguing case arises in connection with certain types of boundary conditions and polygonal approximation of smooth boundaries for plate-bending problems: in the analysis of the deflection of a thin simply supported plate, a sequence of finite element solutions on successively finer meshes may in certain situations converge to the solution of the wrong problem. This problem has been sometimes termed the "Babuška paradox" after the investigations of Babuška [1963]. Our discussion of the subject and related difficulties follows that of Scott [1976] and focuses on a typical boundary-value problem in the classical theory of bending of thin elastic plates.

5.4.2 Plate Problem and Approximation

The problem of determining the small deflection u of a thin flat plate occupying a region $\Omega \subset \mathbb{R}^2$ and under a transverse loading is described by the biharmonic equation

$$\frac{\partial^4 u}{\partial x^4} + 2\frac{\partial^4 u}{\partial x^2\,\partial y^2} + \frac{\partial^4 u}{\partial y^4} = f \quad \text{in} \quad \Omega \tag{5.4.1}$$

together with appropriate boundary conditions on $\partial\Omega$. The present discussion is limited to the case of a smooth boundary and smooth data; $f = p/D$, where p is the prescribed transverse load per unit area on the plate and D is a constant called the flexural rigidity of the plate. In particular, $D = Et^3/12(1 - v^2)$, where t is the plate thickness (a constant), E is Young's modulus, and v is Poisson's ratio for the material, both being constants. We shall assume that $f \in L^2(\Omega)$. The Green's formulas for the biharmonic operator are essential in establishing the variational problem (recall Chapter 1). We note the following relations: for Ω sufficiently smooth and, for instance, $u \in H^4(\Omega)$ and $v \in H^2(\Omega)$,

$$\int_\Omega \Delta u\,\Delta v\,dx = \int_\Omega v\,\Delta^2 u\,dx - \int_{\partial\Omega} v\,\frac{\partial}{\partial n}(\Delta u)\,ds + \int_{\partial\Omega} \Delta u\,\frac{\partial v}{\partial n}\,ds \tag{5.4.2}$$

and

$$\int_\Omega (u_{xx}v_{yy} + u_{yy}v_{xx} - 2u_{xy}v_{xy})\,dx = \int_{\partial\Omega} (u_{tt}v_n - u_{nt}v_t)\,ds \tag{5.4.3}$$

where t and n are tangential-normal coordinates to $\partial\Omega$, $dx \equiv dx\,dy$, and subscripts indicate respective partial derivatives. As the governing equation

(5.4.1) is of order 4, two boundary conditions are required. For example, if the flat plate is clamped, we have the homogeneous boundary conditions

$$u = 0 \quad \text{and} \quad \frac{\partial u}{\partial n} = 0 \quad \text{on} \quad \partial\Omega \qquad (5.4.4)$$

The equivalent variational problem for the clamped plate problem (5.4.1)–(5.4.4) is: Find $u \in H_0^2(\Omega)$ satisfying

$$\int_\Omega \Delta u \, \Delta v \, dx = \int_\Omega fv \, dx \qquad \forall v \in H_0^2(\Omega) \qquad (5.4.5)$$

If u and v in (5.4.5) are sufficiently smooth, we can use (5.4.2) and the clamped conditions (5.4.4) to show that a solution to (5.4.5) also satisfies (5.4.1) and (5.4.4). The case of homogeneous Neumann boundary conditions is also noteworthy. If, instead of (5.4.4), u and $\partial u/\partial n$ are not prescribed on part or all of the boundary, then (5.4.2) implies the natural boundary conditions on $\partial\Omega$

$$\frac{\partial}{\partial n}(\Delta u) = 0 \quad \text{and} \quad \Delta u = 0 \qquad (5.4.6)$$

Of course, when conditions (5.4.6) are applied on $\partial\Omega$, compatibility conditions on the data f must also be satisfied.

Now consider the finite element approximation of the clamped plate problem (5.4.5). The approach is quite standard: First, we discretize the domain as a collection Ω_h of finite elements over which we construct a piecewise-polynomial basis for defining the approximation u_h and test functions v_h. Since the problem (5.4.5) is posed in $H_0^2(\Omega)$, a conforming finite element method can be obtained by use of C^1 elements such as the C^1-quintic triangles described in Chapter 2. If these straight-sided triangles are employed, the boundary $\partial\Omega$ is thereby approximated by a straight-sided polygon—a point of particular significance here.

We shall assume for convenience that Ω is convex and that $\Omega_h \subset \Omega$. Since the boundary is clamped, functions $u_h \in H^h(\Omega_h)$ can be extended by zero outside Ω_h to all of Ω, so that $H^h(\Omega)$ is a subspace of the solution space $H_0^2(\Omega)$. Now we let $u \in H_0^2(\Omega)$ be the solution and w_h the interpolant of u on Ω_h in the finite element space $H^h(\Omega_h)$. As the mesh is uniformly refined, the characteristic mesh size h tends to zero, the boundary $\partial\Omega_h$ approaches $\partial\Omega$, and the interpolated exact solution w_h converges to zero on $\partial\Omega_h$. Consequently, as $h \rightarrow 0$,

$$\inf_{v_h \in H^h(\Omega_h)} \| u - v_h \|_{0,\Omega} \longrightarrow 0 \qquad (5.4.7)$$

This implies that the sequence of finite element solutions u_h converges to the exact solution u of the clamped plate problem. The following estimate holds (see Strang and Berger [1971]):

$$\|u - u_h\|_{0,\Omega_h} \le Ch^{3/2} \tag{5.4.8}$$

provided that u is sufficiently regular and the element is at least of quartic degree (i.e., excluding the composite and other low-degree C^1-elements described in Chapter 2). The rate (5.4.8) is the best possible for u irrespective of element degree k [since the distance between $\partial\Omega$ and $\partial\Omega_h$ is $O(h^2)$].* Thus we see that no major problems result due to polygonal approximation for the clamped plate problem. We shall see that the situation is quite different when a simply supported boundary condition applies.

The relevance of the second Green's formula (5.4.3) becomes clear when we examine other boundary conditions. Let us now consider the situation for a simply supported plate. The deflection u and "bending moment" are zero on $\partial\Omega$, which implies that the boundary conditions are

$$u = 0 \quad \text{and} \quad v\,\Delta u + (1 - v)u_{nn} = 0 \quad \text{on} \quad \partial\Omega \tag{5.4.9}$$

where v is the Poisson ratio introduced earlier.

To embed the second condition of (5.4.9) in the variational statement as a natural boundary condition, we must add to the functional (5.4.5) appropriate boundary integrals. Using the second Green's formula (5.4.3) to recast the boundary integrals as interior integrals, the variational problem for the simply supported plate is: Find u satisfying

$$\int_\Omega [\Delta u\,\Delta v + (1 - v)(2u_{xy}v_{xy} - u_{xx}v_{yy} - u_{yy}v_{xx})]\,dx = \int_\Omega fv\,dx \tag{5.4.10}$$

for all admissible v. Here u and v are required to be in $H^2(\Omega) \cap H_0^1(\Omega)$, so that the essential condition $u = 0$ on $\partial\Omega$ is enforced in the construction of the space. The condition that the moment be zero in (5.4.9) is easily verified to be the natural boundary condition for (5.4.10).

The finite element formulation for the plate with simple support follows the same general procedure as for the clamped plate. Again the domain is discretized as a collection Ω_h of C^1 elements and we take the case of elements with straight sides so that $\partial\Omega_h$ is a polygonal interpolant of the actual boundary $\partial\Omega$. We let $H^h(\Omega_h)$ denote the space of conforming finite element approximations so that $u_h, v_h \in H^h(\Omega_h)$ and $u_h = v_h = 0$ on $\partial\Omega_h$. The finite

---

* The condition $\Omega_h \subset \Omega$ can be relaxed and (5.4.7) still holds. If, in addition, the distance separating $\partial\Omega$ and $\partial\Omega_h$ is restricted to be $O(h^2)$, the error estimate (5.4.8) remains valid.

element problem is then to find $u_h \in H^h(\Omega_h)$ such that

$$\int_{\Omega_h} \{\Delta u_h \, \Delta v_h + (1 - v)[2(u_h)_{xy}(v_h)_{xy} - (u_h)_{xx}(v_h)_{yy}$$

$$- (u_h)_{yy}(v_h)_{xx}]\} \, dx = \int_{\Omega_h} f v_h \, dx \tag{5.4.11}$$

for all $v_h \in H^h(\Omega_h)$.

To investigate the effect of the boundary approximation for the clamped plate, we extended the solution by zero outside Ω_h. Since both v_h and $\partial v_h/\partial n$ are zero on $\partial \Omega_h$ for the clamped plate, extension of v_h by zero does not effect the C^1 continuity of the functions. The situation is quite different for the simply supported plate! We now have $v_h = 0$ on $\partial \Omega_h$ by construction but $\partial v_h/\partial n \neq 0$ on $\partial \Omega_h$, so extension of v_h by zero on $\Omega - \Omega_h$ will not retain C^1 continuity.

Let $\bar{U}$ be the exact solution for the simply supported plate on the approximated domain Ω_h. Then $\bar{U}$ satisfies $\bar{U} = 0$ on $\partial \Omega_h$ and the variational statement

$$\int_{\Omega_h} \{\Delta \bar{U} \, \Delta v + (1 - v)[2\bar{U}_{xy}v_{xy} - \bar{U}_{xx}v_{yy} - \bar{U}_{yy}v_{xx}]\} \, dx = \int_{\Omega_h} f v \, dx \tag{5.4.12}$$

for all $v \in H^2(\Omega_h)$ satisfying $v = 0$ on $\partial \Omega_h$. The approximate solution actually computed is thus the finite element function $u_h \in H^h(\Omega_h)$ approximating $\bar{U}$. Hence, the standard error analysis implies that u_h converges to $\bar{U}$ as the mesh is regularly refined.

On an open straight-line segment of $\partial \Omega_h$, since $\bar{U} = 0$ then $\bar{U}_{tt} = 0$ and this implies that $\Delta \bar{U} = \bar{U}_{nn}$. The second boundary condition in (5.4.9) simplifies accordingly to $\bar{U}_{nn} = 0$, or equivalently, $\Delta \bar{U} = 0$. The plate problem on Ω_h can then be restated as the alternative system

$$\Delta \bar{U} = \bar{W} \quad \text{in} \quad \Omega_h$$
$$\Delta \bar{W} = f \quad \text{in} \quad \Omega_h \tag{5.4.13}$$

with $\bar{U} = \bar{W} = 0$ on $\partial \Omega_h$.

Let U and W be the solution to the corresponding problem

$$\Delta U = W \quad \text{in} \quad \Omega$$
$$\Delta W = f \quad \text{in} \quad \Omega \tag{5.4.14}$$

with $U = W = 0$ on $\partial \Omega$. Then $\bar{W}$ converges to W and $\bar{U}$ converges to U. However, from (5.4.14), the pair (U, W) *clearly solve a different problem than that originally set*; namely, they solve the plate problem with $v = 1$ in (5.4.9). Thus, $U \neq u$ and a finite element sequence of approximations $\{u_h\}$ for this problem not only fails to converge to u but will, in fact, converge to the solution of a different problem. As the mesh size is refined, the finite element solutions converge to the solution for $v = 1$ independent of the actual value of v. This independence of v is to be expected since on any open straight-line boundary segment the moment condition simplifies to $u_{nn} = 0$ and *the Poisson ratio is absent from the problem statement.* For this reason the phenomenon has also occasionally been termed the "Poisson-ratio" effect. As indicated earlier in the discussion, this behavior can be interpreted as due to the fact that extension of u_h by zero outside Ω_h violates the C^1-property of the finite element space when the boundary $\partial \Omega_h$ is simply supported. We also note that the tangent direction has a jump discontinuity at a corner so that the corners are singular points.

5.4.3 Isoparametric Elements

Isoparametric curved elements would appear to offer a simple solution, but other problems arise when these elements are employed. In developing isoparametric elements we construct a polynomial map between the straight-sided master element $\hat{\Omega}$ and a typical element Ω_e which has curved sides. The inverse map from Ω_e to $\hat{\Omega}$ does not enter explicitly in the formulation, but under this mapping the variational functional is transformed to one with nonconstant coefficients, and usually involves rational functions (see Volume I, Chapter 5). In effect, the collection of element maps from Ω_e to $\hat{\Omega}$ equivalently transform the domain Ω with curved boundary to one with straight sides. The problem on the new domain is no longer the plate problem because of the variable coefficients introduced by the mapping. The real difficulty now is that the map from a smooth domain to a polygonal domain can no longer be C^1. Hence, under the transformation the element is no longer conforming. Convergence is suspect even in the case of the clamped plate.

The difficulty can be resolved in several ways. A straightforward strategy is to use triangles with a single curved side on the boundary. The other sides are straight and C^1 conformity is maintained. Let us first comment on the case where the nodal parameters for the boundary conditions remain unchanged and the nodes on the curved sides fall fortuitously on the actual boundary $\partial \Omega$. If the boundary arc for an element is a polynomial of degree $k - 1$, where k is the degree of the element, then the L^2 error of the solution is $O(h^{k+1})$ as before. On reducing the degree of the polynomial on the boundary to $k - 1 - j$, the L^2 error estimate is $O(h^{k+1/2-j})$. Now consider the case where the boundary conditions must be interpolated on the curved sides. These side nodes may be positioned at the boundary edge and essential conditions enforced there, but this choice of boundary nodes pollutes the solution (Berger et al. [1972]). The improved estimate $\| e \|_0 \leq Ch^{k-1}$ is optimal and is obtained if the boundary nodes for $u = 0$ and $u_n = 0$ are the zeros of the polynomials $P_k^{(k-3)}(s)$ and $P_k^{(k-2)}(s)$, where $s \in [s_1, s_2]$ is the arc length along the element boundary, $P_k(s) = (s - s_1)^{k-1}(s_2 - s)^{k-1}$, and the superscript implies differentiation of $P_k(s)$. These polynomials are orthogonal to polynomials of degree $k - 4$ and $k - 3$ in s. The zeros of $P_k^{(k-2)}(s)$ are the Lobatto quadrature points and give the optimal location of the interior nodes for the clamped plate.

5.4.4 Modified Functional

Another approach is to modify the variational functional so that the troublesome boundary conditions are included naturally either by multiplier methods or using a penalty functional. Again the triangulation Ω_h is restricted to have elements with at most one curved side on the boundary and straight sides in the interior, so the approximation is C^1. Let $\partial\Omega_1$ and $\partial\Omega_2$ denote those portions of the boundary which are simply supported and clamped, respectively. Define

$$M(u) = \Delta u + (1 - v)u_{tt} \quad \text{and} \quad N(u) = (\Delta u)_n + (1 - v)\frac{\partial}{\partial s}(u_{nt}) \qquad (5.4.15)$$

where (n, t) are normal-tangent directions and s is arc length. Now construct the modified functional (Nitsche [1972])

$$B_\epsilon(u, v) = \int_\Omega \Delta u \, \Delta v \, dx + \int_{\partial\Omega_1} [N(u)v + uN(v)] \, ds$$

$$+ \int_{\partial\Omega_2} [M(u)v_n + u_n M(v)] \, ds \qquad (5.4.16)$$

$$+ \epsilon h^{-3} \int_{\partial\Omega_1} uv \, dx + \epsilon h^{-1} \int_{\partial\Omega_2} u_n v_n \, ds$$

where ϵ is a positive constant. The finite element solution u_h^ϵ satisfies

$$B_\epsilon(u_h^\epsilon, v_h) = (f, v_h) \qquad (5.4.17)$$

for all $v_h \in H^k(\Omega_h)$, where $(\cdot, \cdot)$ denotes the usual L^2 inner product.

Let p_e be the diameter of the inscribed circle in element Ω_e and $d_e = p_e/h$. Let d_{min} denote the minimum of d_e for elements adjacent to the boundary $\partial\Omega$. There are constants ϵ_0 and h_0 dependent only on d_{min}, Ω, and the polynomial degree k such that, if $\epsilon \geq \epsilon_0$ and $h \leq h_0$,

$$B_\epsilon(v_h, v_h) \geq \frac{\alpha}{2} \|v_h\|_0^2 \qquad \text{for all } v_h \in H^k(\Omega_h)$$

where α is a positive constant. Then u_h^ϵ is well defined and we have the L^2 error estimate

$$\| u - u_h^\epsilon \|_0 \leq Ch^{k-2} \| u \|_k \qquad (5.4.18)$$

with C dependent only on Ω, $d_{\min}$, and k. The dependence on $d_{\min}$ implies restrictions on the mesh such as uniformity, which necessitates small boundary triangles and reduces efficiency. Further modifying the functional and using C^3 spaces of finite elements removes the restriction on $d_{\min}$ but the increase in complexity of the basis and matrix structure are computationally prohibitive.

EXERCISES

5.4.1 Derive the Green's formula for the biharmonic operator. Verify that the variational problem (5.4.5) corresponds to the Euler equation (5.4.1) and natural boundary conditions (5.4.6) as stated.

5.4.2 Verify that the variational statement (5.4.10) yields the simple support condition as natural boundary condition. By examining this natural boundary condition on a straight-edge segment of $\partial\Omega_h$, provide an interpretation of the "Poisson ratio" effect.

5.5 NONCONFORMING ELEMENTS

5.5.1 Introduction

The need for continuity of the finite element approximation of the solution of a boundary-value problem and possibly of its derivatives has been frequently mentioned in our discussion of the use of finite element bases for application of Galerkin's method. We generally seek to construct an approximation space H^h which is a subspace of the space of admissible functions H, and in the applications considered thus far, this requires conforming elements. However, conforming elements, particularly for higher-order problems, are usually more complex. The examples of C^1-elements in Chapter 2 clearly illustrate this fact. Hence, it is natural to enquire if one can relax these interelement continuity requirements (thereby departing from a standard Galerkin method) and still obtain a convergent finite element solution.

Our purpose in this section is to examine *nonconforming* elements and to describe error estimates that indicate the effect of nonconformity on accuracy. We shall also identify specific types of nonconforming elements that yield convergent finite element approximations. This leads directly to the problem of formulating an appropriate "patch test" to determine if a given nonconforming element is permissible. We begin the discussion by introducing the standard form of a variational problem and some examples of nonconforming elements.

5.5.2 Examples

Let us consider a general class of elliptic variational boundary-value problems of the type: Find $u \in H$ such that

$$B(u, v) = f(v) \qquad \forall v \in H \tag{5.5.1}$$

where $B(\cdot, \cdot)$ is the fundamental bilinear form of the problem, f is a linear form corresponding to given data, and H is the space of admissible functions. Two standard examples of (5.5.1) are the Dirichlet problem for Poisson's equation in two dimensions, in which case

$$H = H_0^1(\Omega), \qquad B(u, v) = \int_\Omega \nabla u \cdot \nabla v \, dx \tag{5.5.2}$$

and the Dirichlet problem for the biharmonic operator, particularly the problem of bending of a clamped flat plate, in which case

$$H = H_0^2(\Omega), \qquad B(u, v) = \int_\Omega \gamma(u, v) \, dx$$

$$\gamma(u, v) = \Delta u \, \Delta v + (1 - v) \left\{ 2 \frac{\partial^2 u}{\partial x \, \partial y} \frac{\partial^2 v}{\partial x \, \partial y} - \frac{\partial^2 u}{\partial x^2} \frac{\partial^2 v}{\partial y^2} - \frac{\partial^2 u}{\partial y^2} \frac{\partial^2 v}{\partial x^2} \right\} \tag{5.5.3}$$

$$v = \text{constant}, \quad 0 \leq v < \tfrac{1}{2}$$

In either case, we take

$$f(v) = \int_\Omega fv \, dx, \qquad f \in L^2(\Omega), \quad v \in H \tag{5.5.4}$$

and assume that Ω is a regular domain in $\mathbb{R}^2$ (e.g., a convex polygon).

We have discussed Galerkin finite element approximations of problems of the type (5.5.1) earlier in this volume. The standard procedure for constructing the approximation space H^h is to partition Ω into a collection of E finite elements Ω_e and construct over this mesh of elements basis functions ϕ_i which are piecewise polynomials. In the case of the bilinear form in (5.5.2), we have previously demanded that these basis functions be C^0-functions since they then span a finite-dimensional "approximation space" H^h which is a subspace of H (it being assumed here that $\Omega_h = \Omega$). This leads to a conforming finite element approximation of (5.5.2). Similarly, to construct a conforming finite element approximation of the form in (5.5.3), a C^1-finite element must be employed.

There are, however, successful finite element methods which employ elements that do not satisfy these continuity requirements. These nonconforming elements, being simpler in form, exhibit some advantages over conforming elements, particularly for fourth- and higher-order problems where the continuity requirements of conforming elements lead to quite complex elements and large element matrices. However, not all nonconforming elements behave well—they may give solutions on successively refined meshes that diverge or, more frequently, that converge to the wrong function.

Some examples of nonconforming elements for second-order problems are illustrated in Fig. 5.14. The nonconforming linear triangle in Fig. 5.14a

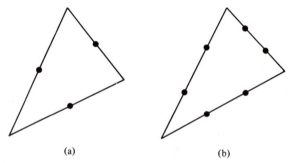

(a)   (b)

FIGURE 5.14 *Some nonconforming linear and quadratic triangles for second-order problems.*

has nodes at the midpoint of each side, and thus the global approximation u_h may be continuous across interelement boundaries only at these midpoints. Similarly, if quadratic shape functions are used with nodes at the trisection point of each side, as shown in Fig. 5.14b, a quadratic nonconforming element is obtained.

For fourth-order problems the elements of Fig. 5.15 have been used. The shape functions for the triangle in Fig. 5.15a are complete quadratics and the degrees of freedom are the function values at the vertices and normal derivatives at the midside nodes, (Morley [1968]). The element in Fig. 5.15b was developed by de Veubeke [1974] and employs a complete cubic: the 10 degrees of freedom are the solution values at the vertices and centroid and the normal derivatives at the Gauss points on each side. The triangle in Fig. 5.15c was proposed by Zienkiewicz [1971]. It has function values and tan-

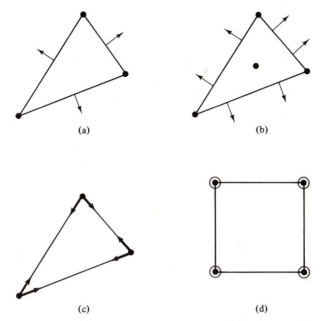

FIGURE 5.15 *Some nonconforming elements for fourth-order problems.*

gential derivatives to give nine degrees of freedom, as shown, and is an incomplete cubic. Finally, the Adini rectangle in Fig. 5.15d (Adini and Clough [1961]) has 12 degrees of freedom, these being the function values and first derivative at each node, and is constructed using a complete cubic plus the two fourth-order terms $\xi^3\eta$ and $\xi\eta^3$.

5.5.3 Nonconforming Approximation of the Variational Problem

Perhaps the first point that we should examine is exactly how a nonconforming approximation of (5.5.1) can even be considered when it is clear that the form $B(\cdot, \cdot)$ is not defined when nonconforming basis functions are introduced. The answer lies in our interpretation of these forms when using finite element methods: namely, that the form $B(\cdot, \cdot)$ is replaced by $B_h(\cdot, \cdot)$, which is obtained by accumulating contributions from each element in the usual manner. Thus, local approximations of $B(\cdot, \cdot)$ and $f(\cdot)$ are made over each element and these are summed over all elements, the possible jumps and discontinuities at element interfaces resulting from nonconformity of u_h being ignored. For example, let u_h^e and v_h^e be local approximations of u and v

over element Ω_e, with

$$u_h^e = \sum_{e=1}^{N_e} u_j^e \psi_j^e(\mathbf{x}), \qquad \mathbf{x} \in \bar{\Omega}_e, \quad 1 \le e \le E, \qquad (5.5.5)$$

where ψ_j^e are the shape functions. We first compute, as usual, the element contributions, $B_e(u_h^e, v_h^e)$ and $f_e(v_h^e)$. These are simply the restrictions of $B(\cdot, \cdot)$ and $f(\cdot)$ to Ω_e. We then construct the sums

$$B_h(u_h, v_h) = \sum_{e=1}^{E} B_e(u_h^e, v_h^e), \qquad f(v_h) = \sum_{e=1}^{E} f_e(v_h^e) \qquad (5.5.6)$$

Our *nonconforming approximation* of (5.5.1) is then to find $u_h \in \hat{H}^h$, such that

$$B_h(u_h, v_h) = f(v_h) \qquad \forall v_h \in \hat{H}^h \qquad (5.5.7)$$

where $\hat{H}^h$ is the finite-dimensional space generated by the nonconforming basis functions and, in general, $\hat{H}^h \not\subset H$.

5.5.4 Error Estimates

Recall that the bilinear form $B(\cdot, \cdot)$ is H-elliptic if a constant $\alpha > 0$ exists such that

$$B(v, v) \ge \alpha \| v \|_H^2 \qquad \forall v \in H \qquad (5.5.8)$$

This condition plays a key role in the stability and well-posedness of elliptic problems and in the determination of error estimates for conforming elements. A similar condition must hold on the discrete space $\hat{H}^h$ if we are to expect nonconforming elements to lead to nonsingular and well-behaved stiffness matrices. We first construct an appropriate norm on $\hat{H}^h$ for analyzing the nonconforming method. We will assume that a functional $\| \cdot \|_h : \hat{H}^h \to \mathbb{R}$ can be defined which is positive definite and satisfies a triangle inequality on $\hat{H}^h$. For example, for the form in (5.5.2) we take

$$\| v_h \|_h = \left(\sum_{e=1}^{E} \int_{\Omega_e} | \nabla v_h |^2 \, dx \right)^{1/2} \qquad (5.5.9)$$

whereas for the form in (5.5.3) we choose

$$\| v_h \|_h = \left\{ \sum_{e=1}^{E} \int_{\Omega_e} \left[\left(\frac{\partial^2 v_h}{\partial x^2} \right)^2 + 2 \left(\frac{\partial^2 v_h}{\partial x\, \partial y} \right)^2 + \frac{\partial^2 v_h}{\partial y^2} \right)^2 \right] dx\, dy \right\}^{1/2} \qquad (5.5.10)$$

We shall confine our discussion to nonconforming finite element approximations for which the following conditions hold:

(a) $\| \cdot \|_h$ is a norm on $\hat{H}^h$.

(b)   There exists a constant $\hat{a} > 0$, independent of h, such that

$$B_h(v_h, v_h) \geq \hat{a} \, \| v_h \|_h^2 \qquad \forall v_h \in \hat{H}^h \tag{5.5.11}$$

(c) $B_h(\cdot, \cdot)$ is continuous on $\hat{H}^h$; that is, there is a constant $\hat{M}$ such that

$$B_h(u_h, v_h) \leq \hat{M} \, \| u_h \|_h \| v_h \|_h \qquad \forall u_h, v_h \in \hat{H}^h$$

Note that under these conditions there exists a unique solution to (5.5.7) for each choice of data f.

An estimate of the error $u - u_h$ in the discrete norm $\| \cdot \|_h$ is easily obtained. If (5.5.11) holds, then

$$
\begin{aligned}
\hat{a} \, \| u_h - v_h \|_h^2 &\leq B_h(u_h - v_h, u_h - v_h) \\
&= f(u_h - v_h) - B_h(u, u_h - v_h) - B_h(v_h - u, u_h - v_h) \\
&\leq \hat{M} \, \| u_h - v_h \|_h \| u - v_h \|_h + f(u_h - v_h) - B_h(u, u_h - v_h)
\end{aligned}
$$

where u_h is the solution of the nonconforming problem (5.5.7), u is the actual solution of (5.5.1), and v_h is an arbitrary element of $\hat{H}^h$. Thus,

$$\| u_h - v_h \|_h \leq C \left[\| u - v_h \|_h + \frac{|f(w_h) - B_h(u, w_h)|}{\| w_h \|_h} \right] \tag{5.5.12}$$

with C an arbitrary constant and $u_h - v_h = w_h \neq 0$. Since $\| \cdot \|_h$ is, by hypothesis, a norm, the triangle inequality holds:

$$
\begin{aligned}
\| e \|_h = \| u - u_h \|_h &= \| u - u_h + v_h - v_h \|_h \\
&\leq \| u - v_h \|_h + \| u_h - v_h \|_h \qquad \forall v_h \in \hat{H}_h
\end{aligned}
\tag{5.5.13}
$$

Combining (5.5.12) and (5.5.13), we arrive at the error bound,

$$\| u - u_h \|_h \leq C \left[\inf_{v_h \in \hat{H}^h} \| u - v_h \|_h + \sup_{w_h \in \hat{H}^h} \frac{|f(w_h) - B_h(u, w_h)|}{\| w_h \|_h} \right] \tag{5.5.14}$$

In most cases, the term $\inf \{ \| u - v_h \|_h, v_h \in \hat{H}^h \}$ can be bounded by the usual interpolation error; for example,

$$\inf_{v_h \in \hat{H}^h} \| u - v_h \|_h = \inf_{v_h \in \hat{H}^h} \left(\sum_{e=1}^{E} \| u - v_h \|_{1, \Omega_e}^2 \right)^{1/2} \leq C(u) h^\sigma \tag{5.5.15}$$

where $\sigma > 0$ depends on the degree k of the element and the regularity r of u.

Clearly, the "sup term" in the bound on the right of (5.5.14) arises from the nonconformity of the method and may be interpreted as a *consistency error*. Obviously, when (5.5.15) holds, a sufficient condition for convergence is the consistency condition

$$\lim_{h \to 0} \sup_{v_h \in \hat{H}^h} \frac{|f(v_h) - B_h(u, v_h)|}{\|v_h\|_h} \longrightarrow 0 \tag{5.5.16}$$

Notice that in the case of conforming elements,

$$\|u - u_h\|_h = \|u - u_h\|_H, \qquad f(w_h) - B_h(u, w_h) = 0$$

and (5.5.14) reduces to our usual error estimate for conforming elements,

$$\|u - u_h\|_H \leq C \inf_{v_h \in H^h} \|u - v_h\|_H \tag{5.5.17}$$

In certain cases, a sharper bound than (5.5.14) can be found. For example, Lascaux and Lesaint [1975] have shown that for the clamped-plate problem a simple application of Green's formula for the form in (5.5.3) yields

$$f(v_h) - B_h(u, v_h) = \begin{cases} E_1(u, v_h) & \text{if } \hat{H}^h \subset C^0(\Omega) \\ E_1(u, v_h) + E_2(u, v_h) + E_3(u, v_h) & \\ & \text{if } \hat{H}^h \not\subset C^0(\Omega) \end{cases} \tag{5.5.18}$$

where

$$\begin{aligned} E_1(u, v_h) &= \sum_{e=1}^{E} \int_{\partial \Omega_e} \left[(1 - v) \frac{\partial^2 u}{\partial t^2} - \Delta u \right] \frac{\partial v_h}{\partial n} \, ds \\ E_2(u, v_h) &= \sum_{e=1}^{E} \int_{\partial \Omega_e} -(1 - v) \frac{\partial^2 u}{\partial n \, \partial t} \frac{\partial v_h}{\partial t} \, ds \\ E_3(u, v_h) &= \sum_{e=1}^{E} \int_{\partial \Omega_e} \frac{\partial (\Delta u)}{\partial n} v_h \, ds \end{aligned} \tag{5.5.19}$$

and (n, t) are outward normal and tangential coordinates on the boundary $\partial \Omega_e$ of element Ω_e. The analysis of certain types of elements, therefore, reduces to a study of the forms E_1, E_2, and E_3 and their behavior as $h \to 0$.

5.5.5 Additional Examples

Second-Order Problems: We now examine the particular case of "Wilson's brick element" following the approach of Ciarlet [1978]. This is a three-dimensional rectangular brick element for second-order problems which is defined by the eight conforming trilinear shape functions $\frac{1}{8}(1 \pm \xi_1)$

$(1 \pm \xi_2)(1 \pm \xi_3)$ and three additional nonconforming quadratic shape functions $\xi_j^2, j = 1, 2, 3$, on the master cube $-1 \leq \xi_i \leq 1, i = 1, 2, 3$. The nodal values at the eight vertices define the trilinear approximation. To complete the description of the element, we can introduce the values of the second derivatives $\partial^2 u / \partial x_j^2$, $j = 1, 2, 3$, at the centroid as nodal degrees of freedom. However, since the second derivatives are constant for any polynomial of quadratic degree, it is better to specify the average value over Ω_e of the second derivatives $\int \partial^2 u / \partial x_j^2 \, dx, j = 1, 2, 3$, as degrees of freedom. For a quadratic polynomial, this average is the value of $\partial^2 u / \partial x_j^2$ at the centroid $\mathbf{x}_c$ scaled by the element volume V_e. This choice admits a broader class of functions for the interpolation since the integral form poses a weaker requirement on the second derivatives. Since the corresponding quadratic shape functions $\hat{\psi}_j(\xi) = \frac{1}{16}(\xi_j^2 - 1), j = 1, 2, 3$, are not C^0 on the surfaces, the element is nonconforming. The global approximation is continuous at the nodes since $\xi_j = \pm 1$ at these points.

The interpolation estimate for this element follows from the Ciarlet–Raviart theory (see Chapter 2 of Volume IV) and is (for $u \in H^l(\Omega)$)

$$|u - \bar{u}_h|_{m, \Omega_e} \leq \frac{C h_e^l}{\rho_e^m} |u|_{l, \Omega_e}, \qquad 0 \leq m \leq l, \quad l = 2, 3 \qquad (5.5.20)$$

where $\bar{u}_h$ is an appropriate projection in the finite element basis, C is a constant independent of the mesh, and h_e and ρ_e are the diameters of the circumscribed and inscribed spheres of Ω_e; $|\cdot|$ denotes the seminorm,

$$|u|_{m, \Omega_e}^2 = \int_{\Omega_e} \sum_{|\alpha| = m} |D^\alpha u|^2 \, dx \qquad (5.5.21)$$

where $\mathbf{x} = (x_1, x_2, \ldots, x_n)$, $\boldsymbol{\alpha} = (\alpha_1, \alpha_2, \ldots, \alpha_n)$, $\alpha_i > 0$, $|\boldsymbol{\alpha}| = \alpha_1 + \alpha_2 + \cdots + \alpha_n$ and

$$D^\alpha u = \frac{\partial^{|\alpha|} u}{\partial x_1^{\alpha_1} \cdots \partial x_n^{\alpha_n}}$$

Using the interpolation result (5.5.20), we obtain for (5.5.15)

$$\inf_{v_h \in \hat{H}_0^h} \|u - v_h\|_h \leq \left(\sum_{e=1}^E |u - \bar{u}_h|_{1, \Omega_e}^2 \right)^{1/2} \leq C h |u|_{2, \Omega} \qquad (5.5.22)$$

where we have taken $u \in H^2(\Omega)$ and assumed a regular family of discretizations and the constant C is again independent of mesh size h. As usual, $v_h \in \hat{H}_0^h$ implies that $v_h \in \hat{H}^h$ and satisfies homogeneous essential boundary conditions. If $u \in H^3(\Omega)$, the estimate is $O(h^2)$, but this gain in the rate will be lost in the estimate of the remaining term. In any event, the corner singularities for convex polygonal domains generally limit the smoothness of the solution.

The estimate for the consistency term in (5.5.14) is obtained from an error analysis of $f(v_h) - B_h(u, v_h)$ for $v_h \in H^h$. One can show that (see Ciarlet [1978])

$$\sup_{v_h \in \mathring{H}^h_0} \frac{|f(v_h) - B_h(u, v_h)|}{\|v_h\|_h} \leq Ch |u|_{2,\Omega} \qquad (5.5.23)$$

Combining the estimates of (5.5.22) and (5.5.23), we see that this nonconforming element converges with a rate that is $O(h)$ in the norm $\| \cdot \|_h$.

Some Nonconforming Plate Elements: As an example of a nonconforming element for fourth-order problems, we mention two triangular "plate" elements proposed by de Veubeke [1974] and shown in Fig. 5.16. The

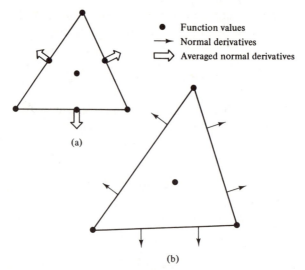

● Function values
⟶ Normal derivatives
⟹ Averaged normal derivatives

(a)

(b)

FIGURE 5.16 *Two nonconforming plate elements of Fraeijs de Veubeke for fourth-order problems.*

element in Fig. 5.16a has as degrees of freedom the values of the function at the vertices, the centroid, and the midpoints of each side, and the mean value of the normal derivative along each edge (10 degrees of freedom) and employs a complete cubic. For these choices, one can show that $P_2(\bar{\Omega}_e) \subset P_e \subset P_3(\bar{\Omega}_e)$, where P_e is the space spanned by the local shape functions for element Ω_e. The element in Fig. 5.16b has as degrees of freedom the values of the function at the vertices and the centroid of each element and the values of the normal derivatives at the Gauss points on each side of the element. Again, complete cubics are employed $[P_e = P_3(\bar{\Omega}_e)]$.

Lascaux and Lesaint [1975] have shown that these elements satisfy the consistency conditions given earlier in the sense that

$$
\left.
\begin{aligned}
&|E_1(u, w - w_h)| + |E_2(u, w - w_h)| \leq Ch\,\|u\|_{3,\Omega}\,\|w - w_h\|_h \\
&\qquad \forall u \in H^3(\Omega), \quad w \in H^2(\Omega), \quad w_h \in \hat{H}^h \\
&\text{and} \\
&|E_3(u, w - w_h)| \leq C(h\,\|u\|_{3,\Omega} + h^2\,\|u\|_{4,\Omega}) \\
&\qquad \forall u \in H^4(\Omega), \quad w \in H^2(\Omega), \quad w_h \in \hat{H}^h
\end{aligned}
\right\}
\qquad (5.5.24)
$$

where E_1, E_2, and E_3 are defined in (5.5.19). Thus, if u is a sufficiently smooth solution of the biharmonic problem with clamped boundary conditions,

$$ f(v_h) - B_h(u, v_h) \longrightarrow 0 \qquad \text{as } h \longrightarrow 0 $$

Of course, the usual assumptions of a quasiuniform mesh and regular refinement are in force. Thus, these elements produce approximations that converge in the $\|\cdot\|_h$-norm. In fact, it follows from (5.5.24) without difficulty that for these elements,

$$ \|u - u_h\|_h \leq C(h\,\|u\|_{3,\Omega} + h^2\,\|u\|_{4,\Omega}) \qquad (5.5.25) $$

whenever the actual solution $u \in H^4(\Omega)$. In fact, under slightly stronger hypotheses, Lascaux and Lesaint [1975] have been able to show that, for these elements,

$$ \|u - u_h\|_h = O(h^2) \qquad \text{for } u \in H^4(\Omega) \qquad (5.5.26) $$

This indicates that these rather simple plate elements may produce very satisfactory results for less computational effort than conforming elements.

5.5.6   Tests for Elements

The mixed success of nonconforming elements throughout the history of finite elements has led to attempts to devise tests to determine if candidate elements are consistent and convergent in appropriate norms. The "patch test" of Irons and Razzaque [1972] represents a simple computational device designed to test a "patch" of nonconforming elements in a given mesh to establish their consistency; the idea has been further refined and analyzed by Strang [1972] and despite its fallability in certain pathological cases remains a useful practical tool for testing nonconforming elements. (Stummel [1980] has shown that the earlier patch test is neither a necessary nor a sufficient condition for convergence and he has proposed a "generalized" patch test.)

The essential idea in the earlier forms of the patch test for elasticity applications due to Irons and Razzaque [1972] was that in the limit of decreasing mesh size the nonconforming finite element solution to the given problem posed on an arbitrary patch of elements should be able to represent the appropriate constant strain states. That is, for second-order problems if the patch test is passed, a linear function is approximated exactly. Similarly, for fourth-order problems, constant curvature on a patch implies that a quadratic function is approximated exactly. Hence, a simple numerical patch test can be devised for, say, the second-order case by simply specifying that the solution be linear on a typical patch, prescribing the solution values on the boundary of the patch and solving the resulting small finite element system to see if in fact the correct linear function is obtained.

Strang's test developed the idea further as follows. Let $B(\cdot, \cdot)$ and $B_h(\cdot, \cdot)$ be the bilinear forms defined earlier in (5.5.1) and (5.5.6). Moreover, let P_m be a polynomial of degree m, where $2m$ is the order of the differential operator and let γ be a nonconforming basis function. Then the bilinear functional is preserved if we satisfy for each such P_m and γ the consistency requirement

$$B(P_m, \gamma) - B_h(P_m, \gamma) = 0 \qquad (5.5.27)$$

For example, the bilinear functional for the Laplacian is given in (5.5.2). Here $m = 1$, so that $P_m(x, y) \equiv P_1(x, y) = a + bx + cy$, where a, b, and c are arbitrary constants. Then, on use of Gauss' divergence theorem on $\Omega_h = \bigcup_e \Omega_e$,

$$B(P_1, \gamma) - B_h(P_1, \gamma) = \oint \gamma \frac{\partial P_1}{\partial n} \, ds \qquad (5.5.28)$$

where the contour integral is performed around all external and internal boundaries, including the interelement boundaries in the interior of the domain. Hence, if the line integral in (5.5.28) vanishes for any arbitrary patch in the mesh, it follows that (5.5.27) holds.

It is worth noting that this sort of patch test is also implicit in the work of Lascaux and Lesaint [1975] on the nonconforming plate elements of de Veubeke in Fig. 5.16, although they do not claim that such a test is necessarily a key to the success of these elements. In particular, for the clamped-plate problem and these elements, they show that

$$f(v_h) - B_h(u, v_h) = 0$$

for any quadratic polynomial $u \in P_2(\Omega)$ and any $v_h \in \hat{H}^h$, which is equivalent to the Irons–Strang patch test for these elements. Since these elements are known to produce convergent approximations in $\| \cdot \|_h$ for solutions u

$\in H^4(\Omega)$, this result suggests that the patch test may be a good indication of the suitability of nonconforming methods.

More recently, Stummel [1980] developed a counterexample to the previous test which we now describe briefly. This counterexample also serves to illustrate the fact that some nonconforming elements such as this, which do not pass the generalized patch test of Stummel, may converge to the solution of a different problem.

Stummel's example is developed here for a special nonconforming basis and our model two-point problem

$$-u''(x) + u(x) = f(x) \quad \text{in} \quad 0 \le x \le 1$$

with

$$u(0) = 0 \quad \text{and} \quad u(1) = 1 \tag{5.5.29}$$

where $f \in L^2(0, 1)$.

The corresponding variational problem is to find $u \in H^1(0, 1)$ with $u(0) = 0$, $u(1) = 1$ such that

$$B(u, v) = f(v) \qquad \forall v \in H_0^1(0, 1) \tag{5.5.30}$$

where

$$B(u, v) = \int_0^1 (u'v' + uv)\, dx$$

and

$$f(v) = \int_0^1 fv\, dx \tag{5.5.31}$$

Consider a uniform discretization of $\Omega = (0, 1)$ as finite elements of length h and a nonconforming basis defining the approximation space $\hat{H}^h$. The finite element problem is then to find $u_h \in \hat{H}^h$ satisfying $u_h(0) = 0$ and $u_h(1) = 1$ such that

$$B_h(u_h, v_h) = f(v_h) \qquad \forall v_h \in \hat{H}^h, \quad v_h(0) = v_h(1) = 0 \tag{5.5.32}$$

where

$$B_h(u_h, v_h) = \sum_{e=1}^{E} \int_{\Omega_e} (u_h'v_h' + u_h v_h)\, dx$$

and

$$f(v_h) = \sum_{e=1}^{E} \int_{\Omega_e} fv_h\, dx \tag{5.5.33}$$

The interesting features of the example pertain to the construction of the space $\hat{H}^h$. Let $\{\chi_j\}$, $j = 0, 1, \ldots, N$, where $N = E$, be the conforming piecewise-linear "hat" functions satisfying $\chi_j(x_i) = \delta_{ij}$ and introduce the discontinuous piecewise-constant nonconforming functions γ_j defined for $j = 1, 2, \ldots, N$ by

$$\gamma_j(x) = \begin{cases} 1, & x_{j-1} < x < x_j \\ 0, & \text{otherwise} \end{cases} \tag{5.5.34}$$

Now let $\hat{H}^h$ be the linear space spanned by $\chi_1, \chi_2, \ldots, \chi_{N-1}$ and $\gamma_2, \gamma_3, \ldots, \gamma_{N-1}$. Any function $v_h \in \hat{H}^h$ can be expressed in the form

$$v_h(x) = \sum_{j=1}^{N-1} a_j \chi_j(x) + \sum_{j=2}^{N-1} b_j \gamma_j(x), \qquad 0 < x < 1 \tag{5.5.35}$$

where the first sum represents the conforming (continuous) part and the second sum represents the nonconforming (discontinuous) part. Let $v_h^1 \in H_0^1(0, 1)$ and $v_h^2 \in L^2(0, 1)$ denote these respective terms. The degrees of freedom may be readily identified as

$$a_j = v_h^1(x_j), \qquad b_j = v_h^2(x_{j-(1/2)}) = \frac{1}{h} \int_{\Omega_e} v_h^2(x) \, dx \tag{5.5.36}$$

where $\Omega_e = (x_{j-1}, x_j)$. Since the γ_j are discontinuous at the interelement nodes, it follows that the trial function v_h in (5.5.35) has jump discontinuities at these points but is smooth in the interior of each element.

With the construction of $\hat{H}^h$ now complete, we can introduce the trial and test functions of (5.5.35) into (5.5.32) and solve for u_h. Note that the derivative γ_j' is zero in an element so that on any element $\Omega_e = (x_{j-1}, x_j)$,

$$\frac{dv_h}{dx} = \frac{dv_h^1}{dx} \quad \text{and} \quad \frac{dv_h^2}{dx} = 0 \tag{5.5.37}$$

As the norm we take [see (5.5.9)]

$$\| v_h \|_h = \sum_{e=1}^{E} \int_{\Omega_e} [v_h^2 + (v_h')^2] \, dx \tag{5.5.38}$$

Using the error analysis of the preceding section it follows that approximability is easily satisfied so that the first term in the bound (5.5.14) approaches zero as h approaches zero at a rate $O(h)$. However, the consistency error in (5.5.14) is not well behaved as $h \longrightarrow 0$, since

$$\sup_{v_h \in \hat{H}^h} \frac{|f(v_h) - B_h(u, v_h)|}{\| v_h \|_h} = \| w_h - \Pi_h w \|_h \tag{5.5.39}$$

where $w = u - g$ for $g(x) = x$, which interpolates the Dirichlet data linearly, $w_h = u_h - g$, and $\Pi_h w$ is the orthogonal projection of w [$w_h - \Pi_h w$ is orthogonal to the nonconforming finite element subspace $\hat{H}^h$ with respect to the inner product $(\cdot, \cdot)_h$ corresponding to the chosen norm (5.5.38)].

One can then show (see Stummel [1980]) that

$$\lim_{h\to 0}\left[\sup_{v_h\in\hat{H}^h}\frac{|f(v_h)-B_h(u, v_h)|}{\|v_h\|_h}\right]\geq\frac{\|w''\|_0^2}{\|w\|_0}>C>0 \qquad (5.5.40)$$

for any solution u such that $u - g \in H^2(0, 1) \cap H_0^1(0, 1)$ and provided that $u - g \neq 0$. It follows then that u_h cannot in general converge to u.

The patch test in (5.5.27) is now considered for Stummel's example. We proceed as in (5.5.28) to write γ_i for γ, so that

$$B(P_1, \gamma_i) - B_h(P_1, \gamma_i) = \int_0^1 (P_1'\gamma_i' + P_1\gamma_i)\, dx$$
$$\qquad\qquad\qquad\qquad - \sum_{e=1}^E \int_{\Omega_e}(P_1'\gamma_i' + P_1\gamma_i)\, dx \qquad (5.5.41)$$

Integrating by parts yields

$$B(P_1, \gamma_i) - B_h(P_1, \gamma_i) = \int_0^1 (-P_1'' + P_1)\gamma_i\, dx$$
$$\qquad\qquad\qquad - \int_0^1 (-P_1'' + P_1)\gamma_i\, dx \qquad (5.5.42)$$
$$\qquad\qquad\qquad + \sum_{j=1}^{N-1}[\gamma_i(x_j^-)P'(x_j^-) - \gamma_i(x_{j-1}^+)P'(x_{j-1}^+)]$$

so that

$$B(P_1, \gamma_i) - B_h(P_1, \gamma_i) = \sum_{j=1}^{N-1}[\gamma_i(x_j^-)P'(x_j^-) - \gamma_i(x_{j-1}^+)P'(x_{j-1}^+)]$$

where x_j^-, x_{j-1}^+ denote the limits as x_j is approached from the left and x_{j-1} from the right in element $\Omega_e \equiv \Omega_j$.

From the construction of γ_i in (5.5.34) it follows that

$$B(P_1, \gamma_i) - B_h(P_1\gamma_i) = P_1'(x_{i-1}) - P_1'(x_i) \qquad (5.5.43)$$

Since P_1' is constant, we have

$$B(P_1, \gamma_i) - B_h(P_1, \gamma_i) = 0 \qquad (5.5.44)$$

so that (5.5.27) is satisfied even though the scheme fails to converge as noted earlier.

The example demonstrates that a more general test is required and Stummel [1980] has developed a generalized patch test which, in the present one-dimensional case, is defined by the requirement that

$$\lim_{h \to 0} \sum_{j=1}^{N-1} \psi(x_j)[\gamma_i(x_j^+) - \gamma_i(x_j^-)] = 0 \qquad (5.5.45)$$

for every smooth test function ψ, $\psi \in C_0^\infty(\mathbb{R}^n)$ and each nonconforming γ_i. On repeating the steps above, we find that (5.5.45) is not satisfied, so this rejects the nonconforming basis as inadmissible. Further details of the generalized test are given in the references cited together with proofs showing its necessity and sufficiency for convergence.

EXERCISES

5.5.1 Verify that $\| \cdot \|_h$ as defined in (5.5.9) satisfies the properties for a norm on $\hat{H}^h$.

5.5.2 Write down the basis functions for the rectangular analogue of "Wilson's brick element" and determine estimates similar to (5.5.23).

5.6 EIGENVALUE PROBLEMS

5.6.1 Introduction

Eigenvalue problems arise frequently in the analysis of periodic oscillations and stability of physical systems. For example, in certain problems of free vibration of structures the solution consists of pairs (λ, u), where $\lambda = \omega^2$, ω being the frequency of vibration, and u is a displacement field representing the corresponding eigenfunction or vibration mode. In turn, these vibration modes can be used to represent the forced motion of the structure. Similarly, the buckling of a column is a familiar example of a stability problem from solid mechanics. Here the buckling load is proportional to the smallest eigenvalue for the corresponding differential operator of beam theory and the buckling mode is the associated eigenfunction. Similar problems arise in nonstructural applications. Finally, eigenvalue problems are also important in some aspects of the numerical analysis of finite element methods, such as the conditioning of the stiffness matrix and the control of numerical oscillations and stability in the numerical integration of certain transient problems.

In the eigenvalue problems considered here we deal with the standard situation in which a boundary-value problem is characterized by a homogeneous governing equation for which a nontrivial solution, satisfying certain homogeneous boundary conditions, exists only if a parameter of the equation has specific values (the eigenvalues). We shall begin with a statement of the linear eigenvalue problem and develop some of the fundamental properties of eigenvalues for linear differential operators. We then construct a variational formulation of a general class of linear eigenvalue problems which, as usual, provides the basis for finite element approximations. Error estimates are established for a standard finite element analysis. We conclude with a brief discussion of some computational aspects.

5.6.2   Linear Eigenvalue Problems

We are principally concerned with the study of linear eigenvalue problems of the form: Find the pair (u, λ) such that

$$
\left.
\begin{aligned}
Au &= \lambda Cu \quad \text{in} \quad \Omega \\[2mm]
Mu &= 0 \quad\quad \text{on} \quad \partial\Omega
\end{aligned}
\right\}
\tag{5.6.1}
$$

with

where A, C, and M are linear operators and Ω is a smooth bounded domain. In many situations of interest C is the identity operator, so that (5.6.1) simplifies to the standard eigenvalue problem. Operators A and M are linear operators and in the case of differential operators for linear boundary-value problems, these operators have the properties described in Chapters 1 and 3. That is, A is an elliptic operator of order $2m$ and M is a compatible boundary operator of order m. Since many of our problems are derived from conservation laws, in keeping with our treatment of elliptic boundary-value problems, we shall assume that A is formally self-adjoint. Using the notation in Chapter 1, (1.3.1)–(1.3.6), we associate with the linear elliptic operator A a bilinear form $B(\cdot, \cdot)$ representing an integral of the type

$$
\begin{aligned}
B(u, v) = \int_{\Omega} \Bigg[& a_{mm}^{11} \frac{\partial^m u}{\partial x_1^m} \frac{\partial^m v}{\partial x_1^m} + a_{mm-1}^{11} \frac{\partial^m u}{\partial x_1^m} \frac{\partial^{m-1} v}{\partial x_1^{m-1}} + \cdots \\
& + a_{mm}^{NN} \frac{\partial^m u}{\partial x_N^m} \frac{\partial^m v}{\partial x_N^m} + \cdots + a_{00}^{NN} uv \Bigg] dx
\end{aligned}
\tag{5.6.2}
$$

where Ω is an open bounded domain in $\mathbb{R}^n$ (typically $n = 1$, 2, or 3).

We can construct a variational statement corresponding to the standard eigenvalue problem [C is the identity in (5.6.1)] of the form: Find (u, λ) with $u \in H$ satisfying the essential conditions $Mu = 0$ on $\partial\Omega$ and such that

$$
B(u, v) = \lambda(u, v)
\tag{5.6.3}
$$

for all test functions $v \in H$, where $H \subset H^m(\Omega)$ is the class of admissible functions and $(\cdot, \cdot)$ denotes the L^2-inner product. The operator A is assumed elliptic and regular, so that the bilinear form $B(\cdot, \cdot)$ is continuous and coercive. That is, we continue to assume that there exist strictly positive constants M and α such that, for all u and v in H,

$$B(u, v) \leq M \| u \|_m \| v \|_m$$

and

$$B(v, v) \geq \alpha \| v \|_m^2$$

$\hspace{2cm}$ (5.6.4)

We now consider some basic properties of the eigensolution of (5.6.3). Under the stated assumptions on A the eigenvalue spectrum may be infinite but is countable; for example, if A is the Laplacian Δ on a unit square $[0, 1] \times [0, 1]$ with $u = 0$ on the boundary, then the eigenvalues are $\{\pi^2(m^2 + n^2)\}$ and eigenfunctions are $\{C \sin m\pi x \sin n\pi y\}$, $m, n = 1, 2, \ldots$, with $C = $ constant $\neq 0$. Let λ^r and λ^s be distinct eigenvalues and u^r, u^s corresponding eigenfunctions satisfying (5.6.3). That is,

$$B(u^r, v) = \lambda^r(u^r, v) \qquad \forall v \in H \hspace{1.5cm} (5.6.5)$$

and

$$B(u^s, v) = \lambda^s(u^s, v) \qquad \forall v \in H \hspace{1.5cm} (5.6.6)$$

Setting $v = u^s$ in (5.6.5) and $v = u^r$ in (5.6.6) and subtracting, we get

$$(\lambda^r - \lambda^s)(u^r, u^s) = 0 \hspace{2cm} (5.6.7)$$

Since λ^r and λ^s are assumed distinct, (5.6.7) implies that the eigenfunctions u^r and u^s are orthogonal with respect to the L^2-inner product. That is,

$$(u^r, u^s) = 0, \qquad r \neq s \hspace{2cm} (5.6.8)$$

It is also easily verified that the eigenfunctions λ^s of (5.6.3) are real: first, we introduce the generalization of the bilinear form and inner product to complex-valued functions with λ belonging to the complex number field; for example, the L^2-inner product becomes $(u, v) = \int u\bar{v} \, dx$, where $\bar{v}$ is the complex conjugate of v. Let us assume now that an eigenvalue λ^r of $Au = \lambda u$, or equivalently of the variational problem, is complex. Taking the complex conjugate of this operator equation, it follows that the complex conjugate $\bar{\lambda}^r$ is also an eigenvalue with eigenfunction $\bar{u}^r$. Then setting $\bar{u}^r$ for u^s in the orthogonality relation (5.6.8), we conclude that $u^r \equiv 0$ in Ω. Hence the eigenvalues are real. We note also that as the eigenvalues are real, we need consider only real eigenfunctions—if u^r were complex, both real and imaginary parts would separately satisfy $Au = \lambda^r u$.

Next we note that if we set $v = u^r \neq 0$ in (5.6.5), we have

$$B(u^r, u^r) = \lambda^r(u^r, u^r)$$

so that the eigenvalue λ^r is defined by the *Rayleigh quotient* as

$$\lambda^r = \frac{B(u^r, u^r)}{(u^r, u^r)} \tag{5.6.9}$$

or

$$\lambda^r = \frac{B(u^r, u^r)}{\| u^r \|_0^2} \tag{5.6.10}$$

We can use the continuity and coercivity properties of $B(\cdot, \cdot)$ in (5.6.4) to develop further properties of the eigenvalues for the class of problems of interest. For instance, since $B(v, v) \geq \alpha \| v \|_m^2$ for all $v \in H$, where α is a strictly positive constant, then it follows from (5.6.10) that

$$\lambda^r = \frac{B(u^r, u^r)}{\| u^r \|_0^2} \geq \frac{\alpha \| u^r \|_m^2}{\| u^r \|_0^2} \geq \alpha > 0 \tag{5.6.11}$$

since $\| u^r \|_0 \leq \| u^r \|_m$, $m \geq 0$.

In fact, setting u and v to be u^r in both continuity and coercivity relations (5.6.4), we have

$$\alpha \| u^r \|_m^2 \leq B(u^r, u^r) \leq M \| u^r \|_m^2 \tag{5.6.12}$$

where α and M are the coercivity and continuity constants for $B(\cdot, \cdot)$. Combining this inequality with the Rayleigh quotient in (5.6.10) gives

$$0 < \alpha \leq \frac{\alpha \| u^r \|_m^2}{\| u^r \|_0^2} \leq \lambda^r \leq \frac{M \| u^r \|_m^2}{\| u^r \|_0^2} \tag{5.6.13}$$

Similar results can be developed for the general eigenvalue problem by constructing a bilinear form associated with operator C to replace the L^2-inner product (u, v) in the variational statement. We leave these extensions and the treatment of more general boundary operators to the exercises.

EXERCISES

5.6.1 Consider the class of two-point eigenvalue problems characterized by

$$-(au')' + bu = \lambda u$$

with

$$u(0) = 0 \quad \text{and} \quad u(1) = 0$$

where $a(x) \geq a_0 > 0$ and $b(x) \geq 0$, a and b being smooth functions. Construct the variational eigenvalue problem and verify the continuity and coercivity properties for the associated bilinear form $B(\cdot, \cdot)$. Write down the Rayleigh quotient.

[*HINT:* Use the Cauchy–Schwarz and Poincaré inequalities.]

5.6.2 Develop the eigenvalue problem for the linear operator in Exercise 5.6.1 with general boundary conditions of the form

$$\alpha_1 u'(0) + \alpha_2 u(0) = 0$$
$$\beta_1 u'(1) + \beta_2 u(1) = 0$$

where at least one of α_1 and α_2 and at least one of β_1 and β_2 are not zero. Consider also the case of periodic end conditions

$$a(0) = a(1), \quad b(0) = b(1), \quad u(0) = u(1), \quad u'(0) = u'(1)$$

Show that the bilinear functionals satisfy the usual coercivity and continuity conditions under the standard assumptions on a and b.

5.6.3 Construct the variational eigenvalue problem for the Laplacian in two dimensions for homogeneous Dirichlet boundary conditions. Using Cauchy–Schwarz and Friedrich's inequalities, show that the bilinear form is continuous and coercive. Write down the Rayleigh quotient.

5.6.4 Construct the variational eigenvalue problem for

$$(au'')'' - (bu')' + cu = \lambda u, \qquad 0 < x < 1$$

with clamped end conditions

$$u(0) = u(1) = 0, \qquad u'(0) = u'(1) = 0$$

and then with simple support end conditions

$$u(0) = u(1) = 0, \qquad u''(0) = u''(1) = 0$$

Determine sufficient conditions on functions a, b, and c for the bilinear form $B(\cdot, \cdot)$ to be continuous and coercive for this problem. Write down the Rayleigh quotient.

5.6.5 Formulate the variational eigenvalue problem for free vibration of a plate that is (a) clamped and (b) simply supported. Develop the expression for the Rayleigh quotient.

5.6.6 Verify that the stationary values of the Rayleigh quotient are the eigenvalues.

5.6.3 Finite Element Approximation

We return now to the standard eigenvalue problems (5.6.1) (with C the identity). The variational form in (5.6.3) is to find $\lambda \in \mathbb{R}$ and $u \in H \subset H^m(\Omega)$ satisfying the homogeneous essential boundary conditions and such that

$$B(u, v) = \lambda(u, v) \qquad \forall v \in H \tag{5.6.14}$$

As in the Galerkin approximation of linear elliptic boundary value problems, the approximate analysis is based on constructing an appropriate finite-dimensional subspace $H^h \subset H$ and posing the variational problem in this subspace. For convenience, we shall take $\Omega_h = \Omega$. The approximate eigenvalue problem for (5.6.14) becomes: find $\lambda_h \in \mathbb{R}$ and $u_h \in H^h$ such that

$$B(u_h, v_h) = \lambda_h(u_h, v_h) \qquad \forall v_h \in H^h \tag{5.6.15}$$

Let us describe a simple illustrative example prior to developing the properties and error estimates in a more general discussion.

Example: The Vibrating String

A simple illustrative example of the linear eigenvalue problem is free vibration of a string of length l fixed at each end and in tension. The corresponding eigenvalue problem is to find λ and u satisfying

$$-u''(x) = \lambda u(x), \qquad u(0) = u(l) = 0 \tag{5.6.16}$$

Integrating by parts in the corresponding weighted-residual expression we obtain the variational form of the eigenvalue problem: find $\lambda \in \mathbb{R}$, $u \in H = H_0^1(0, l)$, such that

$$B(u, v) = \lambda(u, v) \qquad \forall v \in H \tag{5.6.17}$$

where

$$B(u, v) = \int_0^l u'v' \, dx \qquad \text{and} \qquad (u, v) = \int_0^l uv \, dx \tag{5.6.18}$$

The approximate problem is to find $\lambda_h \in \mathbb{R}$, $u_h \in H^h \subset H$, satisfying (5.6.17) for all $v_h \in H^h$. The finite element subspace is constructed in the usual manner. For example, if we use a C^0 basis of piecewise polynomials on a discretization containing N nodes, then

$$u_h(x) = \sum_{j=1}^N u_j \phi_j(x) \tag{5.6.19}$$

273

where $\{\phi_j\}$ are, once again, the global basis functions and $\{u_j\}$ are the nodal degrees of freedom. Substituting (5.6.19) into (5.6.17) and setting $v_h = \phi_i$, $i = 2, 3, \ldots, N - 1$, we have

$$\sum_{j=1}^{N} \left(\int_0^l \phi_i' \phi_j' \, dx \right) u_j = \lambda_h \sum_{j=1}^{N} \left(\int_0^l \phi_i \phi_j \right) dx,$$
$$i = 2, 3, \ldots, N - 1 \tag{5.6.20}$$

or, in matrix form,

$$\mathbf{Ku} = \lambda_h \mathbf{Cu} \tag{5.6.21}$$

where $\mathbf{K} \equiv (k_{ij})$ and $\mathbf{C} \equiv (c_{ij})$ are given in (5.6.20). Equation (5.6.21) is a generalized linear eigenvalue problem for the finite-dimensional discrete problem. This algebraic eigenvalue problem may now be solved to yield the eigenvalues λ_h^s and eigenvectors $\mathbf{u}_h^s$. We observe that the Galerkin finite element method has determined a generalized eigenvalue problem, whereas finite differencing the original problem would lead to a standard eigenvalue problem since $\mathbf{C} = \mathbf{I}$ for the difference system. One can transform (5.6.21) to standard form simply by multiplying through by $\mathbf{C}^{-1}$, but this leads to a system matrix $\mathbf{C}^{-1}\mathbf{K}$ which is generally not sparse and banded, so this strategy is computationally unattractive. Motivated by the form of the difference system, so-called "lumping" techniques have been introduced in which $\mathbf{C}$ is approximated by a diagonal matrix. Analysis of lumping and the eigenvalue properties of the resulting system are important in the solution of transport processes, a topic we consider in some detail in Volume VI.

Turning now to the element calculations for the vibrating string, the element integrals follow from (5.6.20) as

$$k_{ij}^e = \int_{\Omega_e} (\psi_i^e)'(\psi_j^e)' \, dx, \qquad c_{ij}^e = \int_{\Omega_e} \psi_i^e \psi_j^e \, dx \tag{5.6.22}$$

where, as usual, the shape function ψ_i^e is the restriction of ϕ_i to element Ω_e.

If we consider, for example, a uniform mesh of linear elements and set $u_i = \rho^{ih}$ in (5.6.20), the representative discrete equation at node i becomes

$$\beta \rho^{(i-1)h} + \alpha \rho^{ih} + \beta \rho^{(i+1)h} = 0 \tag{5.6.23}$$

with $\beta = 1 + \lambda h^2/6$, and $\alpha = -2 + 4\lambda h^2/6$, where h is the mesh size. Setting $\rho = \exp(\sqrt{-1}\,\theta)$ and solving, we obtain after some algebra

$$u_i^s = \sin \frac{s\pi(ih)}{l}, \qquad \lambda_h^s = \sin^2 \frac{s\pi}{2h} \tag{5.6.24}$$

Observe that the approximate eigenfunctions in this case will interpolate the exact eigenfunctions $\sin s\pi x/l$ at the nodes. Thus we have again a special case of local superconvergence at the interelement nodes (recall Section 5.2).

This max-min principle for the Rayleigh quotient can be used to obtain an upper bound on the approximate eigenvalues and thereby determine error estimates. For our purposes it is better to restate the max-min principle as a min-max principle by noting that if R is maximized over an s-dimensional subspace, the minimum possible value over all such subspaces for this maximum is λ^s.

By applying the foregoing reasoning to the finite-dimensional space H^h in the finite element formulation, we conclude that the approximate eigenvalues also satisfy this principle. Let H_s^h be any s-dimensional subspace of H^h. Then we have

$$\lambda_h^s = \min_{H_s^h} \max_{v_h \in H_s^h} R(v_h) \tag{5.6.34}$$

where the min is taken on all such s-dimensional subspaces H_s^h of H^h.

From (5.6.34) it is evident that

$$\lambda^s \leq \lambda_h^s \tag{5.6.35}$$

since the minimization in (5.6.34) is restricted to s-dimensional subspaces of the finite-dimensional approximation space, whereas the same principle holds for the exact eigenvalues but on the infinite-dimensional space H.

Moreover, since λ_h^s in (5.6.34) is a minimizer on all of $H_s^h \subset H^h$, we can write

$$\lambda_h^s \leq \max_{v_h \in H_s^h} R(v_h) = \max_{v_h \in H_s^h} \frac{B(v_h, v_h)}{\| v_h \|_0^2} \tag{5.6.36}$$

However, $v_h \in H_s^h \subset H^h$, so that v_h is in the approximation space. Hence, v_h can be constructed from functions $w \in H$ by projection: The *elliptic projection* of a function $w \in H$ into H^h with respect to the bilinear form $B(\cdot, \cdot)$ is defined as the function $w_h \in H^h$ satisfying

$$B(w - w_h, v_h) = 0 \qquad \forall v_h \in H^h \tag{5.6.37}$$

On expanding the quadratic functionals $B(w - w_h, w - w_h)$ and using the triangle inequality, we obtain the inequality (Exercise 5.6.15)

$$B(w_h, w_h) \leq B(w, w) \tag{5.6.38}$$

Now define

$$K = 1 - \max_{\substack{w \in H_s \\ \|w\| = 1}} |2(w, w - w_h) - (w - w_h, w - w_h)| \tag{5.6.39}$$

Using these in (5.6.36), we obtain

$$\lambda_h^s \leq \max_{w_h \in H_s^h} \frac{B(w_h, w_h)}{\| w_h \|_0^2} \leq \max_{\substack{w \in H_s \\ \|w\| = 1}} \frac{B(w, w)}{K} \tag{5.6.40}$$

where H_s is the s-dimensional subspace of H spanned by the exact eigen-functions $u^1, u^2, \ldots, u^s$.

We see that (5.6.40) implies that

$$\lambda_h^s \leq \frac{\lambda^s}{K} \tag{5.6.41}$$

We need only estimate $1/K$ to complete the error bound. We recall that

$$K = 1 - \max_{\substack{w \in H_s \\ \|w\| = 1}} |2(w, w - w_h) - (w - w_h, w - w_h)| \tag{5.6.42}$$

Our interpolation estimates in Chapter 1 can be applied directly to the last term to yield

$$\| w - w_h \|_0^2 \leq C_1^2 [h^{k+1} + h^{2(k+1-m)}]^2 \| w \|_{k+1}^2 \tag{5.6.43}$$

where C_1 is a constant, $2m$ is the order of the differential operator A, and k is the degree of the element polynomial basis. The remaining term in (5.6.42) can be bounded by expressing w as a linear combination of the exact eigen-functions $\{u^r\}$ and using

$$(\lambda^s)^{-1} B(u^s, v) = (u^s, v) \tag{5.6.44}$$

with $v = w - w_h$. This yields (Exercise 5.6.17)

$$|(w, w - w_h)| \leq C_2 h^{2(k+1-m)} (\lambda^s)^{[(k+1)/m]-1} \tag{5.6.45}$$

where C_2 is a constant.

Combining the estimates of inequalities (5.6.43) and (5.6.45),

$$K < 1 - |2C_2 h^{2(k+1-m)} (\lambda^s)^{[(k+1)/m]-1} - C_1^2 [h^{k+1} + h^{2(k+1-m)}]^2 \| w \|_{k+1}^2 \tag{5.6.46}$$

and from (5.6.40) and (5.6.35) we have the following bounds for the approxi-mate eigenvalue λ_h^s:

$$\lambda^s \leq \lambda_h^s \leq \lambda^s + C_3 h^{2(k+1-m)} (\lambda^s)^{(k+1)/m} \tag{5.6.47}$$

where h is sufficiently small that $|K| < 1$.

This result can now be applied to bound also the error in the eigen-functions. We assume that both the exact and approximate eigenfunctions have been normalized so that for each r,

$$(u^r, u^r) = (u_h^r, u_h^r) = 1 \tag{5.6.48}$$

Next, we expand the quadratic functional $B(u^s - u_h^s, u^s - u_h^s)$ and use the estimate (5.6.45) to get the bounds for the error in L^2 as

$$\| u^s - u_h^s \|_0 \leq C_1[h^{k+1} + h^{2(k+1-m)}(\lambda^s)^{(k+1)/m}] \tag{5.6.49}$$

and, for the error in H^m, as

$$\| u^s - u_h^s \|_m \leq C_2 h^{2(k+1-m)}(\lambda^s)^{(k+1)/m} \tag{5.6.50}$$

Inequalities (5.6.47), (5.6.49), and (5.6.50) provide the desired error estimates for the approximate eigenvalues λ_h^s and eigenfunctions u_h^s. These inequalities also prove useful in analyzing the stability of finite element schemes for integrating time-dependent problems (Vol. III).

The error analysis for the generalized eigenvalue problem in (5.6.1) is quite analogous to that of the preceeding standard eigenvalue problem. The Rayleigh quotient then becomes

$$R(v) = \frac{B(v, v)}{S(v, v)}, \qquad v \in H \tag{5.6.51}$$

where $S(\cdot, \cdot)$ is the bilinear form associated with operator C in (5.6.1). The previous min-max principle and error analysis are essentially unchanged with the exception that the bound on K now involves $S(\cdot, \cdot)$ rather than the L^2-inner product.

EXERCISES

5.6.7 Construct the element matrices for the vibrating string example in (5.6.16) using a linear element basis. Consider a mesh of four elements of size h and assemble the eigenvalue system. Verify that the computed eigenvalues are

$$\lambda_h = \left\{ \frac{3}{h^2}, \frac{3(10 + 6\sqrt{2})}{7h^2}, \frac{3(10 - 6\sqrt{2})}{7h^2} \right\}$$

Determine the corresponding eigenvectors and compare the results with those in (5.6.24).

5.6.8 Introduce a finite element approximation and construct the discrete eigenvalue problem in integral form for

$$-(au')' + bu = \lambda cu, \qquad 0 < x < 1$$

with mixed-type boundary conditions where a, b, and c are functions of position x. Write down the element matrices for piecewise-linear basis functions.

5.6.9 For the case $c = 1$ and a, b constants in Exercise 5.6.8 with $u(0) = 0$, $u'(1) = 0$, construct the discrete equation at interior node i for a uniform mesh of linear elements. Verify that the nodal values satisfying this discrete equation for eigenfunction u^s are

$$u_i^s = \sqrt{\tfrac{1}{2}} \sin (s - \tfrac{1}{2})\pi i h$$

with eigenvalues

$$\lambda_h^s = \frac{6a[1 - \cos (s - \tfrac{1}{2})h]}{h^2[2 + \cos (s - \tfrac{1}{2})h]} + b$$

5.6.10 Compare the discrete and exact solutions for the problem in Exercise 5.6.9 and write down the error estimates for u_h^s, λ_h^s. Comment on the influence of b on the exact and approximate eigenvalue spectra.

5.6.11 Develop the element matrices for Hermite piecewise cubic elements and the fourth-order problem of Exercise 5.6.4. For $a = 1, b = 0, c = 1$, show that the representative pair of equations at interior node i is

$$-\frac{6}{5h}(u_{i-1} - 2u_i + u_{i+1}) - \frac{1}{10}(u'_{i-1} - u'_{i+1})$$

$$= \lambda_h\Big[\frac{h}{70}(9u_{i-1} + 52u_i + 9u_{i+1}) + \frac{13h^2}{420}(u'_{i-1} - u'_{i+1})\Big]$$

and

$$-\frac{h}{30}(u'_{i-1} - 8u'_i + u'_{i+1}) + \frac{1}{10}(u_{i-1} - u_{i+1})$$

$$= -\lambda_h\Big[\frac{h^3}{420}(3u'_{i-1} - 8u'_i + 3u'_{i+1}) + \frac{13h^2}{420}(u_{i-1} - u_{i+1})\Big]$$

Determine also the equation at the boundary nodes $i = 0, n + 1$.

5.6.12 Compare the finite element eigenfunctions of Exercise 5.6.11 with the exact eigenfunctions. Examine the accuracy of the second eigenvalues λ_h' and discuss the implications to finite element eigenvalue computations.

5.6.13 Approximate the $\mathbf{C}$ matrix in Exercise 5.6.7 by lumping to the diagonal matrix $h\mathbf{I}$ and determine the eigensolution. Compare the solutions for u_h^s and λ_h^s from the lumped system with those of the unlumped (consistent) system in Exercise 5.6.7.

5.6.14 Consider the eigenvalue problem for a vibrating membrane

$$-\Delta u = \lambda u \quad \text{in} \quad \Omega$$

with $u = 0$ on $\partial\Omega$, where Ω is a smooth bounded region in $\mathbb{R}^2$. Write down the variational eigenvalue problem and Rayleigh quotient. Construct a finite element approximation and write down the element contributions in integral form on the master element $\hat{\Omega}$. If C^0-linear approximation on triangles is employed, what are the rates of convergence for u_h^s in the L^2 norm and for λ_h^s?

5.6.15 Expand the quadratic functionals $B(w - w_h, w - w_h)$ and $(w - w_h, w - w_h)$ and use the triangle inequality to establish the intermediate inequalities leading to (5.6.38).

5.6.16 Consider the eigenfunction expansion $w = \sum_{i=1}^s c_i u^i$ and the relation $\lambda^i(u^i, w - w_h) = B(u^i, w - w_h)$, where $w_h = \Pi w$ and Π is the projection operator and satisfies $B(w - \Pi w, v_h) = 0$, $\forall v_h \in H^h$. Expand $B(u^i - \Pi u^i, w - w_h)$ to show that

$$(w, w - w_h) = \sum_{i=1}^s c_i(\lambda^i)^{-1} B(u^i - \Pi u^i, w - w_h)$$

Use this result together with the interpolation error estimate to establish the inequality (5.6.45), (see Strang and Fix [1973]).

5.6.17 Expand the quadratic functional $B(u^s - u_h^s, u^s - u_h^s)$ and use the estimate (5.6.45) to obtain the eigenfunction error estimates in equations (5.6.49) and (5.6.50).

REFERENCES

ADINI, A., and R. W. CLOUGH, "Analysis of Plate Bending by the Finite Element Method," *NSF Report G 7337*, 1961.

AKIN, J. E., "The Generation of Elements with Singularities," *Int. J. Numer. Methods Eng.*, 10, 1249–1259, 1976.

AKIN, J. E., "Elements for the Analysis of Line Singularities," *The Mathematics of Finite Elements with Applications*, Vol. 3, ed. J. R. Whiteman, Academic Press, London, 1979.

ATLURI, S., "A New Assumed Stress Hybrid Finite Element Model for Solid Continua," *AIAA J.*, 9, 1647–1649, 1971.

BABUŠKA, I., "The Theory of Small Changes in the Domain of Existence in the Theory of Partial Differential Equations and Its Applications," *Differential Equations and Their Applications*, Academic Press, New York, 1963.

BABUŠKA, I., "Finite Element Methods for Domains with Corners," *Computing*, 6, 264–273, 1970.

BABUŠKA, I., "The Finite Element Method with Lagrange Multipliers," *Numer. Math.*, 20, 179–192, 1973a.

BABUŠKA, I., "The Finite Element Method with Penalty," *Math. Comput.*, 27 (122), 221–228, 1973b.

BABUŠKA, I., and A. K., AZIZ, "Survey Lectures on the Mathematical Foundations of the Finite Element Method," in *The Mathematical Foundations of the Finite Element Method with Applications to Partial Differential Equations*, ed. A. K. Aziz, Academic Press, New York, 1972.

BABUŠKA, I., J. T., ODEN, and J. K., LEE, "Mixed-Hybrid Finite Element Approximations of Second-Order Elliptic Boundary-Value Problems," *Comput. Methods Appl. Mech. Eng.*, *11*, 175–206, 1977.

BABUŠKA, I., J. T., ODEN, and J. K., LEE, "Mixed-Hybrid Finite Element Approximations of Second-Order Elliptic Boundary-Value Problems, Part II: Weak Hybrid Methods," *Comput. Methods Appl. Mech. Eng.*, *14*, 1–22, 1978.

BABUŠKA, I., and M. B. ROSENWEIG, "A Finite Element Scheme for Domains with Corners," *Numer. Math.*, *20*, 1–21, 1972.

BABUŠKA, I., B. SZABO, and I. KATZ, "The p-Version of the Finite Element Method," *Report WV/CCM-79/1*, Washington University, St. Louis, Mo., 1979.

BARNHILL, R. E., and J. R. WHITEMAN, "Error Analysis of Galerkin Approximations for Dirichlet Problems Containing Boundary Singularities," *J. Inst. Math. Appl.*, *15*, 121–125, 1975.

BARSOUM, R., "On the Use of Isoparametric Finite Elements in Linear Fracture Mechanics," *Int. J. Numer. Methods Eng.*, *10*, 25–37, 1976.

BERCOVIER, M., "Perturbation of Mixed Variational Problems: Application to Mixed Finite Element Methods," *Revue Française d'Automatique Informatique et Recherche Opérationnelle Numer. Anal.*, *12* (3), 211–236, 1978.

BERCOVIER, M., and O. PIRONNEAU, "Estimations d'erreur pour la résolution du problème de Stokes en éléments finis conformes de Lagrange," *C.R. Acad. Sci. Paris, Ser. A*, *285*, 1085–1087, Dec. 19, 1977.

BERGER, A., R. SCOTT, and G. STRANG, "Approximate Boundary Conditions in the Finite Element Method," *Symposia Mathematica*, (Symp. Numer. Anal.), Academic Press, New York, 1972.

BRAMBLE, J., and A. SCHATZ, "On the Numerical Solution of Elliptic Boundary-Value Problems by Least-Square Approximation of the Data," *SYNSPADE*, Academic Press, New York, 107–133, 1970.

BRAMBLE, J., and M. ZLAMAL, "Triangular Elements in the Finite Element Method," *Math. Comput.*, *24*, 809–820, 1970.

BREBBIA, C., *The Boundary Element Method for Engineers*, Pentech Press, London, 1978.

BREBBIA, C., and L. WROBEL, "Applications of Boundary Elements in Water Resources," in *Finite Elements and Water Resources*, Pentech Press, London, 1.3–1.29, 1980.

BREZZI, F., "On the Existence, Uniqueness and Approximation of Saddle-Point Problems Arising from Lagrangian Multipliers," *Revue Française d'Automatique Informatique et Recherche Opérationnelle Numer. Anal. R2*, 129–151, 1974.

BREZZI, F., "Sur une méthode hybride pour l'approximation du problème de la torsion d'une barre élastique," *Lab. Anal. Numer. Con. Naz. Ric. Pub. No. 60*, Pavia, 1974.

BREZZI, F., "Sur la méthode des éléments finis hybrides pour le problème biharmonique," *Numer. Math.*, *24*, 103–131, 1975.

BREZZI, F., and L. D., MARINI, "On the Numerical Solution of Plate Bending Problems by Hybrid Methods," *Revue Française d'Automatique Informatique et Recherche Opérationnelle R3*, 5–50, 1975.

CAREY, G. F., "Derivative Calculation from Finite Element Solutions," *Comp. Meth. Appl. Mech. Eng.*, 1982 (in press).

CAREY, G. F., D. HUMPHREY, and M. F. WHEELER, "Galerkin and Collocation-Galerkin Methods with Superconvergence and Optimal Fluxes," *Int. J. Numer. Meth. Eng. 17*, 937–950, 1981.

CAREY, G. F., and B. A. FINLAYSON, "Orthogonal Collocation on Finite Elements," *Chem. Eng. Sci.*, *30*, 587–596, 1975.

CIARLET, P. G., *The Finite Element Method for Elliptic Problems*, North-Holland, Amsterdam, 1978.

CIARLET, P., and P. RAVIART, "General Lagrange and Hermite Interpolation in $\mathbb{R}^n$ with Applications to Finite Element Methods," *Arch. Ratl. Mech. Anal.*, *46*, 177–199, 1972a.

CIARLET, P., and P. RAVIART, "Interpolation Theory over Curved Elements with Applications to Finite Element Methods," *Comp. Methods Appl. Mech. Eng.*, *1*, 217–249, 1972b.

CLOUGH, R., and C. A. FELIPPA, "A Refined Quadrilateral Element for Analysis of Plate Bending," *Proc. 2nd Conf. Matrix Methods Struct. Mech.*, Wright-Patterson AFB, AFFDL-TR-68-150, 1968.

CLOUGH, R., and J. TOCHER, "Finite Element Stiffness Matrices for Analysis of Plates in Bending," *Proc. Conf. Matrix Methods Struct. Mech.*, Wright-Patterson AFB, AFFDL-TR-66-80, 1965.

COURANT, R., and D. HILBERT, *Methods of Mathematical Physics*, Vol. 2, Interscience, New York, 1962.

CRUSE, T., and F. RIZZO, "Boundary Integral Equation Method: Computational Applications in Applied Mechanics," *Am. Soc. Mech. Eng.*, AMDII, 1975.

DE BOOR, J., and B. SWARTZ, "Collocation at Gaussian Points," *SIAM J. Numer. Anal.*, *10*, 582–606, 1973.

DIAZ, J., "A Collocation-Galerkin Method for the Two-Point Boundary-Value Problem Using Continuous Piecewise Polynomial Spaces," *SIAM J. Numer. Anal.*, *14*(5), 844–858, 1977.

DIAZ, J., "A Collocation-Galerkin Method for Poisson's Equation on Rectangular Regions," *Math. Comp.*, *33*, *145*, 77–84, 1979.

DOUGLAS, J., and T. DUPONT, *Collocation Methods for Parabolic Equations in a Single Space Variable*, Springer-Verlag Lecture Notes in Mathematics No. 385, Springer-Verlag, New York, 1974.

DOUGLAS, J., and T. DUPONT, "Superconvergence for Galerkin Methods for the Two-Point Boundary Problem via Local Projections," *Numer. Math.*, *21*, 270–278, 1973.

DOUGLAS, J., and T. DUPONT, "Galerkin Approximations for the Two-Point Boundary Problem Using Continuous, Piecewise Polynomial Spaces," *Numer. Math.*, *22*, 99–109, 1974.

DUPONT, T, "A Unified Theory of Superconvergence for Galerkin Methods for Two-Point Boundary-Value Problems," *SIAM J. Numer. Anal.*, *13*, 362–368, 1976.

EKELAND, I., and R. TEMAM, *Convex Analysis and Variational Problems*, North-Holland, Amsterdam, 1976.

FAIRWEATHER, G., *Finite Element Galerkin Methods for Differential Equations*, Marcel Dekker, New York, 1978.

FALK, R., and J. E. OSBORN, "Error Estimates for Mixed Methods," *Revue Française d'Automatique Informatique et Recherche Opérationnelle*, Vol. 14, No. 3, 249–277, 1980.

FINLAYSON, B. A., *The Method of Weighted Residuals and Variational Principles*, Academic Press, New York, 1972.

FRAEIJS DE VEUBEKE, B., "Variational Principles and the Patch Test," *Int. J. Numer. Meth. Eng.*, *8*, 783–801, 1974.

FRIED, I., "Shear in C^0 and C^1 Plate Bending Elements," *Int. J. Solids Struct.*, *9*, 449–460, 1973.

FRIED, I., "Finite Element Analyses of Incompressible Materials by Residual Energy Balancing," *Int. J. Solids Struct.*, *10*, 993–1002, 1974.

GALLAGHER, R. H., "Survey and Evaluation of the Finite Element Method in Fracture Mechanics," *Proc. First Int. Conf. Struct. Mech. Reactor Technol.*, Berlin, 637–653, 1971.

GIRAULT, V., and P. A. RAVIART, *Finite Element Approximations of the Navier Stokes Equations*, Lecture Notes in Mathematics No. 749, Springer-Verlag, Berlin, 1979.

GORDON, W. J., "Blending Function Methods of Bivariate and Multivariate Interpolation and Approximation," *SIAM J. Numer. Anal.*, *8*, 158–177, 1971.

GORDON, W. J., and C. A. HALL, "Transfinite Element Methods: Blending Function Interpolation over Arbitrary Curved Element Domains," *Numer. Math.*, *21*, 109–129, 1973.

HAYES, L. J., "An Alternating-Direction Collocation Method for Finite Element Approximations on Rectangles," *Comp. and Math. with Appls.*, *6*, 45–50, 1980.

HELLEN, T. K., "On Special Isoparametric Elements for Linear Elastic Fracture Mechanics," *Int. J. Numer. Meth. Eng.*, *11*, *1*, 200–203, 1977.

HELLINGER, E., "Die allgemeinen ansatze der mechanik der kontinua," *Enz. Math. Wiz.*, *4*, 602–694, 1914.

HENSHELL, D., "Crack Tip Finite Elements Are Unnecessary," *Int. J. Numer. Methods Eng.*, *9*, 495–507, 1975.

HERMANN, L. R., "A Bending Analysis for Plates," *Proc. Conf. Matrix Methods Struct. Mech.*, AFFDL-TR-66-80, Wright-Patterson AFB, 577–602, 1966.

HUGHES, T. J. R., "Equivalence of Finite Elements for Nearly-Incompressible Elasticity," *J. Appl. Mech.*, *44*, 181–183, 1977.

HUGHES, T. J. R., R. L. TAYLOR, and J. F. LEVY, "A Finite Element Method for Incompressible Viscous Flows," *Proc. 2nd Int. Conf. Finite Element Methods Flow Probl.*, Santa Margherita Ligure, Italy, June 1976.

IRONS, B., and A. RAZZAQUE, "Experience with the Patch Test for Convergence of Finite Elements," in *Mathematical Foundations of the Finite Element Method with Applications to Partial Differential Equations*, ed. A. K. Aziz, Academic Press, New York, 557–587, 1972.

JOHNSON, C., "On the Convergence of a Mixed Finite Element Method for Plate Bending Problems," *Numer. Math., 21*, 43–62, 1973.

JOHNSON, C., and J. PITKARANTA, "Analysis of Some Mixed Finite Element Methods Related to Reduced Integration," Research Report, Dept. of Computer Sciences, Chalmers University of Technology, Goteborg, Sweden, 1980.

KIKUCHI, N., "Convergence of a Penalty Method for Variational Inequalities," *TICOM Report 79–16*, Austin, Tex., October 1979.

LADYZHENSKAYA, O. A., *The Mathematical Theory of Viscous Incompressible Flows*, 2nd ed., Gordon and Breach, New York, 1969.

LASCAUX, P., and P. LESAINT, "Some Nonconforming Finite Elements for the Plate Bending Problem," *Revue Française d'Automatique Informatique et Recherche Opérationelle Ser. Rouge Anal. Numer. R-1*, 9–53, 1975.

LEE, J. K., "Convergence of Mixed-Hybrid Finite Element Methods," Ph. D. dissertation, University of Texas at Austin, 1976.

MALKUS, D. S., "Finite Element Analysis of Incompressible Solids," Ph.D. thesis, Boston University, 1975.

MALKUS, D. S., and T. J. R. HUGHES, "Mixed Finite Element Methods—Reduced and Selective Integration Techniques: A Unification of Concepts," *Comput. Methods Appl. Mech. Eng., 15*, 63–81, 1978.

MANSFIELD, L., "Mixed Finite Element Methods for Elliptic Equations," *ICASE Rep. 76–24*, Langley Research Center, Hampton, Va., 1980.

MCLEOD, R., and R. MURPHY, "A Numerical Comparison of High Order Transformation and Isoparametric Transformation Methods," *Comput. Math. Appl., 5*, 241–247, 1979.

MORLEY, L., "The Triangular Equilibrium Element in the Solution of Plate Bending Problems," *Aerosp. Q., 19*, 149–169, 1968.

NITSCHE, J., "On Dirichlet Problems Using Subspaces with Nearly Zero Boundary Conditions," in *Mathematical Foundations of the Finite*

Element Method with Applications to Partial Differential Equations, ed. A. K. Aziz, Academic Press, New York, 409–474, 1972.

NORRIE, D., and G. DE VRIES *Finite Element Bibliography*, Plenum, New York, 1976.

ODEN, J. T., "Some Contributions to the Mathematical Theory of Mixed Finite Element Approximations," in *Theory and Practice in Finite Element Structural Analysis*, University of Tokyo Press, Tokyo, 3–23, 1973.

ODEN, J. T., "A Theory of Mixed Finite Element Approximations of Non-Self-Adjoint Boundary-Value Problems," *Proc. 7th U.S. Natl. Congr. Appl. Mech.*, ASME Publ., 39–51, 1974.

ODEN, J. T., "A Theory of Penalty Methods for Finite Element Approximations of Highly Nonlinear Problems in Continuum Mechanics," *Comput. Struct.*, *8*, 445–449, 1978.

ODEN, J. T., "RIP Methods for Stokesian Flows," *Finite Element Methods in Flow Problems*, Vol. 4, ed. R. H. Gallagher O. C. Zienkiewicz, J. T. Oden, and D. Norrie, Wiley, London, 1982. (Also appeared as *TICOM Rep. 80–11*, Austin, 1980.)

ODEN, J. T., "Penalty Methods for Constrained Problems in Nonlinear Elasticity," *Finite Elasticity*, Maritinus Nijhoff Publishers, Leyden, 1982.

ODEN, J. T., *Qualitative Methods in Nonlinear Mechanics*, Prentice-Hall, Englewood Cliffs, N.J. (to appear, 1983).

ODEN, J. T., and N. KIKUCHI, "Finite Element Methods for Constrained Problems in Elasticity," *Int. J. Numer. Methods Eng.*, 701–725, 1982.

ODEN, J. T., N. KIKUCHI, and Y. J. SONG, "Discrete LBB-Conditions for RIP-Finite Element Methods," *TICOM Report, 80–7*, Austin, Tex., August 1980b.

ODEN, J. T., N. KIKUCHI, and Y. J. SONG, "Penalty Finite-Element Methods for the Analysis of Stokesian Flows," *Comp. Methods Appl. Mech. Eng.*, 1982 (in press).

ODEN, J. T., and J. K. LEE, "Theory and Application of Dual Mixed Hybrid Finite Element Methods to Two-Dimensional Potential Flow Problems," in *Finite Elements in Flow Problems*, Vol. 3, Wiley, London, 123–142, 1978.

ODEN, J. T., and J. N. REDDY, "On Mixed Finite Element Approximations," *SIAM J. Numer. Anal.*, *13*, 393–404, 1976a.

ODEN, J. T., and J. N. REDDY, *Variational Methods in Theoretical Mechanics*, Springer-Verlag, Berlin, 1976b.

ODEN, J. T., N. KIKUCHI, and Y. J. SONG, "Reduced Integration and Exterior Penalty Methods for Finite Element Approximations of Contact Problems in Incompressible Elasticity," *TICOM Report 80–2*, Austin, Tex., August 1980a.

PEARSON, C. E.; *Handbook of Applied Mathematics*, Van Nostrand Reinhold, New York, 1974.

PERCELL, P., and M. F. WHEELER, "A C^1 Finite Element Collocation Method for Elliptic Equations," *SIAM J. Numer. Anal.*, *17*, 605–622, 1980.

PIAN, T. H. H., "Crack Elements," *World Congr. Finite Element Methods Struct. Mech.*, Bournemouth, Dorset, England, 1975.

PIAN, T. H. H., "Derivation of Element Stiffness Matrices by Assumed Stress Distributions," *AIAA J.*, *2*(5), 821–826, 1964.

PIAN, T. H. H., "Element Stiffness Matrices for Boundary Compatibility and for Prescribed Boundary Stresses," *Proc. First Conf. Matrix Methods Struct. Mech.*, Wright-Patterson AFB, 1965, AFDL-TR-66-80, 457–477, 1966.

PIAN, T. H. H., and P. TONG, "Basis of Finite Element Methods for Solid Continua," *Int. J. Numer. Methods Eng.*, *1*, 3–85, 1969.

PRENTER, P. M., and R. D. RUSSELL, "Orthogonal Collocation for Elliptic Partial Differential Equations," *SIAM J. Numer. Anal.*, *13*, 923–939, 1976.

RACHFORD, H., and M. F. WHEELER, "An H^{-1} Galerkin Procedure for the Two-Point Boundary-Value Problem," in *Mathematical Aspects of Finite Elements in Partial Differential Equations*, ed. C. De Boor, Academic Press, New York, 353–382, 1974.

RALSTON, A., *A First Course in Numerical Analysis*, McGraw-Hill, New York, 1965.

RAVIART, P. A., and J. M., THOMAS, "Primal Hybrid Finite Element Methods for 2nd Order Elliptic Equations," *Report 75025*, Université de Paris VI, Laboratoire Analyse Numérique, 1976.

REDDY, J. N., "On the Accuracy and Existence of Primitive Variable Models of Incompressible Viscous Flow," *Int. J. Eng. Sci.*, *16*, 12, 921–930, 1978.

REDDY, J. N., "On the Finite Element Method with Penalty for Incompressible Fluid Flow Problems," *Third Conference on the Mathematics of Finite Elements with Applications*, ed. J. Whiteman, Academic Press, London, 1979a.

REDDY, J. N., "Penalty Finite Element Methods for the Solution of Advection and Free Convection Flows," *3rd Int. Conf. Australia Finite Element Methods*, Sydney, July 1979b.

REDDY, J. N., and J. T. ODEN, "Mathematical Theory of Mixed Finite Element Approximations," *Q. Appl. Math.*, *13*, 255–280, 1975.

REISSNER, E., "Note on the Method of Complementary Energy," *J. Math. Phys.*, *27*, 159–160, 1948.

REISSNER, E., "On a Variational Theorem for Finite Deformations," *J. Math. Phys.*, *32*, 129–135, 1953.

SANI, R. L., P. M. GRESHO, R. L. LEE, D. F. GRIFFITH, and M. ENGLEMAN, "The Cause and Cure (?) of the Spurious Pressures Generated by Certain FEM Solutions of the Incompressible Navier–Stokes Equations; Parts I and II, *Int. J. Numer. Meth. Fluids*, *1*, 17–43 and 171–204, 1981.

SCHATZ, A. H., and L. B. WAHLBIN, "Maximum Norm Estimates in the Finite Element Method on Polygonal Domains," Parts 1 and 2, *Math. Comput.*, *32*, *141*, 73–109, 1978 and *33*, *146*, 465–492, 1979.

SCOTT, R., "The Finite Element Method with Singular Data," *Numer. Math.*, *21*, 317–327, 1973.

SCOTT, R, "A Survey of Displacement Methods for the Plate Bending Problem," *Proc. U.S.-Germany Symp. Formulations Comput. Algorithms Finite Element Anal.*, M.I.T., August 1976.

SHEU, M.-G., "On the Theory of Mixed Finite Element Approximations of Boundary-Value Problems," *Comput. Math. Appl.*, *4*(4), 333–347, 1978.

STERN, M., "Families of Consistent Conforming Elements with Singular Derivative Fields," *Int. J. Numer. Methods Eng.*, *14*, 409–421, 1979.

STRANG, G., "Variational Crimes in the Finite Element Method," in *Mathematical Foundations of the Finite Element Method with Applications to Partial Differential Equations*, ed. A.K. Aziz, Academic Press, New York, 1972.

STRANG, G., and A. E. BERGER, "The Change in Solution Due to

Change in Domain," *Proc. AMS Symp. PDEs*, Berkeley, Calif., 1971.

STRANG, G., and G. FIX, *An Analysis of the Finite Element Method*, Prentice-Hall, Englewood Cliffs, N.J., 1973.

STROUD, A. H., and D. SECREST, *Gaussian Quadrature Formulas*, Prentice-Hall, Englewood Cliffs, N.J., 1966.

STUMMEL, F., "The Limitations of the Patch Test," *Int. J. Numer. Methods Eng.*, *15*, 177–188, 1980.

SZABO, B., "Some Recent Developments in Finite Element Analysis," *Comput. Math. Appl.*, *5*, *2*, 99–116, 1979.

THOMAS, J. M., "Méthodes des éléments finis hybrides duaux pour les problèmes elliptiques du second-order," *Report 75006*, Université de Paris VI et Centre National de la Research Scientifique, 1975.

TONG, P., "New Displacement Hybrid Finite Element Models for Solid Continua," *Int. J. Numer. Methods Eng.*, *2*, 73–85, 1970.

TONG, P., T. H. H. PIAN, and S. J. LASRY, "A Hybrid Element Approach to Crack Problems in Plane Elasticity," *Int. J. Numer. Methods Eng.*, *7*, 297–308, 1973.

VAINBERG, M. M., *Variational Method and Method of Monotone Operators in the Theory of Nonlinear Equations*, Wiley, New York, 1973.

VILLADSEN, J., and M. L. MICHELSEN, *Solution of Differential Equation Models by Polynomial Approximation*, Prentice-Hall, Englewood Cliffs, N.J., 1977.

WACHSPRESS, E., *A Rational Finite Element Basis*, Academic Press, New York, 1975.

WACHSPRESS, E., and R. MCLEOD, eds., "Curved Finite Elements," Special issue of *Comput. Math. Appl.*, *5*(4), 1979.

WHEELER, J. A., "Simulation of Heat Transfer from a Warm Pipeline Buried in Permafrost," *74th Natl. Mtg.*, *AIChE*, New Orleans, 1973.

WHEELER, M. F., "A Galerkin Procedure for Estimating the Flux for Two-Point Boundary Problems," *SIAM J. Numer. Anal.*, *11*, 764–768, 1974.

WHEELER, M. F., "A C^0-Collocation-Finite Element Method for Two-Point Boundary-Value Problems and One Space Dimensional Parabolic Problems," *SIAM J. Numer. Anal.*, *14*(1), 71–90, 1977.

WILSON, E. L., and R. L. TAYLOR, "Incompatible Displacement Models," *Proc. Symp. Numer. Comp. Meth. Struct. Eng.*, O.N.R., University of Illinois, 1971.

ZIENKIEWICZ, O. C., *The Finite Element Method in Engineering Science*, McGraw-Hill, New York, 1971.

ZIENKIEWICZ, O. C., R. L. TAYLOR, and J. M. TOO, "Reduced Integration Techniques in General Analysis of Plates and Shells," *Int. J. Numer. Methods Eng.*, *3*, 275–290, 1971.

INDEX

INDEX